T0190710

Complex Analysis with Applications

RICHARD A. SILVERMAN

Dover Publications, Inc., New York

This Dover edition, first published in 1984, is an unabridged and unaltered republication of the work originally published by Prentice-Hall, Inc., in 1974.

Library of Congress Cataloging in Publication Data

Silverman, Richard A.
Complex analysis with applications.

Reprint. Originally published: Englewood Cliffs, N.J. Prentice-Hall, 1973, c1974.
Bibliography: p.
Includes index.
1. Functions of complex variables. I. Title
QA331.S52 1984 515.9 84-8037
ISBN-13: 978-0-486-64762-3
ISBN-10: 0-486-64762-5

Manufactured in the United States by LSC Communications
64762510 2023
www.doverpublications.com

Contents

Preface

In writing this book, my aim was to give a concise but fully motivated treatment of the rudiments of applied complex analysis, complete enough to contain and amply illustrate all the essential features of the subject, but not so detailed as to overwhelm the beginning student with a surfeit of secondary issues. I believe this aim has been accomplished, with the help of the following pedagogical devices (among others):

1) Extensive problem sets, one at the end of each chapter; these include both abundant exercise material and problems pursuing and amplifying points of theory introduced in the text proper.

2) Hints and answers to many of the problems; these begin on p. 247 and range from numerical answers to detailed solutions in cases where such are called for.

3) Sets of comments, again one at the end of each chapter (before the corresponding problem set); these section-by-section comments are largely of a heuristic nature, and are meant to give the student extra insight into what is "really" going on, what will happen later, etc.

It is hoped that the above features, combined with appropriate clarity and brevity of the main text material, will make the book especially accessible to those encountering complex analysis for the first time. In particular, the approach adopted should help the student keep the forest and the trees well separated in the lush terrain of complex variable theory.

The detailed subject matter of the book is apparent from the table of contents, but there are again several features that warrant special mention:

1) The central topic of integration in the complex plane (the Cauchy theory) is introduced with all deliberate speed, representing as it does the very core of complex analysis.

2) Similarly, the student will encounter complex series and the key topic of power series well within the first half of the book. Thus when he gets to Taylor and Laurent series in Chapters 10 and 11, he will be able to give his undivided attention to the matter at hand, without being distracted by such side issues as the meaning of absolute and uniform convergence, validity of term-by-term integration of series, etc.

3) Both the argument principle and Rouché's theorem are deemed worthy of special attention (see Secs. 12.1 and 12.2). The application of residue theory to the evaluation of improper real integrals is given the detailed treatment it deserves (see Secs. 12.3 and 12.4).

4) Chapter 13 is of a more advanced character than the rest of the book, and might well be omitted in a brief one-semester course, except to the extent that a bit of this material figures in the derivation of the key Schwarz-Christoffel transformation (see Sec. 14.1).

5) Chapter 15 deals with two typical physical applications of complex analysis, namely applications to fluid mechanics and electrostatics. Here the style is appropriately that of mathematical physics rather than of pure mathematics. The Kutta-Joukowski theorem (derived in Sec. 15.26) is a particularly stunning application of complex variable theory. Except for a detail or two, Chapter 15 is independent of Chapters 12–14, and applications-minded instructors may prefer to assign it right after Chapter 11.

Among helpful sources lying on my desk while writing the book, I cite the classic book on functions of a complex variable by I. I. Privalov and my various translations and revisions of the treatise on the same subject by A. I. Markushevich (Prentice-Hall, 1965, 1967). A battery of reviewers, consisting of Profs.Larry Zalcman of the University of Maryland, Raymond Wells, Jr. of Rice University, Kenneth Gross of Dartmouth College, Paul Sally of the University of Chicago, Laurence Hoffmann of Claremont Men's College, and Kenneth Hoffman of the Massachusetts Institute of Technology, subjected earlier drafts of the book to a barrage of valuable criticism, which had the desired effect of leading to substantial improvements in each successive version. Among these reviewers, I found the detailed comments of the last two particularly useful, and the final shape of the manuscript owes much to their constructive advice. I cannot imagine this book arriving at its present form without both the encouragement and the persistent perfectionism of my friend and editor Arthur Wester of Prentice-Hall.

R.A.S.

Complex Analysis with Applications

CHAPTER ONE

Complex Numbers

1.1. Basic Concepts

1.11. By a *complex number* α we mean an ordered pair (a, b) of real numbers a and b:

$$\alpha = (a, b). \tag{1}$$

If $b = 0$ we agree to write $(a, 0) = a$, so that in this case (1) reduces to just

$$\alpha = a.$$

Thus the set R of all real numbers is a proper subset of the set C of all complex numbers.

1.12. Next we define the basic arithmetic operations on complex numbers. Since R is a subset of C, we must require on the one hand that the operations on complex numbers, regardless of how they are defined, give the same results as in the case of ordinary arithmetic when applied to real numbers. On the other hand, if complex numbers are to have general applicability to problems of analysis, we must also require that the arithmetic operations on complex numbers satisfy the usual axioms of "real arithmetic."

a. The *sum* $\alpha + \beta$ of two complex numbers $\alpha = (a, b)$ and $\beta = (c, d)$ is defined by the formula

$$\alpha + \beta = (a + c, b + d). \tag{2}$$

Applying this definition to two real numbers a and c, we get

$$(a, 0) + (c, 0) = (a + c, 0) = a + c,$$

i.e., the operation of addition satisfies the first of the above two requirements.

b. The *product* $\alpha\beta$ of two complex numbers $\alpha = (a, b)$ and $\beta = (c, d)$ is defined by the formula

$$\alpha\beta = (ac - bd, ad + bc). \tag{3}$$

Applying this definition to two real numbers a and c, we get

$$(a, 0)(c, 0) = (ac, 0) = ac,$$

i.e., the operation of multiplication is consistent with the arithmetic of real numbers.

c. Using the definitions (2) and (3), the reader can easily verify that the operations of addition and multiplication of complex numbers obey the following five familiar laws of arithmetic:

1) Addition is commutative: $\alpha + \beta = \beta + \alpha$;
2) Multiplication is commutative: $\alpha\beta = \beta\alpha$;
3) Addition is associative: $\alpha + (\beta + \gamma) = (\alpha + \beta) + \gamma$;
4) Multiplication is associative: $\alpha(\beta\gamma) = (\alpha\beta)\gamma$;
5) Multiplication is distributive with respect to addition:

$$\alpha(\beta + \gamma) = \alpha\beta + \alpha\gamma$$

(the details are left as an exercise).

1.13. In operations on the complex numbers a special role is played by the number represented by the pair (0, 1). This number is denoted by i. Using the definition (3) to square (0, 1), i.e., to multiply (0, 1) by itself, we find that

$$(0, 1)(0, 1) = (-1, 0) = -1.$$

Thus $i^2 = -1$, which leads to the formula

$$i = \sqrt{-1}.$$

This allows us to write an arbitrary complex number $\alpha = (a, b)$ in the form

$$\alpha = (a, b) = (a, 0) + (0, b) = (a, 0) + (b, 0)(0, 1) = a + bi$$

(alternatively $\alpha = a + ib$). The number a is called the *real part* of the complex number α denoted by $\operatorname{Re} \alpha$, while the number b (the coefficient of i in the expression bi) is called the *imaginary part* of α, denoted by $\operatorname{Im} \alpha$. Note that $\alpha = a + bi$ reduces to the real number a if $\operatorname{Im} \alpha = 0$. A complex number $\alpha = a + bi$ is said to be *purely imaginary* if $\operatorname{Re} \alpha = 0$, $\operatorname{Im} \alpha \neq 0$.† Two complex numbers α and β are said to be *equal* if they have the same real

†In definitions the locution "is said to be . . . if" will be used with the meaning "is . . . if and only if" (the dots indicate a missing predicate).

and imaginary parts, i.e., if

$$\text{Re}\,\alpha = \text{Re}\,\beta, \qquad \text{Im}\,\alpha = \text{Im}\,\beta.$$

1.14. Given a complex number $\alpha = a + bi$, the complex number $\bar{\alpha} = a - bi$ with the same real part as α and with an imaginary part of opposite sign is called the *complex conjugate* of α. It follows from (3) that

$$\alpha\bar{\alpha} = a^2 + b^2.$$

1.15. In real arithmetic we have both an *additive unit* 0 and a *multiplicative unit* 1, i.e., (unique) numbers 0 and 1 such that $a + 0 = a$ and $a \cdot 1 = a$ for all real a. The same numbers 0 and 1 serve as additive and multiplicative units in "complex arithmetic." In fact, let δ be the complex additive unit, so that

$$\alpha + \delta = \alpha \tag{4}$$

for every complex number $\alpha = (a, b)$. Then adding the number

$$-\alpha = -1 \cdot \alpha = (-a, -b) \tag{5}$$

(called the *negative* of α) to both sides of (4), we get

$$\delta = 0,$$

since obviously $\alpha + (-\alpha) = (0, 0) = 0$. Similarly, if ϵ is the complex multiplicative unit, i.e., if

$$\alpha\epsilon = \alpha \tag{6}$$

for every complex number $\alpha \neq 0$, then multiplying both sides of (6) by the number

$$\gamma = \frac{1}{a^2 + b^2}\bar{\alpha}, \tag{7}$$

we get

$$\frac{1}{a^2 + b^2}\alpha\bar{\alpha}\epsilon = \frac{1}{a^2 + b^2}\alpha\bar{\alpha}.$$

But $\alpha\bar{\alpha} = a^2 + b^2$ and hence

$$\epsilon = 1.$$

1.16. It follows from the definition (3) that the product of two complex numbers is zero if one or both of the factors is zero. Conversely, if the product of two complex numbers is zero, then at least one of the factors is zero. In fact, suppose

$$\alpha\beta = 0 \tag{8}$$

where $\alpha \neq 0$. Then multiplying both sides of equation (8) by the number (7), we see at once that $\beta = 0$, as required.

1.17. Subtraction is defined as the inverse operation of addition, i.e., by the *difference* $\beta - \alpha$ between two complex numbers $\beta = c + di$ and

$\alpha = a + bi$ (in that order) we mean the number γ such that

$$\alpha + \gamma = \beta. \tag{9}$$

Adding the number (5) to both sides of (9) and solving for γ, we get

$$\gamma = \beta - \alpha = \beta + (-\alpha) = (c - a) + (d - b)i. \tag{10}$$

Note that the negative of α, defined by (5), is just the difference $0 - \alpha$.

1.18. Similarly division is defined as the inverse operation of multiplication. Let

$$\frac{1}{\alpha} \qquad (\alpha \neq 0)$$

(the *reciprocal* of α) be the number γ such that

$$\alpha\gamma = 1.$$

Then γ is just the number given by (7), since

$$\alpha\gamma = \frac{1}{a^2 + b^2}\alpha\bar{\alpha} = 1$$

for this choice of γ. The *quotient* β/α of two complex numbers $\beta = c + di$ and $\alpha = a + bi$ is defined by the formula

$$\frac{\beta}{\alpha} = \frac{1}{\alpha}\beta \qquad (\alpha \neq 0),$$

just as in real arithmetic. Therefore

$$\frac{\beta}{\alpha} = \frac{\bar{\alpha}\beta}{a^2 + b^2} = \frac{(a - bi)(c + di)}{a^2 + b^2} = \frac{(ac + bd) + (ad - bc)i}{a^2 + b^2}. \tag{11}$$

It is clear from (10) and (11) that the numbers $\beta - \alpha$ and β/α are uniquely defined.

1.19. Comparing the formulas

$$(a + c) - (b + d)i = (a - bi) + (c - di),$$
$$(ac - bd) - (ad + bc)i = (a - bi)(c - di)$$

with (2) and (3), we see that the complex conjugate of the sum or product of two complex numbers α and β is just the sum or product of the corresponding complex conjugates $\bar{\alpha}$ and $\bar{\beta}$; more concisely,

$$\overline{\alpha + \beta} = \bar{\alpha} + \bar{\beta}, \qquad \overline{\alpha\beta} = \bar{\alpha}\bar{\beta}.$$

Since subtraction and division are the inverse operations of addition and multiplication, it is easy to see that the same is true of differences and quotients, i.e.,

$$\overline{\beta - \alpha} = \bar{\beta} - \bar{\alpha}, \qquad \overline{\beta/\alpha} = \bar{\beta}/\bar{\alpha}.$$

Consider any equation both sides of which are obtained by applying the operations of addition, subtraction, multiplication and division to various complex numbers. Then the equation clearly remains valid if all complex numbers in both sides are replaced by their complex conjugates. For example,

$$\frac{1+i}{1-i} = i$$

implies

$$\frac{1-i}{1+i} = -i$$

and conversely.

1.2. The Complex Plane

1.21. We can represent any given complex number $\alpha = (a, b)$ by a point in the plane, namely the point with rectangular coordinates a and b (see Figure 1). Each point of this plane, called the *complex plane*,† represents a complex number and conversely. Thinking in terms of such an underlying complex plane, we will use the terms "the complex number α" and "the point α" interchangeably.

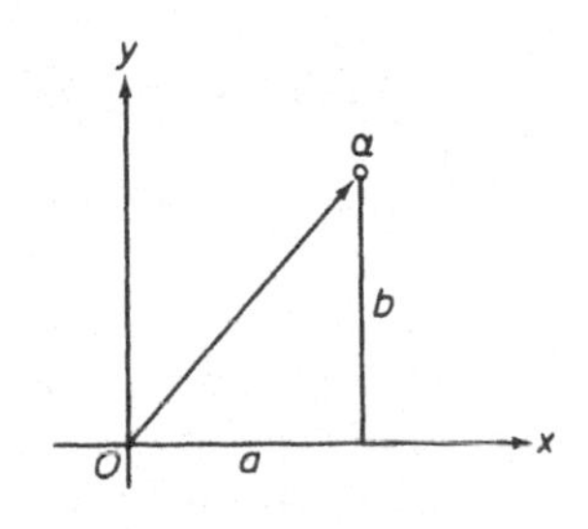

Figure 1

Clearly, every point of the x-axis represents a real number, while every point of the y-axis represents a purely imaginary number (except for the origin of coordinates which represents the complex number 0). For this reason, the x-axis of the complex plane is usually called the *real axis*, while the y-axis is called the *imaginary axis*.

1.22. The complex number $\alpha = (a, b)$ can also be represented by the vector shown in Figure 1, drawn from the origin to the point α. The real part of α is then the projection of the vector α onto the real axis, while the imaginary part is the projection of α onto the imaginary axis.

1.23. To give a geometric construction of the *sum* of two complex numbers α and β, we first represent α and β by appropriate vectors, as in Figure 2. It then follows from the definition (2) that the sum $\alpha + \beta$ is the vector whose components are the sums of the corresponding components of

---

†Or the z-plane, w-plane, . . . , depending on the letter z, w, . . . used to denote a generic complex number.

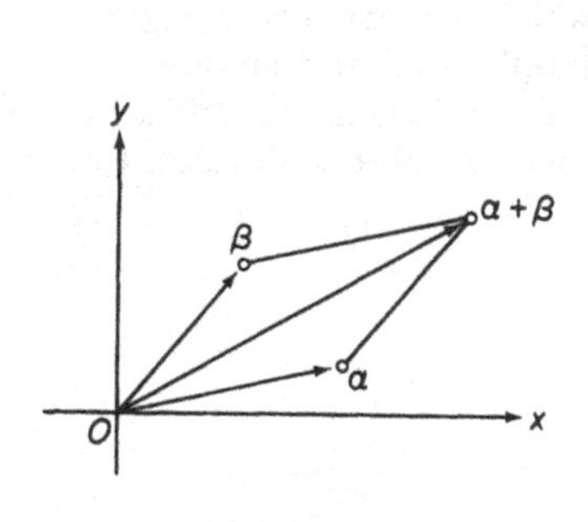

Figure 2

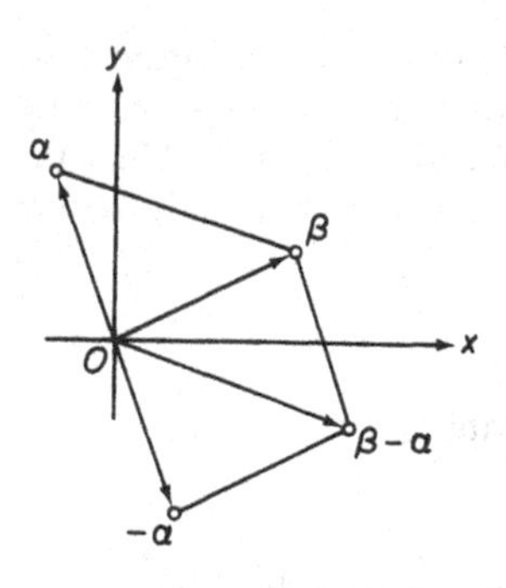

Figure 3

α and β, i.e., the vector $\alpha + \beta$ is the diagonal of the parallelogram constructed on the vectors α and β as sides (see Figure 2).

As shown in Figure 3, the vector representing the *difference* $\beta - \alpha = \beta + (-\alpha)$ is found by using the above "parallelogram rule" to construct the sum of the vectors β and $-\alpha$.

1.3. The Modulus and Argument

1.31. Let $\alpha = (a, b)$ be any point in the complex plane. Then the quantity

$$r = \sqrt{a^2 + b^2} = \sqrt{\alpha\bar{\alpha}},$$

i.e., the distance from the origin to α (alternatively the length of the vector α), is called the *modulus* or *absolute value* of α, denoted by $|\alpha|$. If α is real, the modulus of α obviously reduces to the ordinary absolute value of α. The set of all complex numbers with the same modulus r is clearly represented by the circle of radius r with its center at the origin. The number 0 is the unique complex number of zero modulus.

1.32. Now think of α as a vector in the complex plane. Then the angle θ between α and the positive real axis, more exactly the angle through which the positive real axis must be rotated to give it the same direction as α (regarded as positive if θ is counterclockwise and negative otherwise) is called the *argument* of α, denoted by $\arg \alpha$. Obviously

$$\tan \theta = \frac{b}{a}$$

(see Figure 4). Note that 0 is the only complex number whose argument is undefined.

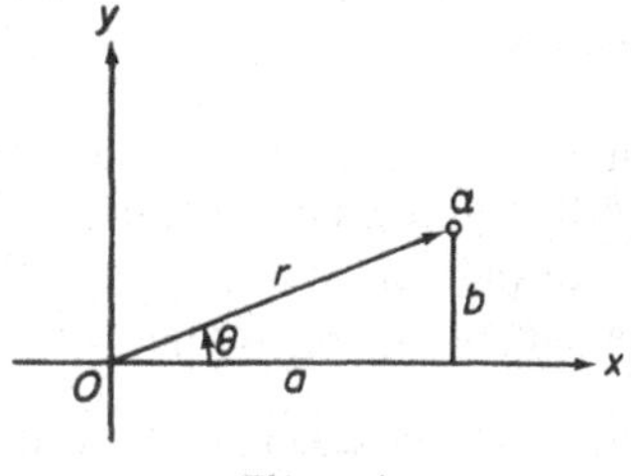

Figure 4

1.33. The argument of every complex number $\alpha \neq 0$ is defined only to within an integral multiple of 2π and hence has infinitely many values. Since r and θ are the polar coordinates of the point $\alpha = (a, b)$, we have

$$a = r\cos\theta, \qquad b = r\sin\theta.$$

It follows that

$$\alpha = r(\cos\theta + i\sin\theta), \tag{12}$$

where (12) is called the *trigonometric* (or *polar*) *form* of the complex number α.

1.34. Forming the product of two complex numbers

$$\alpha = r(\cos\theta + i\sin\theta), \qquad \beta = \rho(\cos\varphi + i\sin\varphi)$$

in trigonometric form, we get

$$\begin{aligned}\alpha\beta &= r\rho[(\cos\theta\cos\varphi - \sin\theta\sin\varphi) + i(\cos\theta\sin\varphi + \sin\theta\cos\varphi)]\\ &= r\rho[\cos(\theta+\varphi) + i\sin(\theta+\varphi)].\end{aligned}$$

It follows that

$$|\alpha\beta| = r\rho = |\alpha||\beta|, \qquad \arg(\alpha\beta) = \theta + \varphi = \arg\alpha + \arg\beta, \tag{13}$$

i.e., *the modulus of the product of two complex numbers equals the product of their moduli, while the argument of the product of two complex numbers equals the sum of their arguments.*

Thus the vector representing $\alpha\beta$ is obtained from the vector representing α by rotating it through the angle $\arg\beta$ in the counterclockwise direction and expanding the resulting vector $|\beta|$ times.† If $|\beta| = 1$ the multiplication reduces to pure rotation. For example, multiplication by i corresponds to rotation through 90° and multiplication by -1 to rotation through 180°. If $\arg\beta = 0$ (so that β is a positive real number), the multiplication reduces to pure expansion.

1.35. We can easily generalize (13) to the case of a product $\alpha\beta\cdots\lambda$ of any number of complex factors. Thus

$$|\alpha\beta\cdots\lambda| = |\alpha||\beta|\cdots|\lambda|,$$
$$\arg(\alpha\beta\cdots\lambda) = \arg\alpha + \arg\beta + \cdots + \arg\lambda,$$

and in particular

$$|\alpha^n| = |\alpha|^n, \qquad \arg(\alpha^n) = n\arg\alpha. \tag{14}$$

We can also write (14) in the form

$$[r(\cos\theta + i\sin\theta)]^n = r^n(\cos n\theta + i\sin n\theta), \tag{14'}$$

a result known as *De Moivre's theorem*.

†Depending on the value of $|\beta|$, the "expansion" corresponds to stretching (if $|\beta| > 1$), shrinking (if $|\beta| < 1$), or neither (if $|\beta| = 1$).

1.36. Complex roots. Given a complex number

$$\alpha = r(\cos\theta + i\sin\theta),$$

we define the *nth root* $\sqrt[n]{\alpha}$ as any complex number which gives α when raised to the nth power. The modulus of $\sqrt[n]{\alpha}$ is obviously just $\sqrt[n]{r}$, while the argument equals

$$\frac{\theta + 2k\pi}{n},$$

where k is any integer. Giving k the values $0, 1, 2, \ldots, n-1$, we get n distinct values of the argument of $\sqrt[n]{\alpha}$. Therefore $\sqrt[n]{\alpha}$ has n and only n (why?) distinct values

$$\sqrt[n]{\alpha} = \sqrt[n]{r}\left(\cos\frac{\theta + 2k\pi}{n} + i\sin\frac{\theta + 2k\pi}{n}\right) \qquad (k = 0, 1, \ldots, n-1).$$

Geometrically these n values of $\sqrt[n]{\alpha}$ correspond to the vertices of a regular n-gon inscribed in the circle of radius $\sqrt[n]{r}$ with its center at the origin (sketch a figure).

1.37. Turning to quotients of complex numbers, we note that

$$\alpha = \frac{\alpha}{\beta}\beta \qquad (\beta \neq 0),$$

and hence, by (13),

$$|\alpha| = \left|\frac{\alpha}{\beta}\right||\beta|, \qquad \arg\alpha = \arg\left(\frac{\alpha}{\beta}\right) + \arg\beta,$$

which implies

$$\left|\frac{\alpha}{\beta}\right| = \frac{|\alpha|}{|\beta|}, \qquad \arg\left(\frac{\alpha}{\beta}\right) = \arg\alpha - \arg\beta,$$

i.e., *the modulus of the quotient of two complex numbers equals the quotient of their moduli, while the argument of the quotient of two complex numbers equals the difference of their arguments.* Thus the vector representing α/β is obtained from the vector representing α by rotating it through the angle $-\arg\beta$ in the counterclockwise direction (i.e., through the angle $\arg\beta$ in the clockwise direction) and expanding the resulting vector $1/|\beta|$ times.

1.38. It is clear from Figure 2 that *the modulus of the sum of two complex numbers cannot exceed the sum of the moduli of the numbers themselves*, i.e.,

$$|\alpha + \beta| \leq |\alpha| + |\beta|. \tag{15}$$

This is an immediate consequence of the fact that the length of one side of a triangle cannot exceed the sum of the lengths of the other two sides (equality occurs in the case where the triangle reduces to a line segment). Replacing β by $\beta - \alpha$ in the "triangle inequality" (15), we get

$$|\beta| = |\alpha + (\beta - \alpha)| \leq |\alpha| + |\beta - \alpha|$$

or equivalently

$$|\beta - \alpha| \geq |\beta| - |\alpha|. \tag{16}$$

Moreover, interchanging α and β in (16) gives

$$|\alpha - \beta| \geq |\alpha| - |\beta|. \tag{16'}$$

Since obviously $|\beta - \alpha| = |\alpha - \beta|$, we can combine (16) and (16′) into the single inequality

$$|\alpha - \beta| \geq ||\alpha| - |\beta||. \tag{17}$$

1.4. Inversion

1.41. Given a point α in the complex plane, we now construct the point representing the number $1/\bar{\alpha}$. If $|\alpha| < 1$ the construction can be accomplished by the following process known as *inversion* (see Figure 5):

1) Draw the circle C of radius 1 centered at the origin O;
2) Draw the straight line L joining O to the point α;
3) Erect the perpendicular to L at the point α, intersecting C in some point P;
4) Draw the tangent to C at P, intersecting L in some point β.

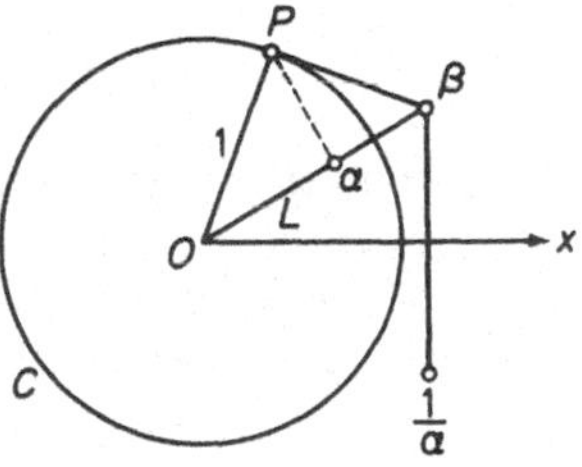

Figure 5

Then β is the required point $1/\bar{\alpha}$. In fact, examination of the triangle $OP\beta$ shows that

$$\frac{|\beta|}{1} = \frac{1}{|\alpha|}$$

in keeping with

$$\left|\frac{1}{\bar{\alpha}}\right| = \frac{1}{|\bar{\alpha}|} = \frac{1}{|\alpha|},$$

while obviously

$$\arg \beta = \arg \alpha$$

in keeping with

$$\arg\left(\frac{1}{\bar{\alpha}}\right) = -\arg \bar{\alpha} = \arg \alpha.$$

The two points α and β are said to be *symmetric with respect to the circle C*.† Note that having found the point

$$\beta = \frac{1}{\bar{\alpha}},$$

†The construction of β in the cases $|\alpha| > 1$ and $|\alpha| = 1$ is left as an exercise (draw the tangent first if $|\alpha| > 1$).

we need only reflect β in the real axis to obtain the *reciprocal*

$$\bar{\beta} = \frac{1}{\alpha}$$

(see Figure 5).

1.42. The same construction with C replaced by a circle C_R of radius R centered at O gives the point

$$\beta = \frac{R^2}{\bar{\alpha}},$$

and the points α and β are now said to be *symmetric with respect to* C_R. Thus two points α and β are symmetric with respect to C_R if and only if they lie on the same ray emanating from O and

$$|\alpha||\beta| = R^2.$$

COMMENTS

1.1. The rules for algebraic operations on complex numbers $\alpha = (a, b)$, $\beta = (c, d), \ldots$ are precisely those obtained if $\alpha, \beta, \ldots$ are thought of as binomials of the form $a + bi$, $c + di, \ldots$ involving an "imaginary unit" $i = \sqrt{-1}$ (Prob. 1). Obviously $\pm i$ is the formal solution of the quadratic equation $x^2 + 1 = 0$, which has no solutions at all for real x. It is indeed remarkable that once complex numbers of the form $z = x + iy$ (x, y real) are admitted, then *every* algebraic equation

$$a_0 + a_1 z + \cdots + a_n z^n = 0$$

with real (or, for that matter, complex) coefficients $a_0, a_1, \ldots, a_n$ has a root, and in fact precisely n roots! This is the content of the celebrated "fundamental theorem of algebra" (Theorem 12.23).

1.2. Note that there is a one-to-one correspondence between the set of all complex numbers C and the set of all points in the plane, or alternatively between C and the set of all vectors in the plane. Note also that the reflection in the real axis of the point (or vector) representing the complex number α is just the point (or vector) representing the complex conjugate $\bar{\alpha}$.

1.3. Just as there are problems in which polar coordinates are more suitable than rectangular coordinates, so it is often more useful to characterize a complex number by its modulus and argument rather than by its real and imaginary parts. Physicists do the same thing in describing a vector in terms of its direction and magnitude rather than in terms of its components.

1.4. The utility of the operation of inversion will be more apparent in Sec. 8.27, when we investigate transformations of the complex plane which carry points symmetric with respect to one circle into points symmetric with respect to another circle.

PROBLEMS

1. Prove that the result of any arithmetic operation on complex numbers $a + bi, c + di, \ldots$ can be obtained by treating $a + bi, c + di, \ldots$, exactly like binomials in an unknown i, with the rule

$$i^2 = -1, \quad i^3 = -i, \quad i^4 = 1, \quad i^5 = i, \ldots$$

being used to eliminate all powers of i higher than the first.

2. Find the complex numbers which are complex conjugates of
a) Their own squares; b) Their own cubes.

3. Calculate the following quantities:

$$\text{a)}\ \frac{1 + i\tan\theta}{1 - i\tan\theta}; \quad \text{b)}\ \frac{(1 + 2i)^3 - (1 - i)^3}{(3 + 2i)^3 - (2 + i)^2}; \quad \text{c)}\ \frac{(1 - i)^5 - 1}{(1 + i)^5 + 1}; \quad \text{d)}\ \frac{(i + i)^9}{(1 - i)^7}.$$

4. What is the locus of the points $z = x + iy$ such that $|z| + \operatorname{Re} z \leq 1$?

5. Find the points $z = x + iy$ such that
a) $|z| \leq 2$; b) $\operatorname{Im} z > 0$; c) $\operatorname{Re} z \leq \frac{1}{2}$; d) $\operatorname{Re}(z^2) = a$;
e) $|z^2 - 1| = a$; f) $\left|\frac{z - 1}{z + 1}\right| \leq 1$; g) $\left|\frac{z - \alpha}{z - \beta}\right| = 1$.

6. Represent the following complex numbers in trigonometric form:
a) $1 + i$; b) $-1 + i$; c) $-1 - i$; d) $1 - i$; e) $1 + \sqrt{3}i$;
f) $-1 + \sqrt{3}i$; g) $-1 - \sqrt{3}i$; h) $1 - \sqrt{3}i$; i) $\sqrt{3} - i$;
j) $2 + \sqrt{3} + i$.

7. By repeated application of the inequality (15), prove that

$$|z_1 + z_2 + \cdots + z_n| \leq |z_1| + |z_2| + \cdots + |z_n|$$

for arbitrary complex numbers $z_1, z_2, \ldots, z_n$.

8. Prove the identity

$$|z_1 + z_2|^2 + |z_1 - z_2|^2 = 2(|z_1|^2 + |z_2|^2),$$

and interpret it geometrically.

9. When do three points z_1, z_2, z_3 lie on a straight line?

10. Let σ be the line segment joining two points z_1 and z_2. Find the point z dividing σ in the ratio $\lambda_1 : \lambda_2$.

11. Let z_1, z_2, z_3 be consecutive vertices of a parallelogram. Find the fourth vertex z_4 (opposite z_2).

12. Show that the center of mass of a system of particles with masses $m_1, m_2, \ldots, m_n$ at the points $z_1, z_2, \ldots, z_n$ lies at the point

$$z = \frac{m_1 z_1 + m_2 z_2 + \cdots + m_n z_n}{m_1 + m_2 + \cdots + m_n}.$$

13. Three points z_1, z_2, z_3 satisfy the conditions

$$z_1 + z_2 + z_3 = 0, \qquad |z_1| = |z_2| = |z_3| = 1.$$

Show that the points lie at the vertices of an equilateral triangle inscribed in the unit circle $|z| = 1$.

14. Four points z_1, z_2, z_3, z_4 satisfy the conditions

$$z_1 + z_2 + z_3 + z_4 = 0, \qquad |z_1| = |z_2| = |z_3| = |z_4| = 1.$$

Show that the points either lie at the vertices of a square inscribed in the unit circle or else coincide in pairs.

15. Calculate the following quantities:

a) $(1 + i)^{25}$; b) $\left(\frac{1+\sqrt{3}i}{1-i}\right)^{30}$; c) $\left(1 - \frac{\sqrt{3}-i}{2}\right)^{24}$;

d) $\frac{(-1+\sqrt{3}i)^{15}}{(1-i)^{20}} + \frac{(-1-\sqrt{3}i)^{15}}{(1+i)^{20}}$.

16. Use De Moivre's theorem to express $\cos nx$ and $\sin nx$ in terms of powers of $\cos x$ and $\sin x$.

17. Express $\tan 6x$ in terms of $\tan x$.

18. Write $\sqrt{1+i}$ in trigonometric form.

19. Find x and y if $x + yi = \sqrt{a + bi}$.

20. Find all the values of the following roots:

a) $\sqrt[3]{1}$; b) $\sqrt[3]{i}$; c) $\sqrt[4]{-1}$; d) $\sqrt[6]{-8}$; e) $\sqrt[8]{1}$; f) $\sqrt{3+4i}$;

g) $\sqrt[3]{-2+2i}$; h) $\sqrt[5]{-4+3i}$; i) $\sqrt[6]{\frac{1-i}{\sqrt{3}+i}}$; j) $\sqrt[8]{\frac{1+i}{\sqrt{3}-i}}$.

21. Prove that the sum of all the distinct nth roots of unity is zero. What geometric fact does this express?

22. Let ϵ be any nth root of unity other than 1. Prove that

$$1 + 2\epsilon + 3\epsilon^2 + \cdots + n\epsilon^{n-1} = \frac{n}{\epsilon - 1}.$$

23. Prove that every complex number $\alpha \neq -1$ of unit modulus can be represented in the form

$$\alpha = \frac{1+it}{1-it},$$

where t is a real number.

24. Prove that

$$|z - 1| \leq ||z| - 1| + |z||\arg z|$$

by a purely geometric argument.

25. Prove that the equation of any circle or straight line in the z-plane can be written in the form

$$Az\bar{z} + \bar{E}z + E\bar{z} + D = 0 \qquad (A, D \text{ real}),$$

where we have a circle if $A \neq 0$, $E\bar{E} - AD > 0$ or a straight line if $A = 0$, $E \neq 0$.

CHAPTER TWO

Limits in the Complex Plane

2.1. The Principle of Nested Rectangles

2.11. Having introduced complex numbers and defined operations on them, we now consider a basic operation of complex analysis, namely passage to the limit in the complex plane. We begin by proving the following familiar proposition of real analysis:

THEOREM ***(Principle of nested intervals).*** *Let $i_1, i_2, \ldots, i_n, \ldots$ be a sequence of closed intervals on the real line such that*

1) *The intervals are "nested," i.e., i_n contains i_{n+1} for every n;*
2) *The length of i_n approaches zero as $n \to \infty$.*

Then there exists one and only one point a belonging to all the intervals.

Proof. Let $i_n = [a_n, b_n]$, and let E be the set of all numbers $a_1, a_2, \ldots, a_n, \ldots$ Since the intervals are nested, given any positive integer k, we have $a_n \leq b_k$ for all $n = 1, 2, \ldots$ ("E is bounded from above by every b_k"). Let a be the least upper bound of E, so that in particular $a_n \leq a$ for all $n = 1, 2, \ldots$† Then a must belong to all the intervals i_n, since otherwise a would exceed some number b_k, which is impossible. The uniqueness of a is trivial, since if there are two points a and a' belonging to all the intervals (see Figure

†The existence of a follows from the completeness of the real number system ("every set of real numbers bounded from above has a least upper bound"). See e.g., R. A. Silverman, *Modern Calculus and Analytic Geometry*, The Macmillan Company, New York (1969), Theorem 2.11.

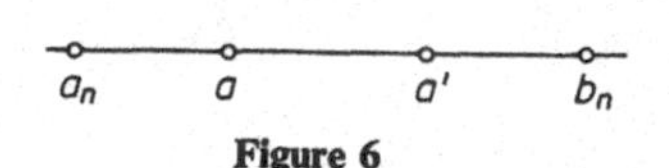

Figure 6

6), the length of every i_n can be no less than the distance between a and a', contrary to condition 2). ▮†

2.12. The above theorem is easily generalized to the case of complex numbers:

THEOREM ***(Principle of nested rectangles).*** *Let $r_1, r_2, \ldots, r_n, \ldots$ be a sequence of rectangles in the complex plane with sides parallel to the coordinate axes such that*

1) *The rectangles are "nested," i.e., r_n contains r_{n+1} for every n;*
2) *The length of the diagonal of r_n approaches zero as $n \longrightarrow \infty$.*

Then there exists one and only one point α belonging to all the rectangles.

Proof. Consider the two sequences of nested closed intervals $i_1, i_2, \ldots, i_n, \ldots$ and $j_1, j_2, \ldots, j_n, \ldots$ obtained by projecting the rectangles $r_1, r_2, \ldots, r_n, \ldots$ onto the real and imaginary axes, respectively (see Figure 7). Since the length of the diagonal r_n approaches zero as $n \longrightarrow \infty$, the same is true of the lengths of the intervals i_n and j_n. By Theorem 2.11,‡ there exists a unique point a on the real axis belonging to all the intervals i_n and a unique point b on the imaginary axis belonging to all the intervals j_n. But then the point $\alpha = a + bi$ obviously belongs to all the rectangles r_n. The uniqueness of α is trivial, since if there are two points α and α' belonging to all the rectangles, the length of the diagonal of every r_n can be no less than the distance between α and α', contrary to condition 2). ▮

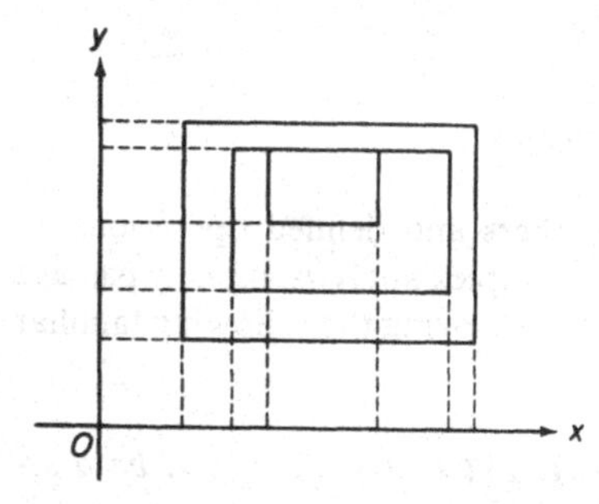

Figure 7

2.2. Limit Points

2.21. Definition. A complex number α is said to be a *limit point* of the infinite sequence of complex numbers

$$z_1, z_2, \ldots, z_n, \ldots \tag{1}$$

---

†The symbol ▮ means Q.E.D. and indicates the end of a proof.

‡Theorem 2.11 refers to the (unique) theorem in Sec. 2.11, Example 2.23b to the example in Sec. 2.23b, etc.

if given any $\epsilon > 0$ (however small), the inequality $|z_n - \alpha| < \epsilon$ holds for infinitely many values of n.†

2.22. By a *neighborhood* of a point α in the complex plane is meant any (circular) disk $|z - \alpha| < \epsilon$ of radius ϵ centered at α. Thus, thinking of the numbers (1) as points in the complex plane, we see that α is a limit point of the sequence (1) if and only if every neighborhood of α contains infinitely many terms of (1).

2.23. Examples

a. The same *point* in the complex plane may well correspond to several or even infinitely many distinct *terms* of the sequence (1). Thus the sequence

$$1, 0, 3, 0, 5, 0, 7, \ldots$$

has the unique limit point 0.

b. The sequence

$$1, 2, 3, \ldots, n, \ldots$$

has no limit points at all.

c. The sequence

$$1, \tfrac{1}{2}, \tfrac{1}{3}, \tfrac{2}{3}, \tfrac{1}{4}, \tfrac{3}{4}, \tfrac{1}{5}, \tfrac{4}{5}, \ldots$$

has two limit points 1 and 0, where the first belongs to the sequence and the second does not.

2.24. Definition. A complex sequence z_n is said to be *bounded* if every term of the sequence is of modulus less than some positive number M, i.e., if $|z_n| < M$ for every n‡; otherwise the sequence is said to be *unbounded*. Geometrically this means that every term of a bounded sequence lies inside some circle of sufficiently large radius M centered at the origin (see Figure 8).

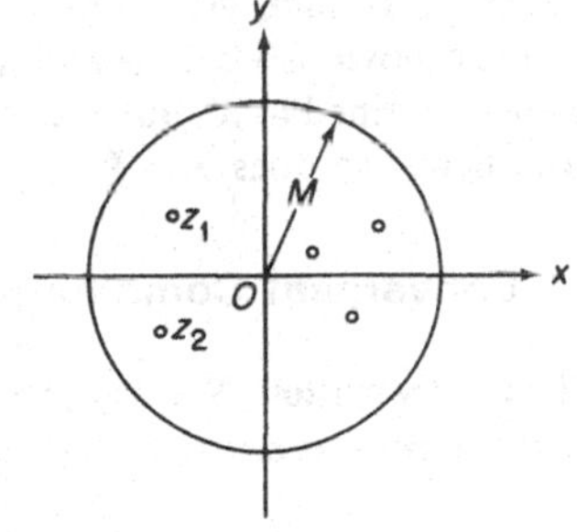

Figure 8

†More concisely, we refer to (1) as a *(complex) sequence*, it being tacitly understood that the sequence is infinite. In the language of Sec. 3.11, such a sequence is just a complex function whose domain is the set of all positive integers $n = 1, 2, \ldots$ We also talk about "the sequence z_n," meaning the sequence (1) with "general term" z_n.

‡We can obviously write $|z_n| \leq M$ instead of $|z_n| < M$ without changing the meaning of boundedness.

2.25. Example 2.23b shows that an unbounded complex sequence need not have a limit point. This cannot happen if the sequence is bounded, as shown by the following

THEOREM ***(Bolzano–Weierstrass).*** *Every bounded complex sequence z_n has at least one limit point.*

Proof. Every term of the sequence z_n lies inside some rectangle r_1 with sides parallel to the coordinate axes (why?). Dividing the sides of r_1 in half and joining midpoints of opposite sides, we partition r_1 into four congruent subrectangles (see Figure 9). Among these four rectangles there is at least one, call it r_2, containing infinitely many terms of the sequence z_n, since otherwise the original rectangle r_1 would contain only a finite number of terms of the sequence, contrary to hypothesis. Next we divide r_2 into four more subrectangles in the same way. At least one of these latter rectangles, call it r_3, must again contain infinitely many terms of the sequence. Continuing this process indefinitely, we get an infinite sequence of rectangles $r_1, r_2, r_3, \ldots, r_n, \ldots$ satisfying the conditions of Theorem 2.12 such that infinitely many terms of the given complex sequence z_n belong to each r_n.† It follows from Theorem 2.12 that there exists a (unique) point α belonging to all the rectangles r_n. Clearly α is a limit point of the sequence z_n. In fact, given any $\epsilon > 0$, let K be the disk of radius ϵ centered at α. Then obviously all the rectangles r_n starting from sufficiently large value of n are contained in K. But r_n contains infinitely many terms of the sequence z_n, and hence so does K. ▮

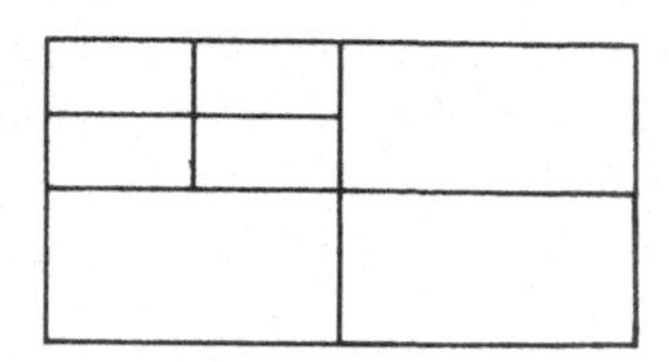

Figure 9

2.3. Convergent Complex Sequences

2.31. Definition. We say that a complex sequence z_n is *convergent with limit* α (or *converges to the limit* α) and write

$$\lim_{n\to\infty} z_n = \alpha \tag{2}$$

or $z_n \to \alpha$ as $n \to \infty$ if given any $\epsilon > 0$ (however small), there exists an integer $N = N(\epsilon) > 0$ such that $|z_n - \alpha| < \epsilon$ for all $n > N$. Geometrically (2) means that every neighborhood of α contains all but a finite number of terms of z_n.

†Again note that some (perhaps infinitely many) terms of the sequence z_n may correspond to the same point in the complex plane.

2.32. THEOREM. *Given two complex sequences z_n and z'_n, suppose*

$$\lim_{n\to\infty} z_n = \alpha, \qquad \lim_{n\to\infty} z'_n = \alpha'.$$

Then

$$\lim_{n\to\infty} (z_n \pm z'_n) = \alpha \pm \alpha',$$

$$\lim_{n\to\infty} z_n z'_n = \alpha\alpha',$$

$$\lim_{n\to\infty} \frac{z_n}{z'_n} = \frac{\alpha}{\alpha'},$$

provided that $\alpha' \neq 0$ in the last formula.

Proof. The exact analogues of the corresponding proofs for real sequences (give the details). ∎

2.33. The merit of the following important test for convergence is that it involves only the terms of the sequence z_n and not the proposed limit of z_n:

THEOREM ***(Cauchy convergence criterion).*** *The complex sequence z_n is convergent if and only if given any $\epsilon > 0$, there exists an integer $N = N(\epsilon) > 0$ such that*

$$|z_m - z_n| < \epsilon$$

for all $m, n > N$.†

Proof. If $z_n \to \alpha$ as $n \to \infty$, then given any $\epsilon > 0$, there exists a positive integer N such that $m, n > N$ implies

$$|z_m - \alpha| < \frac{\epsilon}{2}, \qquad |z_n - \alpha| < \frac{\epsilon}{2}.$$

Therefore

$$|z_m - z_n| = |(z_m - \alpha) + (\alpha - z_n)| \leq |z_m - \alpha| + |z_n - \alpha| < \frac{\epsilon}{2} + \frac{\epsilon}{2} = \epsilon$$

for all $m, n > N$.

Conversely, suppose the sequence z_n satisfies the Cauchy convergence criterion. Then choosing $\epsilon = 1$ (say) and $m_0 > N(1)$, we find that all but finitely many terms of z_n lie in the neighborhood $|z - z_{m_0}| < 1$. Therefore the sequence z_n is bounded, and it follows from the Bolzano–Weierstrass theorem that z_n has a limit point α. Now let $N = N(\epsilon/2)$ and choose $m_1 > N$ such that

$$|z_{m_1} - \alpha| < \frac{\epsilon}{2}$$

(such an integer m_1 exists, since every neighborhood of α must contain infi-

---

†I.e., whenever m and n both exceed N.

nitely many points of the sequence z_n). But then

$$|z_n - \alpha| = |(z_n - z_{m_1}) + (z_{m_1} - \alpha)| \leq |z_n - z_{m_1}| + |z_{m_1} - \alpha|$$
$$< \frac{\epsilon}{2} + \frac{\epsilon}{2} = \epsilon$$

for all $n > N$, i.e., $z_n \to \alpha$ as $n \to \infty$. ∎

2.4. The Riemann Sphere and the Extended Complex Plane

2.41. We now show a way of representing complex numbers by points on a *sphere*. To this end, we draw a sphere Σ tangent to the complex plane Π at the origin O (see Figure 10). The diameter of Σ passing through O is perpendicular to Π and intersects Σ in a second point N, called the (*north*) *pole* for an obvious reason.† Let z be an arbitrary complex number, represented by a point P in the plane Π, and let PN be the straight line joining P to the pole N. Then PN intersects Σ in some point P' (distinct from N), which we regard as representing the given complex number z. Thus every complex number z is represented by a unique point of Σ. Conversely, every point P' of Σ (except N) corresponds to a unique complex number z, namely the number represented by the point P in which the line NP' joining N and P' intersects the plane Π. In this way, we establish a one-to-one correspondence between the points of Σ (other than N) and the points of Π, i.e., between the sphere Σ (with N deleted) and the set of all complex numbers.

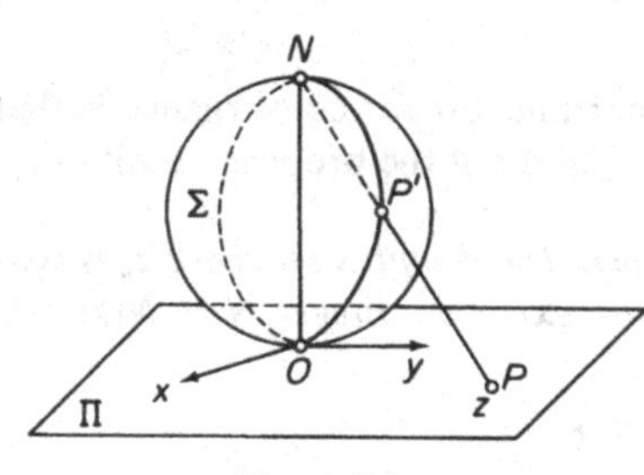

Figure 10

2.42. Definition. We say that a complex sequence z_n *approaches infinity* and write

$$\lim_{n\to\infty} z_n = \infty \tag{3}$$

or $z_n \to \infty$ as $n \to \infty$ if given any $M > 0$ (however large), there exists an integer $\nu = \nu(M) > 0$ such that $|z_n| > M$ for all $n > \nu$. Geometrically (3) means that all but a finite number of terms of z_n lie outside any (arbitrarily large) circle centered at the origin.

2.43. Let z_n be a complex sequence approaching infinity, and let P'_n be the corresponding sequence of points on the sphere Σ constructed above. Then clearly P'_n approaches the pole N (this is made more precise in Prob.

†The south pole of Σ is clearly the origin O itself.

16). Hence it is natural to regard N as corresponding to a unique "ideal point" of the complex plane, which we call the *point at infinity* and denote by ∞. The ordinary complex plane equipped with this extra point ∞ is called the *extended (complex) plane*. To emphasize that ∞ is not a point of the ordinary complex plane, we often refer to the latter as the *finite (complex) plane* and call its points (and the corresponding complex numbers) *finite*. The sphere Σ including the point N is called the *Riemann sphere*, and the mapping just described, which in particular carries ∞ into N and vice versa, is called *stereographic projection*. Thus we see that *stereographic projection establishes a one-to-one correspondence between the extended complex plane and the Riemann sphere*. The points P (or z) and P' in Figure 10 are called *images* of each other under stereographic projection.

2.44. Remark. The exterior E of any (arbitrarily large) circle centered at the origin is called a *neighborhood of infinity* if E is regarded as a set in the extended plane (containing ∞) and a *deleted neighborhood of infinity* if E is regarded as a set in the finite plane (not containing ∞). With this language (3) says that every neighborhood (or deleted neighborhood) of infinity contains all but a finite number of terms of the sequence z_n, in complete analogy with the case of finite limits (see Sec. 2.31).

COMMENTS

2.1. Theorem 2.11 and its complex analogue Theorem 2.12 are immediate consequences of the completeness of the real number system, and will be used to prove the Bolzano–Weierstrass theorem (Theorem 2.25). The latter will be used in turn to prove the key Cauchy convergence criterion (Theorem 2.33).

2.2. A limit point of an infinite set (Prob. 2) can be visualized crudely as a point surrounded by a cloud of infinitely many points of the set, and similarly for the limit point of a sequence except that the "cloud" now consists of points corresponding to infinitely many *terms* of the sequence and hence may consist of only a finite number of points (cf. Example 2.23a). According to the Bolzano–Weierstrass theorem, the terms of a complex sequence must cluster around at least one such limit point unless the sequence has terms of arbitrarily large modulus (in which case they can "escape to infinity"). Put somewhat differently, no finite part of the plane can accommodate infinitely many terms of a sequence without there being at least one cluster of terms.

2.3. We write $N = N(\epsilon)$ to emphasize the dependence of the positive integer N on the preassigned positive number ϵ. A limit is automatically a limit point, since it "attracts" all but finitely many terms of the given sequence

z_n and hence infinitely many terms of z_n. On the other hand, a limit point need not be a limit. In fact, a complex sequence z_n has a (finite) limit α if and only if z_n is bounded and has α as its *unique* limit point (Prob. 4). In particular, a sequence with several limit points cannot be convergent. Loosely speaking, the Cauchy convergence criterion says that the class of sequences whose terms get closer and closer *to some point* (as $n \to \infty$) is precisely the class of sequences whose terms get closer and closer *to each other*.

2.4. The considerations of Sec. 2.4 stem from the desire to put the point at infinity ∞ on the same footing as ordinary (finite) complex numbers. Once this has been done, ∞ is a perfectly acceptable candidate for a limit point or a limit. This leads to a somewhat different version of the Bolzano–Weierstrass theorem (Prob. 18), free of the boundedness required in Theorem 2.25. It can be shown† that stereographic projection carries every circle on the Riemann sphere into a circle (or straight line) in the extended complex plane, and vice versa.

PROBLEMS

1. Find all the limit points of the following sequences:

$$\text{a) } z_n = 1 + (-1)^n \frac{n}{n+1}; \quad \text{b) } z_n = 1 + i^n \frac{n}{n+1}.$$

2. A point α is said to be a *limit point* of a set E of points in the complex plane (as opposed to a sequence z_n) if every neighborhood of α contains infinitely many (distinct) points of E. Show that a sequence z_n can have a limit point α in the sense of Sec. 2.21 without α being a limit point of the set of points corresponding to the terms of z_n. Prove that α is a limit point of a set E if and only if there exists a sequence z_n of distinct points of E converging to α. What is the appropriate version of the Bolzano–Weierstrass theorem for limit points of sets?

3. Find the limit points of the set of all points z such that
   a) $z = \dfrac{1}{m} + \dfrac{i}{n} \quad (m, n = \pm 1, \pm 2, \ldots);$
   b) $z = \dfrac{p}{m} + i\dfrac{q}{n} \quad (m, n, p, q = \pm 1, \pm 2, \ldots);$
   c) $|z| < 1.$

4. Prove that $z_n \to \alpha$ as $n \to \infty$ if and only if
   a) $|z_n - \alpha| \to 0$ as $n \to \infty$;
   b) z_n is bounded and α is its unique limit point.

5. Prove that a complex sequence $z_n = x_n + iy_n$ converges to the limit $\alpha =$

†See e.g., A. I. Markushevich, *Theory of Functions of a Complex Variable*, in three volumes (translated by R. A. Silverman), Prentice-Hall, Inc., Englewood Cliffs, N.J. (1965, 1967), Volume I, Sec. 21.

$a + ib$ if and only if

$$\lim_{n\to\infty} x_n = a, \qquad \lim_{n\to\infty} y_n = b.$$

6. Prove that if $z_n \longrightarrow \alpha$ as $n \longrightarrow \infty$, then $|z_n| \longrightarrow |\alpha|$ as $n \longrightarrow \infty$. Show that the converse is not true.

7. Prove that if $z_n \longrightarrow \alpha \neq 0$ as $n \longrightarrow \infty$ and if θ is any value of arg α, then there is a sequence θ_n, where each θ_n is a value of arg z_n, such that $\theta_n \longrightarrow \theta$ as $n \longrightarrow \infty$ (ignore the finite number of terms z_n which may vanish).

Comment. This fact is indicated by writing

$$\lim_{n\to\infty} \arg z_n = \arg \alpha.$$

8. By the *principal value* of the argument of z, denoted by $\arg_P z$, we mean the (unique) value of $\arg z$ satisfying the inequality $-\pi < \arg z \leq \pi$. Give an example of a complex sequence z_n converging to a limit α such that $\arg_P z_n$ fails to converge. Show that this cannot happen unless $\alpha = 0$ or $\alpha = -1$.

9. Let z_n be a complex sequence such that $|z_n| \longrightarrow r$, $\arg_P z_n \longrightarrow \theta$ as $n \longrightarrow \infty$. Prove that z_n converges to the limit $\alpha = r(\cos\theta + i \sin\theta)$.

10. Prove that if α is a limit point of a sequence z_n, then z_n has a subsequence z_{k_n} converging to α.

11. Prove that the sequence

$$z_n = \frac{q^n}{1 + q^{2n}}$$

converges to the same limit 0 for both $|q| < 1$ and $|q| > 1$.

12. Find the limit, if any, of each of the following sequences:

a) $z_n = \dfrac{2^n}{n!} + \dfrac{i^n}{2^n}$; b) $z_n = \sqrt[n]{n} + inq^n \quad (|q| < 1)$;

c) $z_n = \sqrt[n]{a} + i \sin \dfrac{1}{n} \quad (a > 0)$.

13. Prove that if $z_n \longrightarrow \alpha$ as $n \longrightarrow \infty$, then

$$\lim_{n\to\infty} \frac{z_1 + z_2 + \cdots + z_n}{n} = \alpha.$$

14. Prove the following slight modification of the Cauchy convergence criterion: The complex sequence z_n is convergent if and only if given any $\epsilon > 0$, there exists an integer $N = N(\epsilon) > 0$ such that $|z_{n+p} - z_n| < \epsilon$ for all $n > N$ and all $p = 1, 2, \ldots$

15. Prove that a complex sequence z_n approaches infinity if and only if z_n is an unbounded sequence with no finite limit points.

16. By a *neighborhood of the pole N* on the Riemann sphere Σ we mean the image under stereographic projection of a neighborhood of infinity in the extended complex plane. Describe such neighborhoods geometrically. Given any sequence of points z_n in the finite plane, let P'_n be the corresponding sequence of points on Σ. Prove that $z_n \longrightarrow \infty$ as $n \longrightarrow \infty$ if and only if $P'_n \longrightarrow N$ as $n \longrightarrow \infty$, i.e., if and only if every neighborhood of N contains all but a finite number of terms of the sequence P'_n.

17. Give an example of an unbounded sequence which does not approach infinity.

18. Prove that every infinite set, bounded or not, has at least one limit point in the extended complex plane.

19. Suppose $z_n \longrightarrow \infty$ as $n \longrightarrow \infty$. What does this imply about Re z_n, Im z_n, $|z_n|$ and arg z_n?

20. What sets in the extended plane are the images under stereographic projection of the four hemispheres (northern, southern, eastern and western) of the Riemann sphere?

21. What curve on the Riemann sphere is the image under stereographic projection of a straight line in the extended plane?

22. Does Prob. 13 remain true if $\alpha = \infty$?

CHAPTER THREE

Complex Functions

3.1. Basic Concepts

3.11. Variables and functions. Let E be any set of complex numbers, which we can visualize as a set of points in the complex plane or on the Riemann sphere (in the latter case E may well contain the north pole of the sphere, corresponding to the point at infinity). Then by a *complex variable* we mean a complex number $z = x + iy$ which can take any value in E, and by a *(complex) function* w of the complex variable z we mean a rule assigning a uniquely defined complex number w to every z in E. We then write $w = f(z)$ and call E the *domain (of definition)* of the function.† The set E' of all values of w obtained as z varies over the set E is called the *range* of the function $w = f(z)$, and can again be visualized as a set of points in the complex plane or on the Riemann sphere. Thus w is itself a complex variable, called the *dependent* variable to distinguish it from the *independent* variable z. Note that specifying the function $w = f(z)$ is equivalent to establishing a *mapping* of E onto E', i.e., a correspondence between the sets E and E' such that each point z of E goes into a uniquely defined point w of E' (w is often called the *image* of z under the mapping).

3.12. Remark. If $w = u + iv$ is a function of $z = x + iy$, then u and v are both real functions of the two real variables x and y. From this general point of view, specifying a complex function of a complex variable is equivalent to specifying two real functions of two real variables.

†By the same token, $f(z)$ is said to be *defined in E* (or in any subset of E).

3.13. Single-valued vs. multiple-valued functions. As just defined, the function $w = f(z)$ is *single-valued* in the sense that it assigns one and only one value of w to each value of z in the domain of definition E. We will sometimes find it convenient to enlarge the definition of a function, allowing several (or even infinitely many!) values of w to be assigned to some (or all) values of z. Such functions are said to be *multiple-valued*. However, the term "function" without further qualification will always be understood to mean "single-valued function."

3.14. Inverse functions. The mapping of E onto E' established by a (single-valued) function $w = f(z)$ can always be "read backwards" to give a new function $z = \varphi(w)$, *in general multiple-valued*, carrying each point w in E' into all those points z in E with w as their image under the original mapping $w = f(z)$. The function $z = \varphi(w)$ is called the *inverse* of the function $w = f(z)$. Note that $z = \varphi(w)$ is single-valued if and only if $f(z_1) \neq f(z_2)$ whenever $z_1 \neq z_2$, i.e., if and only if $w = f(z)$ "carries distinct points of E into distinct points of E'." The original function $w = f(z)$ and the corresponding mapping of E onto E' are then said to be *one-to-one*.

3.15. Example. The function

$$w = |z|$$

is single-valued, but its inverse is multiple-valued, carrying the point $w = c$ $(c > 0)$ into the infinitely many points of the circle $|z| = c$.

3.2. Curves and Domains

3.21. a. Let $x(t)$ and $y(t)$ be two continuous real functions of a real variable t taking all values in a closed interval $a \leq t \leq b$. Then the parametric equations

$$x = x(t), \qquad y = y(t) \qquad (a \leq t \leq b) \tag{1}$$

determine a (*continuous*) *curve* C, made up of all points $(x(t), y(t))$ with $a \leq t \leq b$. Setting $z = x + iy$, $z(t) = x(t) + iy(t)$, we can write (1) as a single complex parametric equation

$$z = z(t) \qquad (a \leq t \leq b). \tag{1'}$$

As the parameter t varies from a to b, the point $z = z(t)$ traces out C, starting at the *initial point* $z(a)$ and ending at the *final point* $z(b)$. In this way (1′) equips C with a natural direction of traversal, called the *positive direction* of C. The curve C is said to be *closed* if its initial and final points coincide, i.e., if $z(a) = z(b)$; otherwise C is often called an *arc* (to emphasize that it is not closed).

b. A set E of points in the complex plane is said to be (*arcwise*) *connected* if every pair of points z_1, z_2 in E can be joined by a curve consisting entirely of points of E (with z_1 as its initial point and z_2 as its final point). Thus the state of Kansas (regarded as a point set) is connected, but not the state of Hawaii.

3.22. a. A point z is said to be an *interior point* of a given set E in the complex plane if E contains some neighborhood of z, i.e., if E contains z and every point sufficiently near z. For example, let E be the set of points between two concentric circles excluding the circles themselves. Then every point of E is an interior point (of E). This is no longer true if we let E include points on one or both of the circles. A set E is said to be *open* if it consists entirely of interior points.

b. A set G in the complex plane is said to be a *domain* if it is open and connected.† For example, the set of points between two concentric circles is a domain provided it contains no points of the circles themselves.

c. In keeping with Sec. 2.24, a domain G (or more generally any set E in the complex plane) is said to be *bounded* if its points all lie inside some circle of sufficiently large radius centered at the origin; otherwise G is said to be *unbounded*.

3.23. a. Given any domain G, let G^c be the *complement* of G, i.e., the set of all points of the plane which do not belong to G. If z belongs to G^c, then either some neighborhood of z is entirely contained in G^c or else every neighborhood of z contains both points of G^c and points of G. In the first case we call z an *exterior point* of G, while in the second case we call z a *boundary point* of G. The set of all boundary points of G is called the *boundary* of G. Except for the whole extended complex plane, every domain has a boundary. However, there are domains with no exterior points, e.g., the set of all points of the plane which do not belong to the interval $[-1, 1]$ of the real axis.

b. The set consisting of a given domain G and its boundary is denoted by $\bar{G}$. Such a set is called a *closed domain*. More generally, let $\tilde{G}$ be any set consisting of a domain G and some (possibly all or none) of its boundary points. Then $\tilde{G}$ is called a *region*. Any open domain G or closed domain $\bar{G}$ is a region, but there are regions like the set of all z such that $|z| < 1$ or $z = 1$ (the domain $|z| < 1$ plus the single boundary point $z = 1$) which are

†A domain is often denoted by the letter G (from the German *Gebiet*), just as a general set is often denoted by the letter E (from the French *ensemble*). Do not confuse the present notion of a "domain" with the previous notion of "domain of definition."

neither open domains nor closed domains. Boundary points of $\tilde{G}$ are defined in just the same way as boundary points of G, except that a boundary point of $\tilde{G}$ can now belong to $\tilde{G}$. Thus the boundary of $\tilde{G}$ coincides with that of G.

3.24. a. Given a curve C with equation (1′), suppose distinct values of t in the half-open interval $a \leq t < b$ correspond to distinct points of C. Then C is said to be a *Jordan curve*.† It can be shown that every closed Jordan curve C divides the plane into two distinct domains with C as their common boundary, where one of the domains, called the *interior* of C, is bounded, and the other domain, called the *exterior* of C, is unbounded.‡ There is no loss of generality (why not?) in assuming that the positive direction on C is such that an observer moving along C in the positive direction finds the interior of C on his *left* (this corresponds to traversing C in the counterclockwise direction).

b. If I is the interior of a closed Jordan curve, then the interior of every (other) closed Jordan curve C contained in I is itself contained in I (sketch a figure). An arbitrary domain G is said to be *simply connected* if it has this property, and *multiply connected* otherwise. For example, if G is the exterior of a triangle, then G is multiply connected, since the interior of any closed Jordan curve surrounding the triangle is not entirely contained in G. Similarly, the ring-shaped domain or *annulus*

$$r < |z| < R \tag{2}$$

is multiply connected, since it does not contain the interior of any circle

$$|z| = \rho \qquad (r < \rho < R).$$

c. The above considerations pertain to the finite plane. In the case of the extended plane we modify the definition of a simply connected domain as follows: A domain G in the extended plane is said to be simply connected if given any closed Jordan curve C contained in G, either the interior of C or the exterior of C (including the point at infinity) is itself contained in G. With this definition, the exterior of a triangle is simply connected if it is regarded as containing the point at infinity, and multiply connected otherwise.

d. Let $C_0, C_1, \ldots, C_n$ be $n + 1$ closed Jordan curves such that the curves $C_1, \ldots, C_n$ all lie inside C_0 and do not intersect each other (cf. Figure

†It is to allow for the possibility of C being closed that we resort to a *half-open* interval $a \leq t < b$ in the definition of a Jordan curve.

‡For the proof of this result, known as the *Jordan curve theorem*, see e.g., P. S. Aleksandrov, *Combinatorial Topology, Vol. 1* (translated by H. Komm), Graylock Press, Rochester, N.Y. (1956), Chap. 2. In common parlance, a point is said to lie *inside* a closed Jordan curve C if it belongs to the interior of C and *outside* C if it belongs to the exterior of C.

11). Then the set of points lying inside the curve C_0 and outside the other n curves $C_1, \ldots, C_n$ is a domain G_n (why?). If $n = 0$, so that there are no "inner curves" at all, the domain $G_n = G_0$ is simply connected, being the interior of a closed Jordan curve. If $n > 0$, there are obviously closed Jordan curves contained in G_n (which ones?) whose interiors are not contained in G_n. Therefore G_n is multiply connected. More exactly, the domain G_n is said to be $(n + 1)$-*connected*, since its boundary consists of $n + 1$ disjoint (i.e., nonintersecting) "components," namely the curves $C_0, C_1, \ldots, C_n$. Thus the annulus is doubly connected (its boundary consists of the two circles $|z| = r$ and $|z| = R$), while the shaded domain shown in Figure 11 is triply connected.

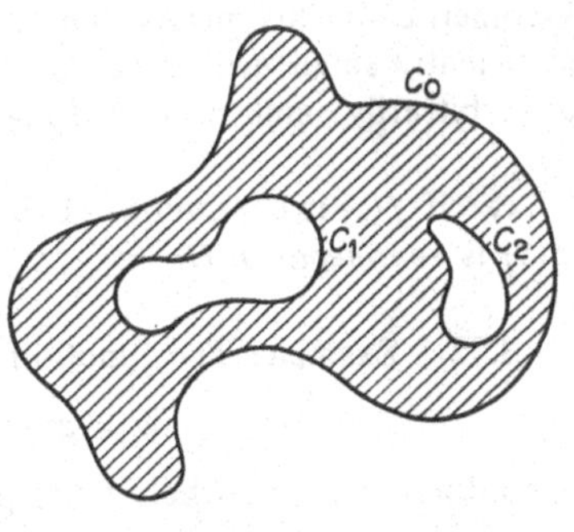

Figure 11

3.3. Continuity of a Complex Function

3.31. a. We say that a complex function $f(z)$ defined in a domain G *approaches the limit A as z approaches a point* z_0 in G and we write

$$\lim_{z \to z_0} f(z) = A$$

or $f(z) \longrightarrow A$ as $z \longrightarrow z_0$ if given any $\epsilon > 0$, there exists a number $\delta = \delta(\epsilon) > 0$ such that

$$|f(z) - A| < \epsilon$$

for all z such that

$$0 < |z - z_0| < \delta.$$

Suppose that in addition $A = f(z_0)$, so that

$$\lim_{z \to z_0} f(z) = f(z_0). \tag{3}$$

Then we say that $f(z)$ is *continuous at* z_0.

b. Thus $f(z)$ is said to be continuous at z_0 if given any $\epsilon > 0$, there exists a number $\delta = \delta(\epsilon) > 0$ such that

$$|f(z) - f(z_0)| < \epsilon$$

for all z such that†

$$|z - z_0| < \delta. \tag{4}$$

†The restriction $0 < |z - z_0|$ or equivalently $z \neq z_0$ is now unnecessary (why?). The points z satisfying (4) can always be assumed to lie in G (if not, pick a smaller δ).

Geometrically this means that the values of $w = f(z)$ at every point of a sufficiently small neighborhood $|z - z_0| < \delta$ of the point z_0 all lie inside an arbitrarily small neighborhood $|w - w_0| < \epsilon$ of the point $w_0 = f(z_0)$.

c. If $f(z)$ is continuous at *every* point z_0 of the domain G, we say that $f(z)$ is *continuous in G*.

3.32. Example. The function

$$w = z^n \qquad (n = 2, 3, \ldots)$$

is continuous in the whole finite plane. In fact, setting $w_0 = z_0^n$ where z_0 is any finite point, we have

$$w - w_0 = z^n - z_0^n = (z - z_0)(z^{n-1} + z^{n-2}z_0 + \cdots + z_0^{n-1}),$$

so that

$$\begin{aligned} |w - w_0| &= |z - z_0||z^{n-1} + z^{n-2}z_0 + \cdots + z_0^{n-1}| \\ &\leq |z - z_0|(r^{n-1} + r^{n-2}r_0 + \cdots + r_0^{n-1}) \end{aligned}$$

(cf. Chap. 1, Prob. 7), where $r = |z|$, $r_0 = |z_0|$. But $|z - z_0| < \delta$ implies

$$r = |z| = |z_0 + (z - z_0)| \leq |z_0| + |z - z_0| < r_0 + \delta$$

and hence

$$|w - w_0| < |z - z_0|[(r_0 + \delta)^{n-1} + (r_0 + \delta)^{n-2}r_0 + \cdots + r_0^{n-1}] < n\delta(r_0 + \delta)^{n-1}.$$

It follows that $|w - w_0|$ can be made less than any preassigned positive number ϵ by suitably choosing δ. Therefore the function $w = z^n$ is continuous at every point $z_0 \neq \infty$.

3.33. Since the definition of continuity for complex functions is formally the same as for real functions, the proofs of the familiar theorems concerning algebraic operations on continuous functions† carry over from the real case to the complex case. Thus if the functions $f(z)$ and $g(z)$ are both continuous at a point z_0, then so are the functions $f(z) \pm g(z)$, $f(z)g(z)$ and $f(z)/g(z)$, provided that $g(z_0) \neq 0$ in the case of the quotient. Moreover, if the function $f(z)$ is continuous at the point z_0 and if the function $\varphi(w)$ is continuous at the point $w_0 = f(z_0)$, then the "composite function" $\varphi(f(z))$ is continuous at the point z_0.

3.34. Remark. Suppose $f(z)$ is defined in a region $\tilde{G}$ containing boundary points of G (in particular, $\tilde{G}$ might be the closed domain $\bar{G}$). Then continuity of $f(z)$ at a boundary point z_0 of $\tilde{G}$ means the same thing as in Sec.

†R.A. Silverman, *op. cit.*, Theorems 4.18–4.20.

3.31b except that (4) is now replaced by†

$$|z - z_0| < \delta, \qquad z \in \tilde{G};$$

i.e., the points z figuring in (4) are required to stay inside $\tilde{G}$ (this is no longer guaranteed merely by the smallness of δ). Correspondingly, we write

$$\lim_{\substack{z \to z_0 \\ z \in \tilde{G}}} f(z) = f(z_0)$$

instead of (3). If $f(z)$ is continuous at every point z_0 of the region $\tilde{G}$, we say that $f(z)$ is *continuous in* $\tilde{G}$ (cf. Sec. 3.31c). In just the same way, we can talk about continuity at a point z_0 of a curve C (in the domain of definition of the given function), this time replacing (3) and (4) by

$$\lim_{\substack{z \to z_0 \\ z \in C}} f(z) = f(z_0)$$

and

$$|z - z_0| < \delta, \qquad z \in C.$$

If $f(z)$ is continuous at every point z_0 of the curve C, we say that $f(z)$ is *continuous on* C.

3.4. Uniform Continuity

3.41. Definition. A complex function $f(z)$ defined in a domain G is said to be *uniformly continuous in* G if given any $\epsilon > 0$, there exists a number $\delta = \delta(\epsilon) > 0$ such that

$$|f(z') - f(z'')| < \epsilon$$

for *all* points z', z'' in G such that

$$|z' - z''| < \delta. \tag{5}$$

The same definition applies to any region $\tilde{G}$ or curve C (with G replaced by $\tilde{G}$ or C everywhere).

3.42. Remark. Unlike the case of ordinary continuity, it is meaningless to talk about uniform continuity at a single point z_0. The key observation here is that the number δ in (5) must be *independent* of the points z', z'' in G. If $f(z)$ is uniformly continuous in a domain G, then $f(z)$ is obviously continuous in G, but a function can be continuous in G without being uniformly continuous in G (see Prob. 13).

3.43. In the next section we will show that a function continuous in a *bounded closed* domain $\bar{G}$ is automatically *uniformly* continuous in $\bar{G}$. The proof will depend on the following proposition, of considerable interest in its own right:

---

†As usual, the symbol $\in$ means "is an element of" or "belongs to."

THEOREM **(Heine–Borel).** *Given a bounded closed domain $\bar{G}$, suppose every point of $\bar{G}$ is the center of a disk K_z. Then $\bar{G}$ can be "covered" by a finite number of disks K_z. More exactly, there exists a finite number of points $z_1, \ldots, z_n$ in $\bar{G}$ such that every point of $\bar{G}$ belongs to at least one of the disks $K_{z_1}, \ldots, K_{z_n}$.*

Proof. Being bounded, the domain $\bar{G}$ lies inside some rectangle r_1 with sides parallel to the coordinate axes. Suppose $\bar{G}$ cannot be covered by a finite number of disks K_z. Then, dividing r_1 into four congruent subrectangles (just as in the proof of Theorem 2.25), we find that the part of $\bar{G}$ lying in at least one of these rectangles, call it r_2, cannot be covered by a finite number of disks K_z. Next, dividing r_2 itself into four more congruent subrectangles, we again find that the part of $\bar{G}$ lying in at least one of these new rectangles, call it r_3, cannot be covered by finitely many K_z. Continuing this process indefinitely, we get an infinite sequence of rectangles $r_1, r_2, r_3, \ldots, r_n, \ldots$ satisfying the conditions of Theorem 2.12 (the principle of nested rectangles) such that the part of $\bar{G}$ in each rectangle can only be covered by infinitely many K_z. It follows from Theorem 2.12 that there is a (unique) point α belonging to all the rectangles r_n. But any given neighborhood of α contains every rectangle r_n with sufficiently large n, and hence contains points of $\bar{G}$. Therefore α belongs to $\bar{G}$ (why?) and is consequently the center of some disk K_α. Let K_α be of radius ρ, and choose n so large that the diagonal of r_n is of length less than ρ. Then all the points of $\bar{G}$ in r_n are covered by the single disk K_α, contrary to the assumption that infinitely many disks are needed to cover these points. This contradiction shows that the original domain $\bar{G}$ can in fact be covered by finitely many K_z. ∎

3.44. THEOREM. *If $f(z)$ is continuous in a bounded closed domain $\bar{G}$, then $f(z)$ is uniformly continuous in $\bar{G}$.*

Proof. Let $f(z)$ be continuous in $\bar{G}$, i.e., at every point of $\bar{G}$. Then given any $z \in \bar{G}$ and $\epsilon > 0$, there exists a disk K_z^* of radius ρ_z centered at z such that $|f(z') - f(z)| < \epsilon/2$ for all z' belonging to K_z^* and $\bar{G}$. But then

$$|f(z') - f(z'')| \leq |f(z') - f(z)| + |f(z) - f(z'')| < \epsilon$$

for all z', z'' belonging to K_z^* and $\bar{G}$. Now replace every K_z^* by the smaller disk K_z of radius $\frac{1}{2}\rho_z$ centered at the same point z. By the Heine–Borel theorem, $\bar{G}$ can be covered by a finite number of these new "half-size" disks, say $K_{z_1}, \ldots, K_{z_n}$. Let δ be the smallest radius of the n disks $K_{z_1}, \ldots, K_{z_n}$. Then $|f(z') - f(z'')| < \epsilon$ for all z', z'' in $\bar{G}$ such that $|z' - z''| < \delta$, thereby proving the uniform continuity of $f(z)$ in $\bar{G}$. In fact, let z', z'' be any two points of $\bar{G}$ such that $|z' - z''| < \delta$. Then, as just shown, z' lies inside some disk K_{z_ν} of radius $\frac{1}{2}\rho_{z_\nu}$ centered at one of the points $z_1, \ldots, z_n$. But $\delta \leq \frac{1}{2}\rho_{z_\nu}$, and hence both z' and z'' lie inside the disk K_z^* of radius ρ_{z_ν} centered at z_ν, so that $|f(z') - f(z'')| < \epsilon$ as asserted. ∎

COMMENTS

3.1. *"Into" vs. "onto" mappings.* Let $f(z)$ be a single-valued function defined in E, with range E', and let E^* be any set containing E', possibly E' itself. Then $f(z)$ is said to map E *into* E^*. If $E^* = E'$ and we want to emphasize this fact, we say that $f(z)$ maps E *onto* E^*, as in Sec. 3.11. Thus every "onto" mapping is an "into" mapping, but not conversely.

3.2. A domain is often called an "open domain" to emphasize that it is an open set. The term "closed domain," although perfectly standard, is somewhat of a misnomer, since a closed domain is not an open set and hence not a domain at all! The term "domain" without further qualification will be understood to mean any (open) domain in the finite plane, whether bounded or unbounded, simply connected or multiply connected.

3.3. The important cases of infinite limits and limits at infinity are considered in Probs. 3 and 4. Once it is known that the product of two continuous functions is itself continuous (Sec. 3.33), Example 3.32 follows at once (by induction) from the obvious continuity of the function $w = z$ at every finite point. In talking about properties valid at every point of a curve, the preposition "on" is preferable to "in," as being more suggestive geometrically. Thus a function defined (continuous, etc.) at every point of a curve C is said to be defined (continuous, etc.) *on* C. Note that in talking about limits or continuity of a function $f(z)$ at a point z_0 it is tacitly assumed that $f(z)$ is defined at points $z \neq z_0$ arbitrarily near z_0. To insure this, we will always assume that the domain of definition of $f(z)$ is a region or a curve.

3.4. The seemingly esoteric Heine–Borel theorem is one of the most important tools of complex analysis, allowing us to replace "infinite coverings" of bounded closed sets (Prob. 11) by "finite subcoverings." The Heine–Borel theorem is used in the proofs of Theorems 3.44, 6.39 and 10.15, as well as in Sec. 5.42.

PROBLEMS

1. Let $\sigma_1, \sigma_2, \ldots, \sigma_n$ be a finite set of straight line segments in the complex plane with definite directions such that the final point of each segment σ_k $(k < n)$ coincides with the initial point of the next segment σ_{k+1}. Then the resulting curve is called a *polygonal curve*, with consecutive *vertices* $P_0, P_1, \ldots, P_n$, where P_0 is the initial point of σ_1 and P_k is the final point of σ_k $(k = 1, 2, \ldots, n)$. Write the parametric equation (1′) of such a curve.
2. Give an example of a domain G such that G and $\bar{G}$ have different boundaries.
3. We say that a function $f(z)$ defined in a domain G *approaches the limit infinity*

(∞) as z approaches a point z_0 in G, and we write

$$\lim_{z \to z_0} f(z) = \infty$$

or $f(z) \longrightarrow \infty$ as $z \longrightarrow z_0$ if given any $M > 0$, there exists a number $\delta = \delta(\epsilon) > 0$ such that $|f(z)| > M$ for all z such that $0 < |z - z_0| < \delta$. Prove that $f(z) \longrightarrow \infty$ as $z \longrightarrow z_0$ if and only if $\varphi(z) = 1/f(z) \longrightarrow 0$ as $z \longrightarrow z_0$.

4. We say that a function $f(z)$ defined in a deleted neighborhood of infinity (see Sec. 2.44) approaches the limit A as *z approaches infinity* (∞) and we write

$$\lim_{z \to \infty} f(z) = A$$

or $f(z) \longrightarrow A$ as $z \longrightarrow \infty$ if given any $\epsilon > 0$, there exists a number $M = M(\epsilon) > 0$ such that $|f(z) - A| < \epsilon$ for all z such that $|z| > M$. Prove that $f(z) \longrightarrow A$ as $z \longrightarrow \infty$ if and only if $\varphi(\zeta) = f(1/\zeta) \longrightarrow A$ as $\zeta \longrightarrow 0$. What is the precise meaning of

$$\lim_{z \to \infty} f(z) = \infty?$$

5. Prove that $f(z) \longrightarrow A$ as $z \longrightarrow z_0$ if and only if the sequence $f(z_n)$ approaches A for every sequence z_n approaching z_0.

6. Prove the following generalization of the Cauchy convergence criterion for sequences (Theorem 2.33): The function $f(z)$ approaches a limit as $z \longrightarrow z_0$ if and only if given any $\epsilon > 0$, there exists a number $\delta = \delta(\epsilon) > 0$ such that $|f(z') - f(z'')| < \epsilon$ whenever both $0 < |z' - z_0| < \delta$ and $0 < |z'' - z_0| < \delta$.

7. Let $f(z)$ be a rational function, i.e., a ratio

$$f(z) = \frac{a_0 + a_1 z + \cdots + a_m z^m}{b_0 + b_1 z + \cdots + b_n z^n} \qquad (a_m \neq 0,\ b_n \neq 0) \tag{6}$$

of two polynomials. Discuss the possible values of $\lim_{z \to \infty} f(z)$.

8. Where is the function (6) continuous?

9. "Every closed Jordan curve is the continuous one-to-one image of a circle." Explain this statement.

10. Prove that if $f(z)$ is continuous in a region $\bar{G}$, then so is $|f(z)|$.

11. A set E in the plane is said to be *bounded* if its points all lie inside some circle of sufficiently large radius centered at the origin, and *closed* if it contains all its limit points. (Note that a closed domain $\bar{G}$ as defined in Sec. 3.23b is also closed in this sense.) Prove that every continuous curve C is bounded and closed. Prove that the boundary of every domain G is closed. Show that the Heine–Borel theorem remains valid for any bounded closed set E, in particular for any continuous curve, if $\bar{G}$ is replaced by E.

12. Let E be a bounded closed domain or a continuous curve, and suppose $f(z)$ is continuous in E. Prove that
   a) $f(z)$ is *bounded in E*, i.e., there exists a number $M > 0$ such that $|f(z)| \leq M$ for all $z \in E$;

b) The *image of E under* $f(z)$, i.e., the set $\mathcal{E}$ of all points $w = f(z)$ with $z \in E$, is itself bounded and closed;

c) E contains points z_0 and Z such that $|f(z_0)| \leq |f(z)| \leq |f(Z)|$ for all $z \in E$.

Comment. We then call $|f(z_0)|$ the *minimum* of $f(z)$ in E, denoted by

$$\min_{z \in E} |f(z)|,$$

and $|f(Z)|$ the *maximum* of $f(z)$ in E, denoted by

$$\max_{z \in E} |f(z)|.$$

13. Is the function

$$f(z) = \frac{1}{1 - z}$$

continuous in the open disk $|z| < 1$? Uniformly continuous?

14. Let K be the unit disk $|z| < 1$, with boundary C (the unit circle $|z| = 1$), and let $f(z)$ be uniformly continuous in K. Prove that the limit

$$\lim_{\substack{z \to z_0 \\ z \in K}} f(z) \tag{7}$$

exists for every point $z_0 \in C$.

15. Let $f(z)$ be the same as in the preceding problem, with "boundary values" defined by (7). Prove that $f(z)$ is continuous on the circle C.

16. The functions

$$\frac{\operatorname{Re} z}{|z|}, \quad \frac{z}{|z|}, \quad \frac{\operatorname{Re} z^2}{|z|^2}, \quad \frac{z \operatorname{Re} z}{|z|}$$

are all defined for $z \neq 0$. Which of them can be defined at the point $z = 0$ in such a way that the "extended" functions are continuous at $z = 0$?

17. Given any domain G in the finite plane other than the plane itself, let C be a curve contained in G, let Γ be the boundary of G and let ρ be the *distance between C and Γ*, i.e., the greatest lower bound of the set of all numbers $|z - \zeta|$ with $z \in C$, $\zeta \in \Gamma$. Prove that $\rho > 0$.

18. Let G, C, Γ and ρ be the same as in the preceding problem, and let D be the set of all points z such that $|z - z_0| < \frac{1}{2}\rho$ for some $z_0 \in C$. Prove that D is a bounded domain (containing C). Prove that the closed domain $\bar{D}$ is contained in G.

CHAPTER FOUR

Differentiation in the Complex Plane

4.1. The Derivative of a Complex Function

4.11. Complex derivatives. We say that a complex function $f(z)$ defined in a domain G is *differentiable* at a point z (in G) if the limit

$$f'(z) = \lim_{\Delta z \to 0} \frac{f(z + \Delta z) - f(z)}{\Delta z} \qquad (z, z + \Delta z \in G) \tag{1}$$

exists and is finite. The limit itself, denoted by $f'(z)$, is then called the *derivative* of $f(z)$ at z.

4.12. Analytic functions. A function $f(z)$ is said to be *analytic in a domain* G if $f(z)$ is differentiable at every point of G and *analytic at a point* z if $f(z)$ is analytic in some neighborhood of z. Note that any function analytic in a domain G is automatically analytic at every point of G.

4.13. Examples

a. The function

$$f(z) = z^2$$

is differentiable in the whole z-plane, since the limit

$$\lim_{\Delta z \to 0} \frac{(z + \Delta z)^2 - z^2}{\Delta z} = \lim_{\Delta z \to 0} \frac{2z\,\Delta z + (\Delta z)^2}{\Delta z} = 2z + \lim_{\Delta z \to 0} \Delta z$$

obviously exists and equals $2z$ at every (finite) point z.

b. The function

$$f(z) = \operatorname{Re} z$$

is continuous in the whole z-plane (why?), but nowhere differentiable! In fact, the limit

$$\lim_{\Delta z \to 0} \frac{\operatorname{Re}(z + \Delta z) - \operatorname{Re} z}{\Delta z} = \lim_{\Delta z \to 0} \frac{\operatorname{Re} \Delta z}{\Delta z} \qquad (\Delta z = \Delta x + i\,\Delta y)$$

fails to exist at every point z. To see this, first let $\Delta z = \Delta x$, $\Delta y = 0$, so that Δz approaches 0 along the real axis, and then let $\Delta z = i\,\Delta y$, $\Delta x = 0$, so that Δz approaches 0 along the imaginary axis. In the first case we get

$$\lim_{\Delta z \to 0} \frac{\operatorname{Re} \Delta z}{\Delta z} = \lim_{\Delta z \to 0} \frac{\Delta x}{\Delta x} = 1,$$

while in the second case

$$\lim_{\Delta z \to 0} \frac{\operatorname{Re} \Delta z}{\Delta z} = \lim_{\Delta z \to 0} \frac{0}{i\,\Delta y} = 0.$$

Since these two values are different, the derivative $f'(z)$ fails to exist.

c. In just the same way, it can be shown that the function

$$f(z) = \bar{z} = x - iy$$

also fails to be differentiable at every point z (give the details).

4.14. Remark. The ease with which we have just constructed nondifferentiable functions is explained by the fact that the requirement of differentiability with respect to a complex variable is much stronger than that of differentiability with respect to a real variable. In fact, differentiability of $f(z)$ at the point z requires that the limit of the "difference quotient"

$$\frac{f(z + \Delta z) - f(z)}{\Delta z}$$

be *independent of the direction* in which the variable point $z + \Delta z$ approaches the fixed point z. The requirement of being differentiable at every point of a domain is even stronger, which explains why functions that are analytic in a domain must have a number of special properties singling them out from the class of all other functions of a complex variable. This book is in large measure devoted to a detailed exploration of the remarkable properties of analytic functions.

4.15. There is a complete analogy between formula (1) defining the derivative of a complex function and the corresponding formula

$$f'(x) = \lim_{\Delta x \to 0} \frac{f(x + \Delta x) - f(x)}{\Delta x} \qquad (a < x < b)$$

defining the derivative of a real function. Hence all the differentiation rules familiar from calculus† carry over at once to the case of complex functions, i.e.,

$$[cf(z)]' = cf'(z)$$

if $f(z)$ is differentiable at z and c is any complex number,

$$[f(z) \pm g(z)]' = f'(z) \pm g'(z),$$

$$[f(z)g(z)]' = f'(z)g(z) + f(z)g'(z)$$

if $f(z)$ and $g(z)$ are differentiable at z,

$$\left[\frac{f(z)}{g(z)}\right]' = \frac{f'(z)g(z) - f(z)g'(z)}{g^2(z)}$$

if $f(z)$ and $g(z)$ are differentiable at z and $g(z) \neq 0$, while

$$[\varphi(f(z))]' = \varphi'(f(z))f'(z)$$

if $f(z)$ is differentiable at z and $\varphi(w)$ is differentiable at $w = f(z)$. Moreover

$$(z^n)' = nz^{n-1}$$

for all $n = 1, 2, \ldots$, just as in the real case. It follows that every polynomial is analytic in the whole complex plane, while every rational function (Chap. 3, Prob. 7) is analytic everywhere except at the points where its denominator vanishes.

4.16. Complex differentials. The concept of the differential of a complex function is formally identical with that of the differential of a real function. Let $w = f(z)$ be *differentiable* at the point z and let $\Delta w = f(z + \Delta z) - f(z)$, so that

$$\lim_{\Delta z \to 0} \frac{\Delta w}{\Delta z} = f'(z).$$

Then

$$\frac{\Delta w}{\Delta z} = f'(z) + \epsilon,$$

where ϵ approaches 0 as $\Delta z \to 0$, or equivalently

$$\Delta w = f'(z)\,\Delta z + \epsilon\,\Delta z. \tag{2}$$

The first term on the right in (2) is called the *differential* of the function w (or the *principal linear part* of the increment Δw) and is denoted by

$$dw = f'(z)\,\Delta z. \tag{3}$$

In particular, choosing $w = z$, we find that

$$dz = 1 \cdot \Delta z = \Delta z,$$

†R. A. Silverman, *op. cit.*, Theorems 5.3–5.6.

i.e., the increment and the differential of the independent variable coincide. Replacing Δz by dz in (3), we get

$$dw = f'(z)\, dz. \tag{3'}$$

This leads to the formula

$$f'(z) = \frac{dw}{dz} = \frac{df(z)}{dz}.$$

The two expressions on the right can be regarded as alternative notations for the derivative $f'(z)$, as well as quotients of differentials.

4.2. The Cauchy–Riemann Equations

4.21. A real function $u(x, y)$ is said to be *differentiable* at the point (x, y) if the increment

$$\Delta u = u(x + \Delta x, y + \Delta y) - u(x, y)$$

can be written in the form

$$\Delta u = A\,\Delta x + B\,\Delta y + \epsilon_1\,\Delta x + \epsilon_2\,\Delta y, \tag{4}$$

where A, B are independent of Δx, Δy, and ϵ_1, ϵ_2 both approach 0 as Δx, $\Delta y \to 0$. It is easy to see that the coefficients A and B are just the partial derivatives $\partial u/\partial x$ and $\partial u/\partial y$ of the function u at the point (x, y). In fact, choosing first $\Delta y = 0$ and then $\Delta x = 0$, we have

$$\frac{\partial u}{\partial x} = \lim_{\Delta x \to 0} \frac{u(x + \Delta x, y) - u(x, y)}{\Delta x} = \lim_{\Delta x \to 0} \frac{A\,\Delta x + \epsilon_1\,\Delta x}{\Delta x} = A + \lim_{\Delta x \to 0} \epsilon_1 = A,$$

$$\frac{\partial u}{\partial y} = \lim_{\Delta y \to 0} \frac{v(x, y + \Delta y) - v(x, y)}{\Delta y} = \lim_{\Delta y \to 0} \frac{B\,\Delta y + \epsilon_2\,\Delta y}{\Delta y} = B + \lim_{\Delta y \to 0} \epsilon_2 = B.$$

4.22. As noted in Sec. 3.12, specifying a function $w = f(z) = u + iv$ of a complex variable $z = x + iy$ is equivalent to specifying two real functions u and v of two real variables x and y. Continuity of u and v obviously implies continuity of w, but differentiability of u and v does not imply differentiability of w, as shown by the continuous nondifferentiable function $w = \text{Re}\, z = x$ considered in Example 4.13b, where $u = x$ and $v = 0$ are obviously differentiable at every point of the plane. Hence the real and imaginary parts of a differentiable function $w = u + iv$ cannot be chosen independently. Instead they must satisfy certain conditions known as the *Cauchy–Riemann equations*, as shown by the following

THEOREM. *The function $w = f(z) = u + iv$ is differentiable at the point $z = x + iy$ if and only if the functions u and v are differentiable at the point*

(x, y) and satisfy the ***Cauchy–Riemann equations***

$$\frac{\partial u}{\partial x} = \frac{\partial v}{\partial y}, \qquad \frac{\partial u}{\partial y} = -\frac{\partial v}{\partial x} \tag{5}$$

at (x, y).

Proof. Suppose $w = f(z)$ is differentiable at z. Then

$$\Delta w = \Delta u + i\,\Delta v = f'(z)\,\Delta z + \epsilon\,\Delta z,$$

where $\epsilon \to 0$ as $\Delta z \to 0$. Writing

$$f'(z) = a + ib, \qquad \epsilon = \epsilon_1 + i\epsilon_2,$$

we have

$$\Delta u + i\,\Delta v = (a + ib)(\Delta x + i\,\Delta y) + (\epsilon_1 + i\epsilon_2)(\Delta x + i\,\Delta y)$$

or

$$\Delta u = a\,\Delta x - b\,\Delta y + \epsilon_1\,\Delta x - \epsilon_2\,\Delta y,$$
$$\Delta v = b\,\Delta x + a\,\Delta y + \epsilon_2\,\Delta x + \epsilon_1\,\Delta y$$

after taking real and imaginary parts, where obviously $\epsilon_1, \epsilon_2 \to 0$ as Δx, $\Delta y \to 0$ since

$$|\Delta z| = \sqrt{(\Delta x)^2 + (\Delta y)^2}, \qquad |\epsilon_1| \le |\epsilon|, \qquad |\epsilon_2| \le |\epsilon|.$$

It follows that u and v are differentiable at (x, y) and

$$\frac{\partial u}{\partial x} = a, \quad \frac{\partial u}{\partial y} = -b, \quad \frac{\partial v}{\partial x} = b, \quad \frac{\partial v}{\partial y} = a. \tag{6}$$

But (6) immediately implies (5).

Conversely, let u and v be differentiable at (x, y) and suppose the Cauchy–Riemann equations (5) hold. Then

$$\begin{aligned}\Delta u &= \frac{\partial u}{\partial x}\Delta x + \frac{\partial u}{\partial y}\Delta y + \alpha_1\,\Delta x + \alpha_2\,\Delta y \\ &= \frac{\partial u}{\partial x}\Delta x - \frac{\partial v}{\partial x}\Delta y + \alpha_1\,\Delta x + \alpha_2\,\Delta y,\end{aligned}$$

$$\begin{aligned}\Delta v &= \frac{\partial v}{\partial x}\Delta x + \frac{\partial v}{\partial y}\Delta y + \beta_1\,\Delta x + \beta_2\,\Delta y \\ &= \frac{\partial v}{\partial x}\Delta x + \frac{\partial u}{\partial x}\Delta y + \beta_1\,\Delta x + \beta_2\,\Delta y,\end{aligned}$$

where $\alpha_1, \alpha_2, \beta_1, \beta_2 \to 0$ as $\Delta x, \Delta y \to 0$. It follows that

$$\begin{aligned}\Delta w &= \Delta u + i\,\Delta v \\ &= \left(\frac{\partial u}{\partial x} + i\frac{\partial v}{\partial x}\right)(\Delta x + i\,\Delta y) + (\alpha_1 + i\beta_1)\,\Delta x + (\alpha_2 + i\beta_2)\,\Delta y \\ &= \left(\frac{\partial u}{\partial x} + i\frac{\partial v}{\partial x}\right)\Delta z + \epsilon\,\Delta z,\end{aligned}$$

where

$$\epsilon = (\alpha_1 + i\beta_1)\frac{\Delta x}{\Delta z} + (\alpha_2 + i\beta_2)\frac{\Delta y}{\Delta z}.$$

But

$$|\epsilon| \leq |\alpha_1 + i\beta_1|\left|\frac{\Delta x}{\Delta z}\right| + |\alpha_2 + i\beta_2|\left|\frac{\Delta y}{\Delta z}\right|$$
$$\leq |\alpha_1 + i\beta_1| + |\alpha_2 + i\beta_2| \leq |\alpha_1| + |\alpha_2| + |\beta_1| + |\beta_2|,$$

and hence $\epsilon \longrightarrow 0$ as $\Delta z \longrightarrow 0$, since $\alpha_1, \alpha_2, \beta_1, \beta_2 \longrightarrow 0$ as Δx, $\Delta y \longrightarrow 0$. Therefore the limit

$$f'(z) = \lim_{\Delta z \to 0} \frac{\Delta w}{\Delta z} = \frac{\partial u}{\partial x} + i\frac{\partial v}{\partial x} + \lim_{\Delta z \to 0} \epsilon = \frac{\partial u}{\partial x} + i\frac{\partial v}{\partial x}$$

exists (and is finite), i.e., $w = f(z)$ is differentiable at z. ∎

4.23. Remark. It follows from Theorem 4.22 that a function $w = f(z) = u + iv$ is analytic in a domain G if and only if its real and imaginary parts u and v are differentiable and satisfy the Cauchy–Riemann equations at every point of G; the derivative $f'(z)$ can then be written in any of the forms

$$f'(z) = \frac{\partial u}{\partial x} + i\frac{\partial v}{\partial x} = \frac{\partial u}{\partial x} - i\frac{\partial u}{\partial y} = \frac{\partial v}{\partial y} + i\frac{\partial v}{\partial x} = \frac{\partial v}{\partial y} - i\frac{\partial u}{\partial y}.$$

As we know from calculus,† a sufficient (but not necessary) condition for differentiability of u and v at a point (x, y) is that u and v have *continuous* partial derivatives at (x, y).

4.3. Conformal Mapping

4.31. Let C be a (continuous) curve with equation

$$z = z(t) \qquad (a \leq t \leq b),$$

and let t_0 be any point of the interval $[a, b]$. Suppose C has a tangent at the point $z_0 = z(t_0)$. This means that the vector

$$\Delta z = z(t_0 + \Delta t) - z(t_0)$$

has a "limiting direction" as $\Delta t \longrightarrow 0$, or more exactly that the limit

$$\psi = \lim_{\Delta t \to 0} \arg \Delta z \tag{7}$$

exists.‡ Since $\Delta t \longrightarrow 0$ implies $\Delta z \longrightarrow 0$, we can just as well write (7) as

$$\psi = \lim_{\Delta z \to 0} \arg \Delta z. \tag{7'}$$

---

†R. A. Silverman, *op. cit.*, Theorem 12.3 (also Prob. 10, p. 716).

‡If z_0 is an end point of C, so that $t = a$ or $t = b$, then $\Delta t \longrightarrow 0$ while staying positive or negative.

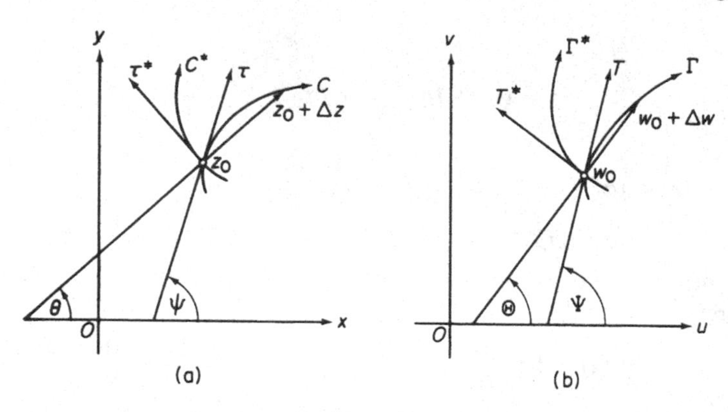

Figure 12

Geometrically the tangent to C at z_0 is represented by the ray τ through z_0 making the angle ψ with the positive real axis (see Figure 12a where $\theta = \arg \Delta z$).

Now let $f(z)$ be continuous in a domain G containing the curve C. Then $f(z)$ maps C into a curve Γ in the w-plane with equation

$$w = f(z(t)) \qquad (a \leq t \leq b).$$

Let

$$\Delta w = f(z_0 + \Delta z) - f(z_0),$$

so that in particular $\Delta w \longrightarrow 0$ as $\Delta z \longrightarrow 0$. Suppose further that $f(z)$ has a *nonzero* derivative $f'(z_0)$ at the point $z_0 \in C$. Then since

$$\lim_{\Delta z \to 0} \frac{\Delta w}{\Delta z} = f'(z_0), \tag{8}$$

we have

$$\lim_{\Delta z \to 0} \arg \frac{\Delta w}{\Delta z} = \arg f'(z_0)$$

(cf. Chap. 2, Prob. 7), where the fact that $f'(z_0) \neq 0$ is crucial since arg 0 is undefined. But

$$\arg \frac{\Delta w}{\Delta z} = \arg \Delta w - \arg \Delta z$$

(see Sec. 1.37), and hence

$$\lim_{\Delta z \to 0} \arg \Delta w = \lim_{\Delta z \to 0} \arg \Delta z + \arg f'(z_0).$$

Since the limit on the right exists, so does the limit on the left, i.e., Γ has a tangent T at the point $w_0 = f(z_0)$, with inclination

$$\Psi = \psi + \arg f'(z_0) \tag{9}$$

(see Figure 12b where $\Theta = \arg \Delta w$). In other words, *the inclination of T exceeds the inclination of τ by the angle* $\arg f'(z_0)$.

4.32. Next let $f(z)$ be the same as before, and let C and C^* be two curves in G which intersect at a point z_0, where they have tangents τ and τ^* (see Figure 12a).† Then the *angle between C and C** (in that order) is defined as the angle between τ and τ^*, measured from τ and τ^*. Let Γ and Γ^* be the "images of C and C^* under the mapping $f(z)$," i.e., the two curves in the w-plane into which $f(z)$ maps C and C^*. Then, as we have seen, Γ and Γ^* have tangents T and T^* at the point $w_0 = f(z_0)$, where T and T^* are obtained by rotating τ and τ^* through the *same* angle $\arg f'(z_0)$. Therefore the angle between Γ and Γ^* equals the angle between C and C^*, both in magnitude and direction (i.e., both in absolute value and sign).

4.33. A mapping by a continuous function which preserves magnitudes of angles between curves passing through a given point z_0 is said to be *isogonal* at z_0. If $f(z)$ is isogonal at z_0 and in addition preserves directions of angles between curves passing through z_0, then $f(z)$ is said to be *conformal* at z_0. Thus we have just shown that if $f(z)$ is continuous in a domain G and has a nonzero derivative $f'(z_0)$ at a point $z_0 \in G$, then $f(z)$ is conformal at z_0. By the same token, if $f(z)$ is analytic in a domain G, then $f(z)$ is conformal at every point of G where $f'(z) \neq 0$.‡

4.34. Examples

a. The mapping $w = z^2$ is conformal at every point $z \neq 0$, since the derivative $w' = 2z$ is nonvanishing for $z \neq 0$. However $w = z^2$ fails to be conformal at the point $z = 0$ where the derivative w' vanishes. In fact, since

$$\arg w = \arg z^2 = 2 \arg z,$$

the mapping doubles every angle with its vertex at the origin.

b. The mapping $w = \bar{z}$ is isogonal but not conformal at every point z. In fact, this mapping reduces simply to reflection in the real axis, and hence carries any pair of rays intersecting at angle θ into a pair of rays intersecting at angle $-\theta$, as shown in Figure 13 (where the z- and w-planes are combined).

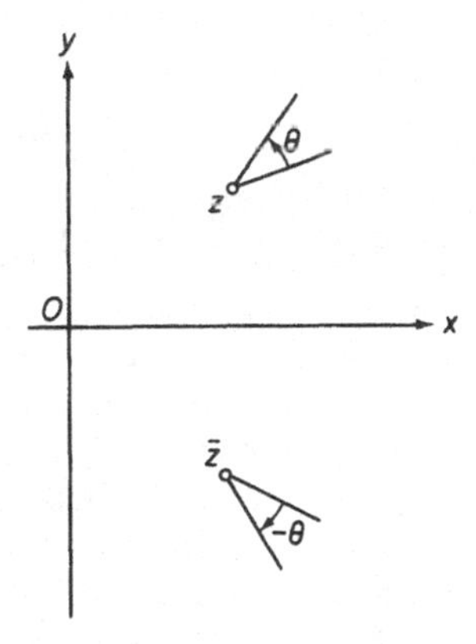

Figure 13

†This situation is often described by saying that "C and C^* form an angle with its vertex at z_0."

‡Conversely, it can be shown (see Chap. 10, Prob. 24) that if $f(z)$ is analytic in a domain G, then $f(z)$ fails to be conformal at every point of G where $f'(z) = 0$.

4.35. Having found a simple geometric interpretation for the argument of the derivative $f'(z_0)$, we now interpret the modulus of the derivative, namely the quantity $|f'(z_0)|$. To this end, we note that (8) immediately implies

$$|f'(z_0)| = \lim_{\Delta z \to 0} \left|\frac{\Delta w}{\Delta z}\right| = \lim_{\Delta z \to 0} \frac{|\Delta w|}{|\Delta z|}$$

(cf. Chap. 2, Prob. 6). But $|\Delta z|$ is the distance between two neighboring points z_0 and $z_0 + \Delta z$ in the z-plane, while $|\Delta w|$ is the distance between their images $w_0 = f(z_0)$ and $w_0 + \Delta w = f(z_0 + \Delta z)$ in the w-plane. Hence the ratio

$$\frac{|\Delta w|}{|\Delta z|}$$

is in effect the *magnification* (or expansion) of the infinitesimal vector Δz produced by the mapping $w = f(z)$, and $\mu = |f'(z_0)|$ is the "limiting" magnification (as $\Delta z \longrightarrow 0$) at the point z_0. Here the word "magnification" is used in the same general sense as in Sec. 1.34, corresponding to stretching (if $\mu > 1$), shrinking (if $\mu < 1$), or neither (if $\mu = 1$).

COMMENTS

4.1. According to Sec. 3.31a, (1) means that given any $\epsilon > 0$, there exists a number $\delta = \delta(\epsilon) > 0$ such that

$$\left| f'(z) - \frac{f(z + \Delta z) + f(z)}{\Delta z} \right| < \epsilon$$

for all Δz such that $0 < |\Delta z| < \delta$. In connection with (2), note that if

$$\Delta w = A\,\Delta z + \epsilon\,\Delta z$$

where A is independent of Δz and $\epsilon \longrightarrow 0$ as $\Delta z \longrightarrow 0$, then $f'(z)$ exists and equals A (divide by Δz and take the limit as $\Delta z \longrightarrow 0$).

4.2. In connection with Theorem 4.22, it is essential to note that the class of differentiable functions of *two* variables $u = u(x, y)$ is *smaller* than the class of functions for which the partial derivatives $\partial u/\partial x$, $\partial u/\partial y$ exist (cf. Prob. 5) and *larger* than the class of functions for which $\partial u/\partial x$, $\partial u/\partial y$ exist and are continuous (cf. Sec. 4.23). For functions of one variable, whether real or complex, there is no distinction between a differentiable function and a function for which the derivative exists (why not?), as implicit in the definitions of Sec. 4.11. It is a trivial matter to prove that a complex function $w = u + iv$ with a derivative at a point z satisfies the Cauchy–Riemann equations at z (Prob. 4). Theorem 4.22 says much more than this, namely that the class of complex functions $w = u + iv$ with a derivative at z is *precisely the same* as the class of functions w whose real and imaginary parts u and v are differentiable at z and satisfy the Cauchy–Riemann equations at z.

4.3. As in Chap. 2, Prob. 7, formula (7) means that given any $\epsilon > 0$ there is a number $\delta = \delta(\epsilon) > 0$ and a function $\theta(\Delta t)$ equal to a value of $\arg \Delta z$ for every $\Delta t \neq 0$ such that $|\theta(\Delta t) - \psi| < \epsilon$ whenever $0 < |\Delta t| < \delta$. Clearly ψ is defined only to within an integral multiple of 2π. It can be shown that the stereographic projection of Sec. 2.43 is a conformal mapping, with angles between curves on the Riemann sphere being defined in the natural way (see e.g., A. I. Markushevich, *op. cit.*, Volume I, Sec. 23).

PROBLEMS

1. Where is the function $f(z) = z \operatorname{Re} z$ differentiable? How about the function $f(z) = |z|$?

2. Prove that a function $w = f(z)$ which is differentiable at a point z is automatically continuous at z.

3. Prove that if $f'(z) = 0$ at every point of a domain G, then $f(z)$ is constant in G.

4. By requiring that

$$f'(z) = \lim_{\Delta z \to 0} \frac{f(z + \Delta z) - f(z)}{\Delta z} = \lim_{\Delta z \to 0} \frac{\Delta u + i\,\Delta v}{\Delta x + i\,\Delta y}$$

have the same value for $\Delta z = \Delta x$ ($z + \Delta z$ approaching z along a line parallel to the real axis) and for $\Delta z = i\,\Delta y$ ($z + \Delta z$ approaching z along a line parallel to the imaginary axis), give a quick proof of the Cauchy–Riemann equations.

5. Prove that the function

$$u(x, y) = \begin{cases} x & \text{if } |y| > |x|, \\ -x & \text{if } |y| \leq |x| \end{cases}$$

is continuous and has partial derivatives $\partial u/\partial x$ and $\partial u/\partial y$ at the origin but is not differentiable there.

6. Show that the function $f(z) = \sqrt{|xy|}$ is continuous and satisfies the Cauchy–Riemann equations at the origin but is not differentiable there.

7. Prove that the Cauchy–Riemann equations (5) take the form

$$\frac{\partial u}{\partial r} = \frac{1}{r}\frac{\partial v}{\partial \theta}, \qquad \frac{\partial v}{\partial r} = -\frac{1}{r}\frac{\partial u}{\partial \theta} \tag{5'}$$

in polar coordinates ($x = r\cos\theta$, $y = r\sin\theta$).

8. Use (5′) to verify the analyticity of $f(z) = z^n$ ($n = 1, 2, \ldots$) in the whole plane.

9. A function $f(z)$ is said to be *infinitely differentiable* in a domain G if it has derivatives

$$f'(z), \qquad f''(z) = \frac{df'(z)}{dz}, \qquad f'''(z) = \frac{df''(z)}{dz}, \ldots$$

of all orders at every point of G. Give examples of infinitely differentiable functions.

10. What is the image in the w-plane of the line segment

$$z = 1 + it \qquad (-1 \leq t \leq 1)$$

under the mapping $w = z^2$?

11. Find the angle α through which a curve drawn from the point z_0 is rotated under the mapping $w = z^2$ if

a) $z_0 = i$; b) $z_0 = -\frac{1}{4}$; c) $z_0 = 1 + i$; d) $z_0 = -3 + 4i$.

Find the corresponding values of the magnification μ.

12. Answer the same question for the mapping $w = z^3$.

13. Prove that if $f(z)$ is analytic in a domain G, then $\overline{f(z)}$ is isogonal but not conformal at every point of G where $f'(z) \neq 0$. How about $f(\bar{z})$?

14. Which part of the complex plane is stretched and which part shrunk under the following mappings:
a) $w = z^2$; b) $w = z^2 + z$; c) $w = 1/z$?

15. Find an analytic function which maps angles at the point z_0 into angles five times larger.

16. The mapping

$$w = f(z) = az + b \qquad (a \neq 0), \tag{10}$$

where a and b are arbitrary complex numbers (except for the condition $a \neq 0$), is called an *entire linear transformation.* Prove that
a) $f(z)$ is one-to-one in the extended complex plane (mapping ∞ into ∞);
b) $f(z)$ is conformal at every point of the finite plane;
c) Under the mapping $f(z)$ the tangents to all curves in the finite plane are rotated through the same angle $\arg a$ and the magnification at every point equals $|a|$;
d) If $a = 1$, then $f(z)$ reduces to a shift of the whole plane by the vector b.

17. Suppose $a \neq 1$ in (10) as well as $a \neq 0$. Prove that (10) can be written in the form

$$w - z_0 = a(z - z_0), \tag{10'}$$

where z_0 is determined from the equation

$$z_0 = az_0 + b.$$

Comment. The point z_0 is called a *fixed point* of the transformation (10), since (10) carries z_0 into itself, i.e., leaves z_0 unchanged.† The point at infinity (∞) is always a fixed point of the transformation (10) and the only fixed point if $a = 1$, $b \neq 0$.

18. Using (10′), show that the general transformation (10) with $a \neq 1$ is equivalent to a rotation of the whole plane through the angle $\arg a$ about the (finite) fixed point

$$z_0 = \frac{b}{1 - a},$$

†Synonymously, z_0 is said to be *invariant under* (10).

together with a uniform magnification $|a|$ relative to the point z_0 (synonymously, a *similarity transformation* with *center* z_0 and *ratio* $|a|$).

19. Find the rotation, magnitude and finite fixed point (if any) corresponding to each of the following entire linear transformations:
a) $w = 2z + 1 - 3i$; b) $w = iz + 4$; c) $w = z + 1 - 2i$.

20. Find the entire linear transformation with fixed point $1 + 2i$ carrying the point i into the point $-i$.

21. Find the entire linear transformation carrying the triangle with vertices at the points $0, 1, i$ into the similar triangle with vertices at the points $0, 2, 1 + i$.

22. The mapping

$$w = f(z) = \frac{az + b}{cz + d}, \tag{11}$$

where a, b, c and d are arbitrary complex numbers (except that c and d are not both zero), is called a *fractional linear transformation* (or *Möbius transformation*). Prove that
a) $f(z)$ reduces to an entire linear transformation if $c = 0$;
b) $f(z)$ reduces to a constant if $ad - bc = 0$;
c) If $c \neq 0$, $ad - bc \neq 0$, then $f(z)$ has a nonzero derivative $f'(z)$ for all z except $z = \delta = -d/c$;
d) If $c \neq 0$, $ad - bc \neq 0$, then $f(z)$ is conformal at all finite points except possibly at δ (however see Prob. 26), where the angle α through which tangents to curves are rotated has the same value

$$\alpha = \arg f'(z) = \arg \frac{ad - bc}{c^2} - 2 \arg (z - \delta)$$

along any given ray emanating from δ, and the magnification has the same value

$$\mu = |f'(z)| = \left|\frac{ad - bc}{c^2}\right| \frac{1}{|z - \delta|^2}$$

along any given circle with center δ.

23. Let μ be the same as in the preceding problem. Show that
a) $\mu = 1$ at every point of the circle with equation

$$|z - \delta| = \frac{1}{|c|}\sqrt{|ad - bc|},$$

called the *isometric circle* of the transformation (11);
b) $\mu > 1$ inside γ and approaches ∞ as $z \longrightarrow \delta$;
c) $\mu < 1$ outside γ and approaches 0 as $z \longrightarrow \infty$.†

24. Let

$$w = f(z) = \frac{az + b}{cz + d} \qquad (c \neq 0, ad - bc \neq 0). \tag{11'}$$

Then clearly

$$\lim_{z \to \delta} = f(z) = \infty, \qquad \lim_{z \to \infty} f(z) = \frac{a}{c} = A$$

---

†Cf. Chap. 3, Probs. 3 and 4.

($\delta = -d/c$). Suppose we complete the definition of $f(z)$ by setting

$$f(\delta) = \infty, \qquad f(\infty) = A.$$

Show that $f(z)$ is a one-to-one mapping of the extended complex plane onto itself, with inverse

$$z = \frac{dw - b}{-cw + a}.$$

25. Two continuous curves C and C^* in the extended z-plane are said to form an angle of α radians with its *vertex at infinity* if their images L and L^* under the transformation $\zeta = 1/z$ form an angle of α radians with its vertex at the origin (of the ζ-plane). Show that the real and imaginary axes form an angle of $\pi/2$ radians with its vertex at infinity.

26. Prove that the mapping (11′) is conformal at the point $\delta = -d/c$, i.e., that it carries any two curves C and C^* in the z-plane forming an angle of α radians with its vertex at δ into two curves Γ and Γ^* in the w-plane forming an angle of α radians with its vertex at infinity.

27. Prove that the mapping (11′) is *conformal at infinity*, i.e., that it carries any two curves C and C^* in the z-plane forming an angle of α radians with its vertex at infinity into two curves Γ and Γ^* in the w-plane forming an angle of α radians with its vertex at $A = a/c$.

Comment. Thus finally we have shown that the fractional linear transformation (11′) is conformal at every point of the extended z-plane.

28. Prove that the entire linear transformation (10) is conformal at infinity (and hence at every point of the extended z-plane) if $a \neq 0$.

29. A function $f(z)$ is said to be *analytic at infinity* if the function $\varphi(\zeta) = f(1/\zeta)$ is analytic at $\zeta = 0$. Suppose $f(z)$ is analytic at infinity. Then $f(\infty)$, the *value of $f(z)$ at infinity*, is defined as

$$f(\infty) = \varphi(0).$$

Prove that

$$\lim_{z \to \infty} f(z) = f(\infty),$$

while

$$\lim_{z \to \infty} f'(z) = 0.$$

30. Suppose $f(z)$ is analytic at infinity. Then $f'(\infty)$, the *derivative of $f(z)$ at infinity*, is defined as

$$f'(\infty) = \varphi'(0),$$

where $\varphi(\zeta) = f(1/\zeta)$. Show that in general

$$f'(\infty) \neq \lim_{z \to \infty} f'(z).$$

31. Calculate $f'(\infty)$ for the fractional linear transformation (11′). How does the result explain the conformality of (11′) at infinity?

32. Suppose

$$\lim_{z \to z_0} f(z) = \infty$$

(z_0 may be infinite), and suppose the function

$$\varphi(z) = \frac{1}{f(z)}$$

is analytic at the point z_0, with derivative $\varphi'(z_0) \neq 0$. Prove that $f(z)$ is conformal at z_0.

CHAPTER FIVE

Integration in the Complex Plane

5.1. The Integral of a Complex Function

5.11. A curve C with parametric equation

$$z = z(t) \qquad (a \leq t \leq b)$$

is said to be *smooth* if $z(t)$ has a continuous nonvanishing derivative $z'(t) \neq 0$ at every point of the interval $a \leq t \leq b$.† Let $f(z)$ be any function of a complex variable defined in a domain G of the z-plane, and let C be any smooth curve contained in G with initial point z_0 and final point Z. Choosing points $z_0, z_1, z_2, \ldots, z_{n-1}, z_n = Z$ arranged consecutively along C in the positive direction (the direction of increasing t), we form the sum

$$\sum_{k=1}^{n} f(\zeta_k)\,\Delta z_k, \tag{1}$$

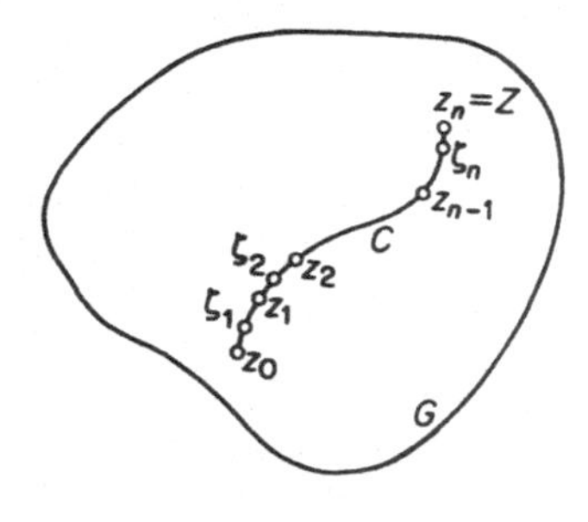

Figure 14

where $\Delta z_k = z_k - z_{k-1}$ and ζ_k is an arbitrary point of the arc $\widehat{z_{k-1}z_k}$ (see Figure 14). Let l_k be the length of $\widehat{z_{k-1}z_k}$ (see Prob. 2), and let

$$\lambda = \max\{l_1, l_2, \ldots, l_n\}$$

Suppose the limit

$$\lim_{\lambda \to 0} \sum_{k=1}^{n} f(\zeta_k)\,\Delta z_k \tag{2}$$

†At $t = a$, $z'(t)$ denotes the right-hand derivative and is continuous from the right, while at $t = b$, $z'(t)$ denotes the left-hand derivative and is continuous from the left.

exists, regardless of the choice of the points z_k, ζ_k. Then $f(z)$ is said to be *integrable along C*, and the limit (2), denoted by

$$\int_C f(z)\,dz,$$

is called the *integral of $f(z)$ along C*.

5.12. THEOREM. *If $f(z)$ is continuous in a domain G containing a smooth curve C, then $f(z)$ is integrable along C.*

Proof. Let

$$z_k = x_k + iy_k, \quad \zeta_k = \xi_k + i\eta_k, \quad \Delta z_k = \Delta x_k + i\,\Delta y_k \qquad (k = 1, 2, \ldots, n)$$

and

$$f(z) = u(x, y) + iv(x, y).$$

Then (1) can be written in the form

$$\sum_{k=1}^{n} f(\zeta_k)\,\Delta z_k = \sum_{k=1}^{n} (u_k + iv_k)(\Delta x_k + i\,\Delta y_k)$$
$$= \sum_{k=1}^{n} (u_k\,\Delta x_k - v_k\,\Delta y_k) + i\sum_{k=1}^{n} (v_k\,\Delta x_k + u_k\,\Delta y_k),$$

where

$$u_k = u(\xi_k, \eta_k), \qquad v_k = v(\xi_k, \eta_k).$$

But as $\lambda \longrightarrow 0$, the first sum on the right approaches the line integral

$$\int_C u\,dx - v\,dy,$$

while the second sum approaches the line integral

$$\int_C v\,dx + u\,dy.$$

It follows that (2) exists† and equals

$$\int_C f(z)\,dz = \int_C u\,dx - v\,dy + i\int_C v\,dx + u\,dy. \tag{3}$$

Note that the theorem remains valid if $f(z)$ is continuous along C itself (see Sec. 3.34) rather than in some domain containing C. ∎

5.13. We can write (3) in the particularly concise form

$$\int_C f(z)\,dz = \int_C (u + iv)(dx + i\,dy). \tag{3'}$$

Suppose C has the parametric equation

$$z = z(t) = x(t) + iy(t) \qquad (a \le t \le b). \tag{4}$$

---

†The role played by the continuity of $f(z)$ and the smoothness of C in guaranteeing the existence of the line integrals is particularly clear from formulas (5) and (5′).

Then clearly

$$\int_C f(z)\,dz = \int_a^b f(z(t))z'(t)\,dt$$
$$= \int_a^b [u(z(t))x'(t) - v(z(t))y'(t)]\,dt$$
$$+ i\int_a^b [v(z(t))x'(t) + u(z(t))y'(t)]\,dt,$$

i.e.,

$$\int_C f(z)\,dz = \int_a^b R(t)\,dt + i\int_a^b I(t)\,dt, \tag{5}$$

where

$$\begin{aligned} R(t) &= \operatorname{Re} f(z(t))z'(t), \\ I(t) &= \operatorname{Im} f(z(t))z'(t). \end{aligned} \tag{5'}$$

Thus use of (5) reduces the calculation of a complex integral to the calculation of two real integrals.

5.14. Suppose C is a smooth curve made up of (smooth) arcs $C_1, C_2, \ldots, C_n$ "joined end to end,"† and let $f(z)$ be continuous on C (i.e., continuous in a domain containing C or possibly only along C itself). Then clearly

$$\int_C f(z)\,dz = \int_{C_1} f(z)\,dz + \int_{C_2} f(z)\,dz + \cdots + \int_{C_n} f(z)\,dz, \tag{6}$$

as we see at once by choosing the end points of the arcs $C_1, C_2, \ldots, C_n$ among the points z_k ($k = 0, 1, \ldots, n$) figuring in the sum (1). A curve C made up of smooth arcs joined end to end may not be smooth itself; in this case we say that C is *piecewise smooth* and use (6) to *define* the integral of $f(z)$ along C (however see Prob. 6). Clearly (6) continues to hold if any of the arcs C_k is only piecewise smooth rather than smooth. To see this, we need only decompose each piecewise smooth arc C_k into further smooth subarcs.

5.15. Example. Let C be any piecewise smooth curve joining two points z_0 and Z. Then

$$\int_C z^n\,dz = \frac{1}{n+1}(Z^{n+1} - z_0^{n+1}), \tag{7}$$

where n is an integer other than -1 and C does not go through the point $z = 0$ if n is negative. In fact, if C has the parametric equation (4), then

†More exactly, the final point of C_k coincides with the initial point of C_{k+1} for $k = 1, 2, \ldots, n-1$.

$$\int_C z^n\,dz = \int_a^b z^n(t)z'(t)\,dt = \int_a^b \frac{1}{n+1}\frac{d}{dt}z^{n+1}(t)\,dt$$
$$= \frac{1}{n+1}z^{n+1}(t)\Big|_{t=a}^{t=b} = \frac{1}{n+1}(Z^{n+1} - z_0^{n+1}) \qquad (n \neq -1).$$

Note that this integral does not depend on the particular curve C joining the points z_0 and Z. If C is a closed curve, then $z_0 = Z$ and (7) reduces to

$$\int_C z^n\,dz = 0.$$

Formula (7) is obvious for $n = 0$, since then

$$\int_C dz = \lim_{\lambda\to 0}\sum_{k=1}^{n} \Delta z_k = \lim_{\lambda\to 0}\sum_{k=1}^{n}(z_k - z_{k-1}) = z_n - z_0 = Z - z_0.$$

5.2. Basic Properties of the Integral

5.21. THEOREM. *Let $f(z)$ be continuous on a piecewise smooth curve C. Then*

$$\int_{C^-} f(z)\,dz = -\int_C f(z)\,dz,$$

where C^- denotes the curve C traversed in the negative direction.†

Proof. We need only note that

$$\int_{C^-} f(z)\,dz = \lim_{\lambda\to 0}\sum_{k=1}^{n} f(\zeta_{n-k+1})(z_{n-k} - z_{n-k+1})$$
$$= -\lim_{\lambda\to 0}\sum_{k=1}^{n} f(\zeta_k)(z_k - z_{k-1}) = -\int_C f(z)\,dz,$$

where λ, z_k and ζ_k have the same meaning as in Sec. 5.11. ∎

5.22. THEOREM. *Let $f(z)$ and $g(z)$ be continuous on a piecewise smooth curve C. Then*

$$\int_C [\alpha f(z) + \beta g(z)]\,dz = \alpha\int_C f(z)\,dz + \beta\int_C f(z)\,dz,$$

where α and β are arbitrary complex numbers.

Proof. This time we have

$$\int_C [\alpha f(z) + \beta g(z)]\,dz = \lim_{\lambda\to 0}\sum_{k=1}^{n}[\alpha f(\zeta_k) + \beta g(\zeta_k)]\,\Delta z_k$$
$$= \alpha\lim_{\lambda\to 0}\sum_{k=1}^{n} f(\zeta_k)\,\Delta z_k + \beta\lim_{\lambda\to 0}\sum_{k=1}^{n} g(\zeta_k)\,\Delta z_k$$
$$= \alpha\int_C f(z)\,dz + \beta\int_C g(z)\,dz. \quad ∎$$

†I.e., in the reverse of the positive direction (see Sec. 3.21a).

5.23. THEOREM. *Let $f(z)$ be continuous on a piecewise smooth curve C, and suppose $|f(z)| \leq M$ for all $z \in C$.*† *Then*

$$\left|\int_C f(z)\,dz\right| \leq Ml,$$

where l is the length of C.

Proof. Note that

$$\left|\sum_{k=1}^{n} f(\zeta_k)\,\Delta z_k\right| \leq \sum_{k=1}^{n} |f(\zeta_k)||\Delta z_k| \leq M \sum_{k=1}^{n} |\Delta z_k| \leq Ml,$$

where the last step follows from the obvious fact that the length of any polygonal curve inscribed in C is less than the length of C itself.‡ ∎

5.3. Integrals Along Polygonal Curves

The following two propositions involving integrals along polygonal curves may appear a bit specialized at this point, but they will be needed in the next section to prove one of the key theorems of complex analysis (Cauchy's integral theorem).

5.31. LEMMA. *Let $f(z)$ be continuous in a domain G containing a piecewise smooth curve C. Then given any $\epsilon > 0$, there exists a polygonal curve L inscribed in C and contained in G such that*

$$\left|\int_C f(z)\,dz - \int_L f(z)\,dz\right| < \epsilon.$$

Proof. Let D be a bounded domain containing C such that the closed domain $\bar{D}$ is contained in G, let Γ be the boundary of D, and let ρ be the distance between C and Γ, as defined in Chap. 3, Prob. 17.§ Being continuous in G, the function $f(z)$ is continuous in $\bar{D}$ and hence uniformly continuous in $\bar{D}$, by Theorem 3.44. Hence, given any $\epsilon > 0$, there is a number $\delta = \delta(\epsilon) > 0$ such that

$$|f(z') - f(z'')| < \frac{\epsilon}{2l}$$

for all points $z', z'' \in \bar{D}$ such that $|z' - z''| < \delta$, where l is the length of C. Suppose we divide C into n arcs $\gamma_1, \gamma_2, \ldots, \gamma_n$, each of length less than

$$\delta^* = \min\{\delta, \rho\},$$

---

†Note that $\max_{z \in C} |f(z)|$ is a possible choice of the constant M.

‡Given a curve C with parametric equation (4), any polygonal curve (cf. Chap. 3, Prob. 1) with consecutive vertices at the points $z_k = z(t_k)$ where $a = t_0 < t_1 < t_2 < \cdots < t_{n-1} < t_n = b$ is said to be *inscribed in C*, for an obvious geometric reason.

§Concerning the existence of D, see Chap. 3, Prob. 18, noting that if G is the whole plane we can choose D to be any disk $|z| < R$ containing C.

by choosing points $z_0, z_1, z_2, \ldots, z_{n-1}, z_n$ along C in the positive direction (with z_0 the initial point of C and z_n the final point). Let L be the polygonal curve inscribed in C with consecutive vertices $z_0, z_1, z_2, \ldots, z_{n-1}, z_n$ and segments $\sigma_1, \sigma_2, \ldots, \sigma_n$ (see Figure 15, where the domain D is not shown).

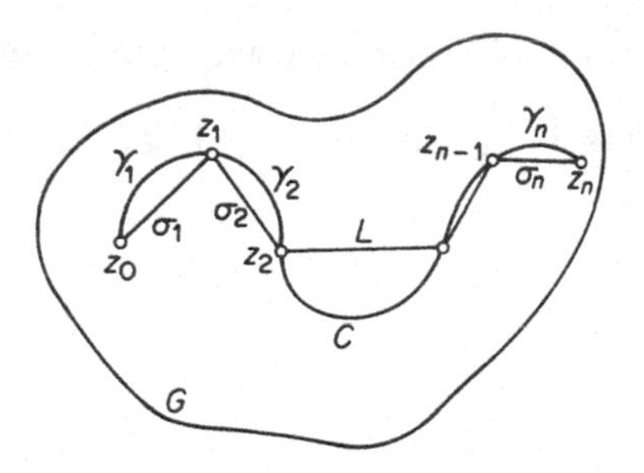

Figure 15

Since the length of every arc γ_k is less than δ^*, the distance between any two points of γ_k or of σ_k is certainly less than δ^*. In particular, every segment σ_k of the polygonal curve L is of length less than δ^*. But $\delta^* \leq \rho$, and hence L must be contained in $\bar{D}$ and therefore in G.

Now consider the sum

$$S = \sum_{k=1}^{n} f(z_k)\,\Delta z_k \qquad (\Delta z_k = z_k - z_{k-1}),$$

approximating the integral $\int_C f(z)\,dz$. Clearly

$$S = \int_{\gamma_1} f(z_1)\,dz + \cdots + \int_{\gamma_n} f(z_n)\,dz, \tag{8}$$

since

$$\Delta z_k = \int_{\gamma_k} dz$$

(recall Example 5.15). On the other hand,

$$\int_C f(z)\,dz = \int_{\gamma_1} f(z)\,dz + \cdots + \int_{\gamma_n} f(z)\,dz, \tag{9}$$

and hence, subtracting (8) from (9), we get

$$\int_C f(z)\,dz - S = \int_{\gamma_1} [f(z) - f(z_1)]\,dz + \cdots + \int_{\gamma_n} [f(z) - f(z_n)]\,dz.$$

But $|f(z) - f(z_k)| < \epsilon/2l$ on every arc γ_k and hence, by Theorem 5.23,

$$\left|\int_C f(z)\,dz - S\right| \leq \left|\int_{\gamma_1} [f(z) - f(z_1)]\,dz\right| + \cdots + \left|\int_{\gamma_n} [f(z) - f(z_n)]\,dz\right|$$
$$< l_1 \frac{\epsilon}{2l} + \cdots + l_n \frac{\epsilon}{2l},$$

where l_k is the length of γ_k. Therefore

$$\left|\int_C f(z)\,dz - S\right| < \frac{\epsilon}{2}, \tag{10}$$

since $l_1 + \cdots + l_n = l$.

In just the same way, replacing C by L and γ_k by σ_k, we have

$$S = \int_{\sigma_1} f(z_1)\,dz + \cdots + \int_{\sigma_n} f(z_n)\,dz \tag{8'}$$

instead of (8), since

$$\Delta z_k = \int_{\sigma_k} dz,$$

and

$$\int_L f(z)\,dz = \int_{\sigma_1} f(z)\,dz + \cdots + \int_{\sigma_n} f(z)\,dz \tag{9'}$$

instead of (9). It follows that

$$\int_L f(z)\,dz - S = \int_{\sigma_1} [f(z) - f(z_1)]\,dz + \cdots + \int_{\sigma_n} [f(z) - f(z_n)]\,dz.$$

But $|f(z) - f(z_k)| < \epsilon/2l$ on every segment σ_k and hence

$$\left|\int_L f(z)\,dz - S\right| < \lambda_1 \frac{\epsilon}{2l} + \cdots + \lambda_n \frac{\epsilon}{2l},$$

where λ_k is the length of σ_k. Therefore

$$\left|\int_L f(z)\,dz - S\right| < \frac{\epsilon}{2}, \tag{10'}$$

since $\lambda_1 + \cdots + \lambda_n \leq l$ (the length of the inscribed polygonal curve L cannot exceed the length of C itself). Combining (10) and (10'), we finally find that

$$\left|\int_C f(z)\,dz - \int_L f(z)\,dz\right| \leq \left|\int_C f(z)\,dz - S\right| + \left|S - \int_L f(z)\,dz\right|$$
$$< \frac{\epsilon}{2} + \frac{\epsilon}{2} = \epsilon. \quad ∎$$

5.32. LEMMA. *Let $f(z)$ be continuous in a simply connected domain G containing a closed polygonal curve L. Then*

$$\int_L f(z)\,dz = \int_{\Delta_1} f(z)\,dz + \int_{\Delta_2} f(z)\,dz + \cdots + \int_{\Delta_n} f(z)\,dz, \tag{11}$$

where the curves $\Delta_1, \Delta_2, \ldots, \Delta_n$ are all "triangular contours"† contained in G.

---

†I.e., perimeters of triangles. The term *contour* is used as a synonym for *curve*, mainly when the curve is closed.

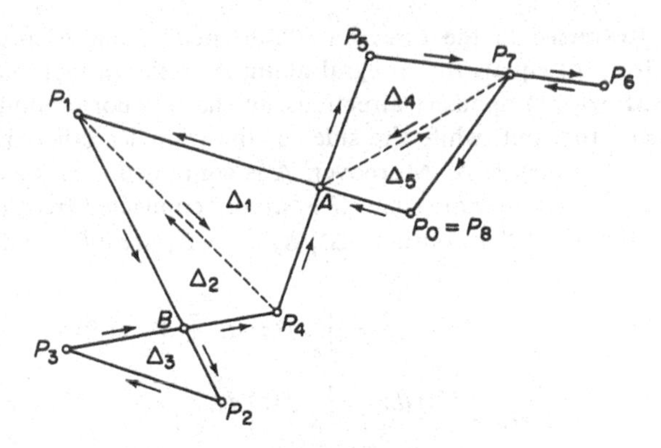

Figure 16

Proof. Let L be the closed polygonal curve $P_0P_1P_2 \cdots P_6P_7P_8$ shown in Figure 16, traversed in the direction indicated by the arrows. This curve exhibits the typical behavior of the general closed polygonal curve,† namely "self-intersections" at the points A and B (not among the vertices of L) and a segment P_6P_7 traversed twice in opposite directions. It is clear from the figure and formula (6) that

$$\begin{aligned}\int_L f(z)\,dz &= \int_{AP_1P_2P_3P_4A} f(z)\,dz + \int_{P_0AP_5P_6P_7P_0} f(z)\,dz \\ &= \int_{AP_1BP_4A} f(z)\,dz + \int_{BP_2P_3B} f(z)\,dz + \int_{P_0AP_5P_6P_7P_0} f(z)\,dz \\ &= \int_{AP_1BP_4A} f(z)\,dz + \int_{BP_2P_3B} f(z)\,dz + \int_{P_0AP_5P_7P_0} f(z)\,dz \\ &\quad + \int_{P_7P_6} f(z)\,dz + \int_{P_6P_7} f(z)\,dz.\end{aligned}$$

But the last two integrals cancel each other out, by Theorem 5.21, and hence

$$\int_L f(z)\,dz = \int_{AP_1BP_4A} f(z)\,dz + \int_{BP_2P_3B} f(z)\,dz + \int_{P_0AP_5P_7P_0} f(z)\,dz, \qquad (12)$$

i.e., we have reduced the integral of $f(z)$ along the original closed polygonal curve L to a sum of integrals along three curves (one already a triangular contour) which are *Jordan curves* as well as closed polygonal curves. The interior of any such curve Λ can be partitioned into a finite number of triangles by drawing suitable "cross cuts" (the dashed lines in the figure) connecting certain vertices of Λ. But the sum of the integrals of $f(z)$ along the triangular contours $\Delta_\alpha, \Delta_\beta, \ldots, \Delta_\nu$ equal to the perimeters of these

†For further justification of this assertion, see A. I. Markushevich, *op. cit.*, Volume I, pp. 266–268.

triangles (traversed in the directions "inherited" from Λ and in the first instance from L) equals the integral along Λ itself; in fact, each cross cut is traversed twice in opposite directions, so that the corresponding integrals cancel each other out, while the sides of the triangles other than the cross cuts make up the curve Λ. Moreover, Λ is contained in G, being part of L, and hence so is its interior, since G is simply connected (recall Sec. 3.24b). Therefore the triangular contours $\Delta_\alpha, \Delta_\beta, \ldots, \Delta_\nu$ are all contained in G. In our case,

$$\int_{AP_1BP_4A} f(z)\,dz = \int_{\Delta_1} f(z)\,dz + \int_{\Delta_2} f(z)\,dz,$$

$$\int_{BP_3P_2B} f(z)\,dz = \int_{\Delta_3} f(z)\,dz,$$

$$\int_{P_6AP_5P_7P_6} f(z)\,dz = \int_{\Delta_4} f(z)\,dz + \int_{\Delta_5} f(z)\,dz,$$

where $\Delta_1, \Delta_2, \Delta_3, \Delta_4, \Delta_5$ are the triangular contours shown in the figure (for simplicity, we use the same symbols for the triangles and their perimeters). Substituting these formulas into (12), we finally get

$$\int_L f(z)\,dz = \int_{\Delta_1} f(z)\,dz + \int_{\Delta_2} f(z)\,dz + \int_{\Delta_3} f(z)\,dz + \int_{\Delta_4} f(z)\,dz + \int_{\Delta_5} f(z)\,dz,$$

in keeping with (11).

The proof in the case of a general closed polygonal curve is identical in principle. First we successively "split off" closed polygonal *Jordan* curves $\Lambda_1, \Lambda_2, \ldots, \Lambda_N$ from L, dropping all segments of L traversed twice in opposite directions. We then partition the interiors of $\Lambda_1, \Lambda_2, \ldots, \Lambda_N$ into triangles with perimeters $\Delta_1, \Delta_2, \ldots, \Delta_n$ satisfying (11). ∎

5.4. Cauchy's Integral Theorem

5.41. Let $f(z)$ be continuous in a domain G, and let C_1, C_2 be two curves in G with the same initial and final points. Then we may have

$$\int_{C_1} f(z)\,dz = \int_{C_2} f(z)\,dz \tag{13}$$

for every such pair of curves C_1 and C_2, as in Example 5.15 where $f(z) = z^n$, or else (13) may fail to hold, as in Prob. 9 where $f(z) = \operatorname{Re} z$. Thus we are led to the problem of finding conditions under which (13) holds for all curves C_1, C_2 joining the same two points, or equivalently conditions under which

$$\int_C f(z)\,dz = 0 \tag{14}$$

for every closed curve C lying in G (see Prob. 8). In the first of the two cases just considered, the function $f(z)$ is analytic, while in the second it is not (recall Example 4.13b). This suggests that the analyticity of $f(z)$ is somehow responsible for the validity of (14), a conjecture confirmed by the following key theorem of complex analysis:

THEOREM *(Cauchy's integral theorem).*† *Let $f(z)$ be analytic in a simply connected domain G. Then*

$$\int_C f(z)\,dz = 0$$

for every piecewise smooth closed curve C contained in G.

Proof. Suppose

$$\int_\Delta f(z)\,dz = 0 \tag{15}$$

for every triangular contour Δ contained in G. Then, by Lemma 5.32,

$$\int_L f(z)\,dz = 0 \tag{16}$$

for every closed polygonal curve L contained in G, since the integral on the left can always be reduced to a finite sum of integrals along triangular contours. Moreover, by Lemma 5.31, given any $\epsilon > 0$, there is a closed polygonal curve L such that

$$\left|\int_C f(z)\,dz - \int_L f(z)\,dz\right| < \epsilon,$$

and hence

$$\left|\int_C f(z)\,dz\right| < \epsilon$$

because of (16). But then

$$\int_C f(z)\,dz = 0,$$

since ϵ is arbitrarily small.

Thus the whole proof reduces to proving (15), i.e., to showing that the integral of $f(z)$ along any triangular contour Δ contained in G vanishes. Hence, writing

$$\left|\int_\Delta f(z)\,dz\right| = M, \tag{17}$$

our goal is to show that $M = 0$. To this end, we draw the line segments joining the midpoints of the sides of Δ, thereby dividing Δ into four congruent subtriangles Δ_1, Δ_2, Δ_3 and Δ_4, as shown in Figure 17, all traversed in the

†For a weaker version of Cauchy's integral theorem, which is correspondingly easier to prove, see Prob. 14.

counterclockwise direction.† Clearly

$$\int_{\Delta} f(z)\,dz = \int_{\Delta_1} f(z)\,dz + \int_{\Delta_2} f(z)\,dz + \int_{\Delta_3} f(z)\,dz + \int_{\Delta_4} f(z)\,dz,$$

since each of the four segments joining midpoints of Δ is traversed twice in opposite directions, and hence gives rise to a pair of integrals which cancel each other out, while the remaining sides of $\Delta_1, \Delta_2, \Delta_3, \Delta_4$ make up the perimeter of Δ. It follows that the integral along at least one of the contours $\Delta_1, \Delta_2, \Delta_3, \Delta_4$, call it $\Delta^{(1)}$, can be no smaller than $M/4$ in absolute value, i.e.,

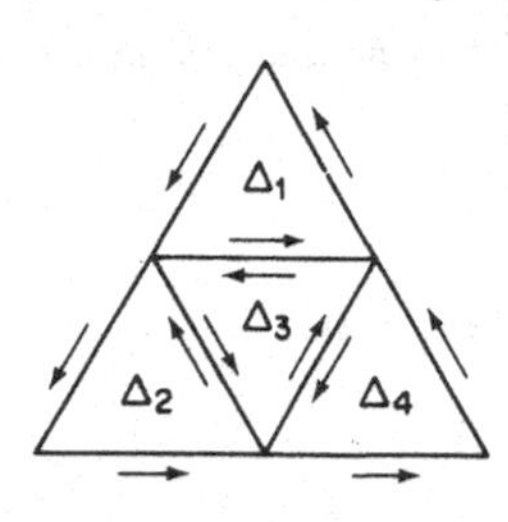

Figure 17

$$\left|\int_{\Delta^{(1)}} f(z)\,dz\right| \geq \frac{M}{4}, \tag{18}$$

since otherwise

$$\left|\int_{\Delta} f(z)\,dz\right| \leq \left|\int_{\Delta_1} f(z)\,dz\right| + \cdots + \left|\int_{\Delta_4} f(z)\,dz\right| < M,$$

contrary to (17). Next we divide the triangle $\Delta^{(1)}$ itself into four congruent subtriangles. Then the same argument shows that the integral along one of these new triangles, call it $\Delta^{(2)}$, can be no smaller than $M/4^2$ in absolute value, i.e.,

$$\left|\int_{\Delta^{(2)}} f(z)\,dz\right| \geq \frac{M}{4^2},$$

if (18) is to hold. Continuing this process indefinitely, we get an infinite sequence of triangles $\Delta^{(0)} = \Delta, \Delta^{(1)}, \ldots, \Delta^{(n)}, \ldots$, each containing the next and contained in G (here we use the fact that G is simply connected), such that

$$\left|\int_{\Delta^{(n)}} f(z)\,dz\right| \geq \frac{M}{4^n} \qquad (n = 0, 1, 2, \ldots). \tag{19}$$

Note that if l is the perimeter of Δ and l_n the perimeter of $\Delta^{(n)}$, then

$$l_n = \frac{l}{2^n}.$$

But the conclusion of Theorem 2.12 obviously continues to hold for nested triangles instead of nested rectangles with l_n playing the role of r_n (why?). Hence there is a unique point $z_0 \in G$ belonging to all the infinitely many triangles $\Delta^{(0)}, \Delta^{(1)}, \ldots, \Delta^{(n)}, \ldots$

Next, using the assumed analyticity of $f(z)$ in G, we observe that $f(z)$

†For simplicity, we use the same symbols $\Delta, \Delta_1, \Delta_2$, etc. (and the same word "triangle") to denote both triangular contours and the closed domains bounded by such contours. Whether we are talking about contours or domains will always be clear from the context.

has a finite derivative $f'(z_0)$ at z_0. Hence given any $\epsilon > 0$, there is a $\delta > 0$ such that $0 < |z - z_0| < \delta$ implies

$$\left|\frac{f(z) - f(z_0)}{z - z_0} - f'(z_0)\right| < \epsilon$$

or equivalently

$$|f(z) - f(z_0) - (z - z_0)f'(z_0)| < \epsilon|z - z_0|. \tag{20}$$

Moreover

$$\begin{aligned}\int_{\Delta^{(n)}} [f(z) - f(z_0) - (z - z_0)f'(z_0)]\, dz \\ &= \int_{\Delta^{(n)}} f(z)\, dz - f(z_0)\int_{\Delta^{(n)}} dz - f'(z_0)\int_{\Delta^{(n)}} z\, dz + z_0 f'(z_0)\int_{\Delta^{(n)}} dz \\ &= \int_{\Delta^{(n)}} f(z)\, dz,\end{aligned}$$

since

$$\int_{\Delta^{(n)}} dz = \int_{\Delta^{(n)}} z\, dz = 0$$

by Sec. 5.15 and Prob. 11, while (20) holds for all $z \in \Delta^{(n)}$ if n is sufficiently large, say $n > N$, since then $\Delta^{(n)}$ is contained in the disk $|z - z_0| < \delta$. But $z \in \Delta^{(n)}$ implies $|z - z_0| < l_n$ (why?), and hence

$$|f(z) - f(z_0) - (z - z_0)f'(z_0)| < \epsilon l_n$$

for all $z \in \Delta^{(n)}$ if $n > N$. Therefore, by Theorem 5.23,

$$\left|\int_{\Delta^{(n)}} f(z)\, dz\right| = \left|\int_{\Delta^{(n)}} [f(z) - f(z_0) - (z - z_0)f'(z_0)]\, dz\right| < \epsilon l_n^2 = \frac{\epsilon l^2}{4^n}.$$

Comparing this with (19), we get

$$\frac{M}{4^n} \le \left|\int_{\Delta^{(n)}} f(z)\, dz\right| < \frac{\epsilon l^2}{4^n} \qquad (n > N),$$

so that $M < \epsilon l^2$. But then $M = 0$, since M is intrinsically nonnegative and ϵ is arbitrarily small. ∎

5.42. Let C be a piecewise smooth closed *Jordan* curve with interior I, and let $f(z)$ be analytic "inside and on" C, i.e., analytic in the closed domain $\bar{I}$. Then $f(z)$ is analytic in a simply connected domain G containing $\bar{I}$ and in particular containing C, so that

$$\int_C f(z)\, dz = 0 \tag{21}$$

by Cauchy's integral theorem. In fact, given any point $z \in \bar{I}$, $f(z)$ is analytic in some open disk K_z centered at z (see Sec. 4.12). But then $\bar{I}$ can be covered by a finite number of these disks, say $K_{z_1}, \ldots, K_{z_n}$, by the Heine–Borel theorem. The set of all points belonging to at least one of the disks $K_{z_1}, \ldots, K_{z_n}$, being obviously open and connected, is a domain G containing $\bar{I}$. Moreover, G is simply connected, being the interior of a closed Jordan curve (which one?).

Actually, it can be shown that (21) remains valid if $f(z)$ is analytic in I and only continuous in $\bar{I}$, a result known as the *generalized Cauchy integral theorem.*†

5.43. Next, as in Sec. 3.24d, let $C_0, C_1, \ldots, C_n$ be $n + 1$ piecewise smooth closed Jordan curves such that the curves $C_1, \ldots, C_n$ all lie inside C_0 and do not intersect each other. Then the set of points lying inside the curve C_0 and outside the other n curves $C_1, \ldots, C_n$ is an $(n + 1)$-connected domain D, whose boundary consists of the $n + 1$ curves $C_0, C_1, \ldots, C_n$. Suppose $f(z)$ is analytic in $\bar{D}$. Then we have

$$\int_{C_0} f(z)\,dz = \int_{C_1} f(z)\,dz + \int_{C_2} f(z)\,dz + \cdots + \int_{C_n} f(z)\,dz. \tag{22}$$

To see this, we join each of the curves $C_0, C_1, \ldots, C_n$ to the next and C_n back to C_0 by drawing $n + 1$ nonintersecting auxiliary arcs $\gamma_0, \gamma_1, \ldots, \gamma_n$, thereby dividing $\bar{D}$ into two closed domains bounded by two closed Jordan curves Γ and Γ' made up of the arcs $\gamma_0, \gamma_1, \ldots, \gamma_n$ and parts of the curves $C_0, C_1, \ldots, C_n$, as illustrated in Figure 18 for the case $n = 2$ (here $\gamma_0 = ab$, $\gamma_1 = cd$, $\gamma_2 = ef$, $\Gamma = amfendcpba$, $\Gamma' = abp'cdn'efm'a$). But $f(z)$ is analytic inside and on both Γ and Γ', so that

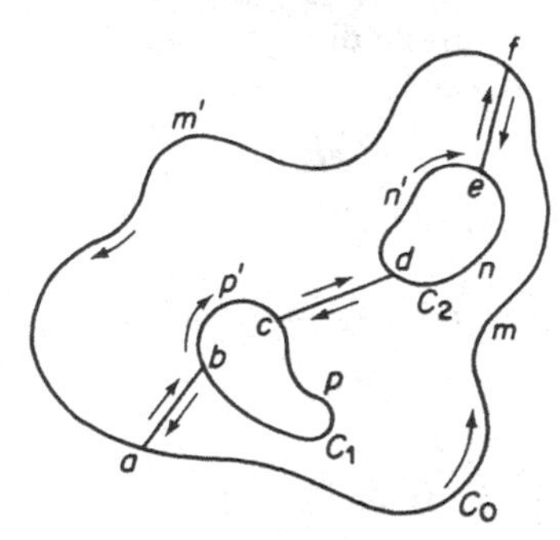

Figure 18

$$\int_{\Gamma} f(z)\,dz = 0, \qquad \int_{\Gamma'} f(z)\,dz = 0$$

by Cauchy's integral theorem. Adding these integrals and noting that each of the arcs $\gamma_0, \gamma_1, \ldots, \gamma_n$ is traversed twice in opposite directions, so that the corresponding integrals cancel each other out, we get

$$\int_{C_0} f(z)\,dz + \int_{C_1^-} f(z)\,dz + \int_{C_2^-} f(z)\,dz = \int_{\Gamma} f(z)\,dz + \int_{\Gamma'} f(z)\,dz = 0$$

in the case shown in the figure, and more generally

$$\int_{C_0} f(z)\,dz + \int_{C_1^-} f(z)\,dz + \int_{C_2^-} f(z)\,dz + \cdots + \int_{C_n^-} f(z)\,dz = 0. \tag{23}$$

But (23) is equivalent to (22), by Theorem 5.21. If there are only two curves

†See e.g., A. I. Markushevich, *op. cit.*, Volume III, Theorem 3.10.

C_0 and C_1 (where C_1 lies inside C_0), then (22) reduces to

$$\int_{C_0} f(z)\,dz = \int_{C_1} f(z)\,dz, \tag{22'}$$

provided of course that $f(z)$ is analytic in $\bar{D}$, i.e., on the curves C_0, C_1 and in the domain between them.

5.44. Example. Let C be any piecewise smooth closed Jordan curve, and suppose the point $z = 0$ lies outside C. Then, since the function $1/z$ is analytic everywhere except at the point $z = 0$, the integral

$$\int_C \frac{dz}{z}$$

vanishes, by Cauchy's integral theorem. However, if C encloses the point $z = 0$ (i.e., if the point $z = 0$ lies inside C), then, by (22′),

$$\int_C \frac{dz}{z} = \int_\gamma \frac{dz}{z},$$

where γ is any circle centered at $z = 0$ lying inside C. Suppose γ is of radius R. Then†

$$z = R(\cos\theta + i\sin\theta),$$

$$dz = R(-\sin\theta + i\cos\theta)\,d\theta = iR(\cos\theta + i\sin\theta)\,d\theta,$$

$$\frac{dz}{z} = i\,d\theta$$

if $z \in \gamma$, so that

$$\int_\gamma \frac{dz}{z} = \int_0^{2\pi} i\,d\theta = 2\pi i.$$

Therefore

$$\int_C \frac{dz}{z} = 2\pi i$$

for every piecewise smooth closed Jordan curve C enclosing the point $z = 0$.

Somewhat more generally, consider the integral

$$\int_C \frac{dz}{z - z_0},$$

where C is any piecewise smooth closed Jordan curve enclosing the point $z = z_0$. Making the substitution $z - z_0 = \zeta$, $dz = d\zeta$, we see that

$$\int_C \frac{dz}{z - z_0} = \int_{C'} \frac{d\zeta}{\zeta},$$

---

†Alternatively, we can write

$$z = Re^{i\theta}, \qquad dz = iRe^{i\theta}\,d\theta, \qquad \frac{dz}{z} = i\,d\theta,$$

anticipating Sec. 8.13a.

where the new curve C' now encloses the point $\zeta = 0$ (describe C'). It follows that

$$\int_C \frac{dz}{z - z_0} = 2\pi i. \tag{24}$$

On the other hand, if the point $z = z_0$ lies outside C, we have

$$\int_C \frac{dz}{z - z_0} = 0,$$

since the function $1/(z - z_0)$ is then analytic inside and on C. ·

5.5. Indefinite Complex Integrals

5.51. THEOREM. *Let $f(z)$ be analytic in a simply connected domain G. Then the integral*†

$$F(z) = \int_{z_0}^{z} f(\zeta)\, d\zeta, \tag{25}$$

taken along any piecewise smooth curve in G with fixed initial point z_0 and variable final point z, defines a single-valued analytic function in G with derivative $F'(z) = f(z)$.

Proof. The fact that $F(z)$ is independent of the curve joining z_0 to z (and hence is single-valued) follows at once from Cauchy's integral theorem and Prob. 8. Given any point $z \in G$, let K be any neighborhood of z contained in G, and let $z + h$ be any point of K. Then

$$F(z + h) - F(z) = \int_{z_0}^{z+h} f(\zeta)\, d\zeta - \int_{z_0}^{z} f(\zeta)\, d\zeta = \int_{z}^{z+h} f(\zeta)\, d\zeta, \tag{26}$$

where we can choose the "path of integration" in the last integral to be the line segment joining z to $z + h$ (see Figure 19). Dividing (26) by h and noting that

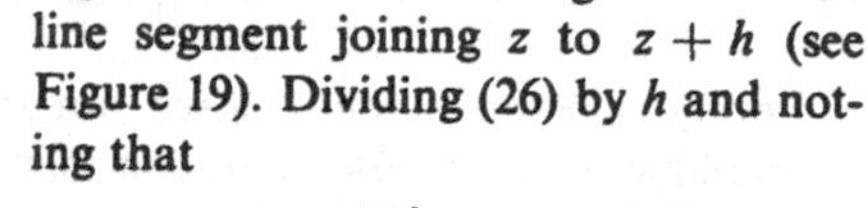

$$f(z) = f(z)\frac{1}{h}\int_{z}^{z+h} d\zeta = \frac{1}{h}\int_{z}^{z+h} f(z)\, d\zeta,$$

we get

$$\frac{F(z + h) - F(z)}{h} - f(z) = \frac{1}{h}\int_{z}^{z+h} [f(\zeta) - f(z)]\, d\zeta. \tag{27}$$

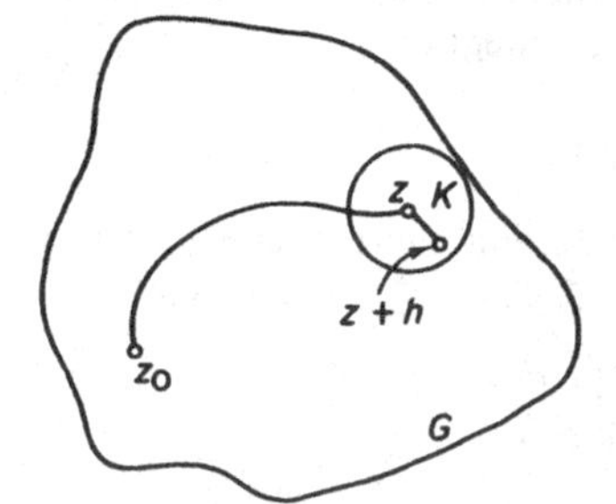

Figure 19

Since $f(z)$ is continuous at z, given any $\epsilon > 0$, there is a number $\delta = \delta(\epsilon) > 0$

†Note that $F(z)$ vanishes automatically if $z = z_0$.

such that $|z - \zeta| < \delta$ implies

$$|f(\zeta) - f(z)| < \epsilon.$$

Hence, applying Theorem 5.23 to (27), we find that

$$\left|\frac{F(z+h) - F(z)}{h} - f(z)\right| < \epsilon\frac{|h|}{|h|} = \epsilon$$

if $|h| < \delta$. But then

$$\lim_{h\to 0} \frac{F(z+h) - F(z)}{h} = f(z),$$

i.e., $F(z)$ is differentiable at z (and hence analytic in G since $z \in G$ is arbitrary), with derivative $F'(z) = f(z)$. ∎

5.52. Remark. It should be noted that we need not assume analyticity of $f(z)$ in proving Theorem 5.51, since it is clearly sufficient to assume that $f(z)$ is continuous in G and that the integral of $f(z)$ along every piecewise smooth closed curve contained in G vanishes. In fact, with these assumptions, G need no longer be simply connected (why not?).

5.53. Every single-valued function $\Phi(z)$ defined in a domain G such that $\Phi'(z) = f(z)$ for all $z \in G$ is called an *indefinite integral* (or *antiderivative*) of $f(z)$. Thus, according to Theorem 5.51, the function (25) is an indefinite integral of $f(z)$. In fact, (25) is essentially the "most general" indefinite integral of $f(z)$, as shown by the following

THEOREM. *Every indefinite integral of $f(z)$ is of the form*

$$\Phi(z) = F(z) + c = \int_{z_0}^{z} f(\zeta)\, d\zeta + c \qquad (z_0, z \in G), \tag{28}$$

where c is a complex constant.

Proof. Let

$$\Phi(z) - F(z) = \psi(z) = u(x, y) + iv(x, y).$$

Then clearly $\psi(z)$ is analytic in the underlying domain G, with derivative

$$\psi'(z) = \Phi'(z) - F'(z) = f(z) - f(z) = 0.$$

But

$$\psi'(z) = \frac{\partial u}{\partial x} + i\frac{\partial v}{\partial x} = \frac{\partial v}{\partial y} - i\frac{\partial u}{\partial y}$$

(see Sec. 4.23), and hence

$$\frac{\partial u}{\partial x} = \frac{\partial v}{\partial x} = \frac{\partial u}{\partial y} = \frac{\partial v}{\partial y} = 0$$

at every point of G. Therefore u and v are constant in G, i.e.,

$$\psi(z) = u + iv = c \qquad (c = \text{constant}),$$

which is equivalent to (28). ∎

5.54. Choosing $z = z_0$ in (28), we get $c = \Phi(z_0)$. This immediately leads to the formula

$$\int_{z_0}^{z} f(\zeta)\, d\zeta = \Phi(z) - \Phi(z_0)$$

expressing the "definite integral" on the left as a difference between the values of the indefinite integral at the end points of the path of integration. Thus, confining ourselves to the case of functions analytic in a simply connected domain, we see that complex integration (just like real integration) can be regarded both as a process of summation (cf. Sec. 5.11) and as the operation inverse to differentiation.

5.6. Cauchy's Integral Formula

5.61. THEOREM. *Let $f(z)$ be analytic in a domain G containing a piecewise smooth closed Jordan curve C and its interior. Then*

$$f(z_0) = \frac{1}{2\pi i}\int_C \frac{f(z)}{z - z_0}\, dz \tag{29}$$

if z_0 lies inside C.

Proof. If z_0 lies inside C, the function

$$\frac{f(z)}{z - z_0} \tag{30}$$

is analytic everywhere except at the point z_0. Let γ_R be a circle of radius R centered at z_0 which is small enough to lie inside C. Then by (22′),

$$\int_C \frac{f(z)}{z - z_0}\, dz = \int_{\gamma_R} \frac{f(z)}{z - z_0}\, dz.$$

But the value of the integral on the right is independent of the radius of γ_R, and hence

$$\int_C \frac{f(z)}{z - z_0} = \lim_{R\to 0}\int_{\gamma_R} \frac{f(z)}{z - z_0}\, dz.$$

Thus, to prove (29), we need only show that

$$\lim_{R\to 0}\int_{\gamma_R} \frac{f(z)}{z - z_0}\, dz = 2\pi i f(z_0), \tag{31}$$

i.e., that given any $\epsilon > 0$, there is a number $\delta = \delta(\epsilon) > 0$ such that

$$\left|\int_{\gamma_R} \frac{f(z)}{z - z_0}\, dz - 2\pi i f(z_0)\right| < \epsilon \tag{32}$$

whenever $R < \delta$. The left-hand side of (32) can be written in the form

$$\left|\int_{\gamma_R} \frac{f(z)}{z - z_0}\, dz - f(z_0)\int_{\gamma_R} \frac{dz}{z - z_0}\right| = \left|\int_{\gamma_R} \frac{f(z) - f(z_0)}{z - z_0}\, dz\right|,$$

because of (24). But $f(z)$ is continuous at z_0 (cf. Chap. 4, Prob. 2), and hence given any $\epsilon > 0$, there is a $\delta = \delta(\epsilon) > 0$ such that

$$|f(z) - f(z_0)| < \frac{\epsilon}{2\pi}$$

whenever $|z - z_0| < \delta$. Therefore, by Theorem 5.23, we have

$$\left|\int_{\gamma_R} \frac{f(z)}{z - z_0}\,dz - 2\pi i f(z_0)\right| < \frac{1}{R}\frac{\epsilon}{2\pi}2\pi R = \epsilon$$

whenever $R < \delta$, thereby proving (31). ∎

5.62. Formula (29), known as *Cauchy's integral formula*, relates the values of $f(z)$ inside the contour C to the values of $f(z)$ on C itself. Note that if z_0 lies outside C, then (30) is analytic inside and on C, and hence

$$\frac{1}{2\pi i}\int_C \frac{f(z)}{z - z_0} = 0$$

by Cauchy's integral theorem.

5.7. Infinite Differentiability of Analytic Functions

5.71. THEOREM. *If $f(z)$ is analytic in a domain G, then $f(z)$ is infinitely differentiable in G, i.e., $f(z)$ has derivatives of all orders in G. In fact, the nth derivative of $f(z)$ is given by the formula*†

$$f^{(n)}(z_0) = \frac{n!}{2\pi i}\int_C \frac{f(z)}{(z - z_0)^{n+1}}\,dz \qquad (z_0 \in G,\ n = 0, 1, 2, \ldots), \qquad (33)$$

where C is any piecewise smooth closed Jordan curve enclosing z_0 such that G contains both C and its interior.

Proof. We will prove (33) by induction, noting first that (33) reduces to Cauchy's integral formula (29) if $n = 0$. Assuming that (33) holds for a nonnegative integer $n - 1$, we now show that (33) also holds for n, thereby completing the induction. This will be accomplished by direct calculation of the quantity

$$f^{(n)}(z_0) = \lim_{h \to 0} \frac{f^{(n-1)}(z_0 + h) - f^{(n-1)}(z_0)}{h} \qquad (z_0 \in G).$$

Just as in the proof of Theorem 5.61, the right-hand side of (33) can be replaced by

$$\frac{n!}{2\pi i}\int_{\gamma_R} \frac{f(z)}{(z - z_0)^{n+1}}\,dz,$$

†By definition, $f^{(0)}(z) = f(z)$, $0! = 1$.

where γ_R is a circle of radius R centered at z_0 which is small enough to lie inside C. Therefore, choosing $|h| < R$, so that $z_0 + h$ lies inside γ_R, and assuming that (33) holds for $n - 1$, we have

$$\frac{f^{(n-1)}(z_0+h)-f^{(n-1)}(z_0)}{h}$$
$$=\frac{(n-1)!}{2\pi ih}\int_{\gamma_R} f(z)\left[\frac{1}{(z-z_0-h)^n}-\frac{1}{(z-z_0)^n}\right]dz$$
$$=\frac{(n-1)!}{2\pi ih}\int_{\gamma_R} f(z)\frac{(z-z_0)^n-(z-z_0-h)^n}{(z-z_0-h)^n(z-z_0)^n}dz$$
$$=\frac{(n-1)!}{2\pi i}\int_{\gamma_R} f(z)\frac{(z-z_0)^{n-1}+(z-z_0)^{n-2}(z-z_0-h)+\cdots+(z-z_0-h)^{n-1}}{(z-z_0-h)^n(z-z_0)^n}dz, \tag{34}$$

where in the last step we use the algebraic identity

$$a^n - b^n = (a-b)(a^{n-1} + a^{n-2}b + \cdots + b^{n-1}).$$

Let

$$M = \max_{z\in\gamma_R} |f(z)|, \tag{35}$$

as in Chap. 3, Prob. 12. Then it follows from (34) and (35) with the help of Theorem 5.23 that

$$\left|\frac{f^{(n-1)}(z_0+h)-f^{(n-1)}(z_0)}{h}-\frac{n!}{2\pi i}\int_{\gamma_R}\frac{f(z)}{(z-z_0)^{n+1}}dz\right|$$
$$=\left|\frac{(n-1)!}{2\pi i}\int_{\gamma_R} f(z)\frac{(z-z_0)^n+(z-z_0)^{n-1}(z-z_0-h)+\cdots+(z-z_0)(z-z_0-h)^{n-1}-n(z-z_0-h)^n}{(z-z_0-h)^n(z-z_0)^{n+1}}dz\right|$$
$$\le\frac{(n-1)!}{2\pi}2\pi RM|h|\frac{(2R)^{n-1}+2(2R)^{n-1}+\cdots+n(2R)^{n-1}}{(R-|h|)^nR^{n+1}} \tag{36}$$

where in the last step we use the estimate

$$R - |h| = ||z - z_0| - |h|| \le |z - z_0 - h| \le |z - z_0| + |h| < 2R$$

(cf. Sec. 1.38), together with the fact that

$$\begin{aligned}(z-z_0)^n &+ (z-z_0)^{n-1}(z-z_0-h) + \cdots + (z-z_0)(z-z_0-h)^{n-1} - n(z-z_0-h)^n\\
&= [(z-z_0)^n - (z-z_0-h)^n] + \cdots + [(z-z_0)^2(z-z_0-h)^{n-2} - (z-z_0-h)^n]\\
&\quad + [(z-z_0)(z-z_0-h)^{n-1} - (z-z_0-h)^n]\\
&= h[(z-z_0)^{n-1} + (z-z_0)^{n-2}(z-z_0-h) + \cdots + (z-z_0-h)^{n-1}]\\
&\quad + \cdots + h(z-z_0-h)^{n-2}[(z-z_0) + (z-z_0-h)] + h(z-z_0-h)^{n-1}.\end{aligned}$$

But the right-hand side of (36) approaches zero as $h \to 0$, and hence the same is true of the left-hand side, i.e.,

$$\begin{aligned}f^{(n)}(z_0) &= \lim_{h\to 0}\frac{f^{(n-1)}(z_0+h) - f^{(n-1)}(z_0)}{h}\\
&= \frac{n!}{2\pi i}\int_{\gamma_R}\frac{f(z)}{(z-z_0)^{n+1}}dz = \frac{n!}{2\pi i}\int_C\frac{f(z)}{(z-z_0)^{n+1}}dz. \quad \blacksquare\end{aligned}$$

5.72. Remark. It is an immediate consequence of Theorem 5.71 that if $f(z)$ is analytic in a domain G, then so are all the derivatives

$$f'(z), f''(z), \ldots, f^{(n)}(z), \ldots \tag{37}$$

In particular, the derivatives (37) are all automatically continuous in G (cf. Chap. 4, Prob. 2).

5.73. It is now an easy matter to prove a result which is essentially the converse of Cauchy's integral theorem:

THEOREM *(Morera). Let $f(z)$ be continuous in a domain G, and suppose*

$$\int_C f(z)\,dz = 0 \tag{38}$$

for every piecewise smooth closed curve C contained in G. Then $f(z)$ is analytic in G.

Proof. According to Theorem 5.51 and Sec. 5.52, the integral

$$F(z) = \int_{z_0}^{z} f(\zeta)\,d\zeta,$$

taken along any piecewise smooth curve in G with initial point z_0 and final point z, defines a single-valued analytic function in G, with derivative $F'(z) = f(z)$. But, as just noted, $F'(z)$ is itself analytic in G, being the derivative of a function analytic in G. Therefore $f(z)$ is analytic in G. ∎

5.8. Harmonic Functions

5.81. Definitions. A real function $u = u(x, y)$ of two real variables x and y is said to be *harmonic* in a domain G if it has continuous second partial derivatives

$$\frac{\partial^2 u}{\partial x^2}, \quad \frac{\partial^2 u}{\partial x\,\partial y}, \quad \frac{\partial^2 u}{\partial y\,\partial x}, \quad \frac{\partial^2 u}{\partial y^2}$$

at every point of G† and satisfies *Laplace's equation*

$$\frac{\partial^2 u}{\partial x^2} + \frac{\partial^2 u}{\partial y^2} = 0$$

everywhere in G. Let $u = u(x, y)$ and $v = v(x, y)$ be two functions harmonic in a domain G, which satisfy the Cauchy–Riemann equations

$$\frac{\partial u}{\partial x} = \frac{\partial v}{\partial y}, \quad \frac{\partial u}{\partial y} = -\frac{\partial v}{\partial x} \tag{39}$$

†The function u and its first partial derivatives $\partial u/\partial x$, $\partial u/\partial y$ are then automatically continuous in G and

$$\frac{\partial^2 u}{\partial x\,\partial y} = \frac{\partial^2 u}{\partial y\,\partial x}$$

(R. A. Silverman, *op. cit.*, Theorems 12.1–12.3).

(Sec. 4.22) at every point of G. Then u and v are said to be *conjugate harmonic functions* (in G), and each of the functions u and v is said to be the *conjugate harmonic function* (or simply the *harmonic conjugate*) of the other.

5.82. As shown by the following theorem, there is an intimate connection between harmonic functions and analytic functions:

THEOREM. *Let $f(z) = u(x, y) + iv(x, y)$ be a complex function defined in a domain G. Then $f(z)$ is analytic in G if and only if its real part $u = u(x, y)$ and imaginary part $v = v(x, y)$ are conjugate harmonic functions in G.*

Proof. If $f(z)$ is analytic in G, then u and v are differentiable and satisfy the Cauchy–Riemann equations (39) at every point of G, by Theorem 4.22. Moreover, the function $f'(z)$ is also analytic in G, being the derivative of a function analytic in G (Sec. 5.72). But

$$f'(z) = \frac{\partial u}{\partial x} - i\frac{\partial u}{\partial y} = \frac{\partial v}{\partial y} + i\frac{\partial v}{\partial x}$$

(Sec. 4.23), and hence both pairs of functions

$$\frac{\partial u}{\partial x}, \qquad -\frac{\partial u}{\partial y} \tag{40}$$

and

$$\frac{\partial v}{\partial y}, \qquad \frac{\partial v}{\partial x} \tag{40'}$$

are differentiable in G and satisfy the Cauchy–Riemann equations at every point of G. Writing the first Cauchy–Riemann equation for (40) and the second for (40′), we get

$$\frac{\partial}{\partial x}\left(\frac{\partial u}{\partial x}\right) = \frac{\partial}{\partial y}\left(-\frac{\partial u}{\partial y}\right), \qquad \frac{\partial}{\partial y}\left(\frac{\partial v}{\partial y}\right) = -\frac{\partial}{\partial x}\left(\frac{\partial v}{\partial x}\right)$$

or

$$\frac{\partial^2 u}{\partial x^2} + \frac{\partial^2 u}{\partial y^2} = 0, \qquad \frac{\partial^2 v}{\partial x^2} + \frac{\partial^2 v}{\partial y^2} = 0.$$

Hence both functions u and v satisfy Laplace's equation everywhere in G, and to complete the proof that u and v are conjugate harmonic functions in G we need only show that u and v have continuous second partial derivatives at every point of G. But this follows at once from the fact that $f''(z)$ is analytic and hence continuous in G (Sec. 5.72), since $f''(z)$ can be written in any of the forms

$$\begin{aligned} f''(z) &= \frac{\partial^2 u}{\partial x^2} + i\frac{\partial^2 v}{\partial x^2} = -\frac{\partial^2 u}{\partial y^2} - i\frac{\partial^2 v}{\partial y^2} = -i\frac{\partial^2 u}{\partial x\,\partial y} + \frac{\partial^2 v}{\partial x\,\partial y} \\ &= -i\frac{\partial^2 u}{\partial y\,\partial x} + \frac{\partial^2 v}{\partial y\,\partial x}. \end{aligned}$$

Conversely, if u and v are conjugate harmonic functions in G, then in particular u and v have continuous first partial derivatives at every point of G, and hence are differentiable in G (recall Sec. 4.23). Since u and v also satisfy the Cauchy–Riemann equations at every point of G, it follows from Theorem 4.22 that $f(z) = u + iv$ is analytic in G. ▮

5.83. Example. The function

$$f(z) = z^3$$

is analytic in the whole plane, and hence its real and imaginary parts

$$u = x^3 - 3xy^2, \qquad v = 3x^2y - y^3 \tag{41}$$

are a pair of conjugate harmonic functions in the whole plane. It is easily verified that both functions (41) satisfy Laplace's equation.

5.84. THEOREM. *If $u = u(x, y)$ is harmonic in a simply connected domain G, then the harmonic conjugate of u is given by*

$$v = v(x, y) = \int_{(x_0, y_0)}^{(x, y)} -\frac{\partial u}{\partial y}\,dx + \frac{\partial u}{\partial x}\,dy + c, \tag{42}$$

where c is an arbitrary real constant and the line integral is along any piecewise smooth curve in G with initial point $(x_0, y_0) \in G$ and final point $(x, y) \in G$.

Proof. If v exists, then

$$dv = \frac{\partial v}{\partial x}\,dx + \frac{\partial v}{\partial y}\,dy = -\frac{\partial u}{\partial y}\,dx + \frac{\partial u}{\partial x}\,dy, \tag{43}$$

since v must satisfy the Cauchy–Riemann equations (39). The existence of v is guaranteed if the right-hand side of (43) is an exact differential in G, which in turn is implied by the condition†

$$\frac{\partial}{\partial x}\left(\frac{\partial u}{\partial x}\right) = \frac{\partial}{\partial y}\left(-\frac{\partial u}{\partial y}\right). \tag{44}$$

But (44) is automatically satisfied, being just another way of writing Laplace's equation for the harmonic function u. Substituting (43) into the obvious identity

$$v = \int_{(x_0, y_0)}^{(x, y)} dv + c,$$

we get (42). ▮

5.85. It follows from Theorems 5.82 and 5.84 that, to within an arbitrary purely imaginary constant, the function

$$f(z) = u(x, y) + iv(x, y) = u(x, y) + i\int_{(x_0, y_0)}^{(x, y)} -\frac{\partial u}{\partial y}\,dx + \frac{\partial u}{\partial x}\,dy$$

is the unique analytic function in G with $u(x, y)$ as its real part.

---

†See the comment to Sec. 5.8.

COMMENTS

5.1. The integral of $f(z)$ along C can be defined for a general "rectifiable" curve C, i.e., for a general curve C with a well-defined length (see Prob. 2), but the case of piecewise smooth C is sufficient for our purposes. By the same token, we will confine ourselves to the case of continuous $f(z)$, although the integral of $f(z)$ along C exists for more general $f(z)$.

5.2. Each of Theorems 5.21–5.23 obviously remains valid if the word "continuous" is replaced by "integrable."

5.3. The key idea of Lemma 5.31 is that the integral of $f(z)$ along a piecewise smooth curve C can be approximated *arbitrarily closely* by the integral of $f(z)$ along a polygonal curve inscribed in C and still contained in the underlying domain G. In particular, if the integral of $f(z)$ along all polygonal curves contained in C can be shown to vanish, the same must be true of the integral of $f(z)$ along the given curve C. (The only number arbitrarily close to 0 is the number 0 itself.) The key idea of Lemma 5.32 is that the integral of $f(z)$ along a polygonal curve contained in G can be reduced to a finite sum of integrals of $f(z)$ along triangular contours contained in G, provided G is simply connected. In particular, if the integral of $f(z)$ along all triangular contours contained in G can be shown to vanish, the same must be true of the integral of $f(z)$ along any polygonal curve contained in G and hence along any piecewise smooth curve contained in G. The proof of the fundamental Cauchy integral theorem (Theorem 5.41) is now well under way.

5.4. In connection with Cauchy's integral theorem, it is crucial to note that $f(z)$ is assumed merely to be *analytic* in G, i.e., to have a derivative $f'(z)$ at every point of G. If $f(z)$ were known to have a *continuous* derivative in G, then the proof of Cauchy's integral theorem given in Sec. 5.41 would be unnecessarily complicated, and in particular there would be no need for Lemmas 5.31 and 5.32, since a simpler proof based on the use of Green's theorem would then suffice, at least in the case where C is a Jordan curve (see Prob. 14). Thus a remarkable aspect of Theorem 5.41 is precisely that no requirements other than analyticity are imposed on $f(z)$. The continuity of $f'(z)$ and, remarkably enough, the infinite differentiability of $f(z)$ then follow as *implications* of Theorem 5.41 (see Sec. 5.72). This fact cannot be emphasized too strongly.

5.5. In Theorem 5.51 it is crucial that G be simply connected. Otherwise $F(z)$ may well be multiple-valued, as in Chap. 9, Prob. 20.

5.6. According to Cauchy's integral formula (Theorem 5.61), if the values of $f(z)$ are changed inside the curve C, then the values of $f(z)$ must change on C if $f(z)$ is to remain analytic. This is further evidence of the very special character of analytic functions (cf. Secs. 4.14 and 10.24).

5.7. The algebraic details of the proof of Theorem 5.71 are a little tedious and need not be worked through. Note that (33) is precisely the result of n-fold differentiation of Cauchy's integral formula (29), under the assumption that the differentiation behind the integral sign is justified. The proof of (33) for $n = 1$ is particularly simple. We need only note that

$$\frac{f(z_0 + h) - f(z_0)}{h} = \frac{1}{2\pi i h} \int_{\gamma_R} f(z) \left[\frac{1}{z - z_0 - h} - \frac{1}{z - z_0} \right] dz$$
$$= \frac{1}{2\pi i} \int_{\gamma_R} \frac{f(z)}{(z - z_0 - h)(z - z_0)} dz,$$

and hence

$$\left| \frac{f(z_0 + h) - f(z_0)}{h} - \frac{1}{2\pi i} \int_{\gamma_R} \frac{f(z)}{(z - z_0)^2} dz \right| \leq \frac{1}{2\pi} 2\pi R M |h| \frac{1}{(R - |h|)R^2},$$

where the right-hand side obviously approaches zero as $h \to 0$, implying

$$f'(z_0) = \frac{1}{2\pi i} \int_{\gamma_R} \frac{f(z)}{(z - z_0)^2} dz.$$

5.8. According to *Green's theorem*, proved in advanced calculus, if the functions $P = P(x, y)$, $Q = Q(x, y)$ and their partial derivatives are continuous on a piecewise smooth closed Jordan curve C and its interior I, then

$$\int_C P\,dx + Q\,dy = \iint_I \left(\frac{\partial Q}{\partial x} - \frac{\partial P}{\partial y} \right) dx\,dy.$$

Therefore, if $\partial Q/\partial x = \partial P/\partial y$ everywhere in a simply connected domain G, we have

$$\int_C P\,dx + Q\,dy = 0$$

for every such curve C contained in G, and hence the integral

$$\Phi = \Phi(x, y) = \int_{(x_0, y_0)}^{(x, y)} P\,dx + Q\,dy$$

is "path-independent" (by an obvious modification of Prob. 8), or equivalently, $P\,dx + Q\,dy$ is an exact differential (of the function Φ).

PROBLEMS

1. "A smooth curve has a continuously varying tangent." Explain this statement.

2. Given a continuous curve C with parametric equation

$$z = z(t) = x(t) + iy(t) \qquad (a \leq t \leq b), \tag{45}$$

let L be a polygonal curve inscribed in C with vertices at the points $z_k = z(t_k)$, where $a = t_0 < t_1 < t_2 < \cdots < t_{n-1} < t_n = b$. Then

$$\sum_{k=1}^{n} |\Delta z_k| = \sum_{k=1}^{n} |z_k - z_{k-1}| \tag{46}$$

is clearly the length of L. Suppose the set of numbers (46) corresponding to all possible polygonal curves inscribed in C has a finite least upper bound l. Then l is called the *length* of C, and C itself is said to be *rectifiable* (more concisely, C is said to be of length l). Prove that if C is smooth, then C is of length l, where l satisfies the estimate

$$\sqrt{m_x^2 + m_y^2}\,(b-a) \leq l \leq \sqrt{M_x^2 + M_y^2}\,(b-a),$$

involving the quantities†

$$m_x = \min_{a \leq t \leq b} |x'(t)|, \qquad m_y = \min_{a \leq t \leq b} |y'(t)|,$$

$$M_x = \max_{a \leq t \leq b} |x'(t)|, \qquad M_y = \max_{a \leq t \leq b} |y'(t)|.$$

3. Given a continuous curve $\widehat{AB}$ with end points A and B, let P be any other point of $\widehat{AB}$. Prove that $\widehat{AB}$ is rectifiable if and only if the arcs $\widehat{AP}$ and $\widehat{PB}$ are rectifiable. Prove that the length of $\widehat{AB}$ is then the sum of the lengths of $\widehat{AP}$ and $\widehat{PB}$. Prove that every piecewise smooth curve is rectifiable.

4. Prove that the length l of a smooth curve with parametric equation (45) is given by

$$l = \int_a^b |z'(t)|\,dt = \int_a^b \sqrt{[x'(t)]^2 + [y'(t)]^2}\,dt.$$

Prove that the same is true for a piecewise smooth curve, despite the fact that in this case $z'(t)$ may fail to exist at a finite number of points of the interval $a \leq t \leq b$.

5. Let C be a piecewise smooth curve made up of smooth arcs $C_1, C_2, \ldots, C_n$, and let $f(z)$ be continuous on C. Prove that

$$\int_C f(z)\,dz = \int_{C_1} f(z)\,dz + \int_{C_2} f(z)\,dz + \cdots + \int_{C_n} f(z)\,dz$$

directly from the definition

$$\int_C f(z)\,dz = \lim_{\lambda \to 0} \sum_{k=1}^{n} f(\zeta_k)\,\Delta z_k$$

(Sec. 5.11).

6. Let C be a piecewise smooth curve of length l, with parametric equation (45), and let $s(t)$ be the length of the variable arc of C with initial point $z(a)$ and final point $z(t)$. Prove that $s = s(t)$ is continuous and strictly increasing in the interval $a \leq t \leq b$, with a continuous strictly increasing inverse $t = t(s)$ in the interval $0 \leq s \leq l$. Show that C has a parametric representation of the form

$$z = \tilde{z}(s) \qquad (0 \leq s \leq l) \tag{45'}$$

with the arc length s as parameter. Prove that the derivative $\tilde{z}'(s)$ exists and is of modulus 1 at all but a finite number of points of the interval $0 \leq s \leq l$.

---

†Thus m_x is the minimum and M_x the maximum of $|x'(t)|$ in the interval $a \leq t \leq b$, and similarly for m_y and M_y (cf. Chap. 3, Prob. 12).

7. Let C be a closed curve with parametric equation (45), and let t_0 be any point in the open interval $a < t < b$. Let C_1 be the curve

$$z = z(t) \qquad (a \leq t \leq t_0),$$

and let γ be the curve

$$z = z(t) \qquad (t_0 \leq t \leq b),$$

so that $C = C_1 + \gamma$, i.e., C is obtained by first traversing C_1 and then traversing γ. Prove that the curves C_1 and $C_2 = \gamma^-$ have the same initial and final points.

8. Prove that

$$\int_{C_1} f(z)\,dz = \int_{C_2} f(z)\,dz \tag{47}$$

for all curves C_1 and C_2 contained in a given domain G with the same initial and final points if and only if

$$\int_C f(z)\,dz = 0 \tag{48}$$

for all *closed* curves C contained in G.

9. Let σ_1 be the segment joining the points 0 and $2 + i$, σ_2 the segment joining the points 0 and 2, and σ_3 the segment joining the points 2 and $2 + i$. Moreover, let $C_1 = \sigma_1$, $C_2 = \sigma_2 + \sigma_3$, $C = C_1 + C_2^-$, so that C_1 and C_2 have the same initial and final points, while C is closed. Prove that

$$\int_{C_1} \operatorname{Re} z\,dz \neq \int_{C_2} \operatorname{Re} z\,dz,$$

while

$$\int_C \operatorname{Re} z\,dz \neq 0.$$

Comment. Thus the kind of "path-independent" behavior of complex integrals exhibited in Example 5.15 and analyzed in Prob. 8 is the exception rather than the rule.

10. Evaluate the integral

$$\int_C |z|\,dz$$

where C is
a) The line segment with initial point -1 and final point 1;
b) The semicircle $|z| = 1$, $\operatorname{Im} z \geq 0$ with initial point -1 and final point 1;
c) The circle $|z| = r$ with arbitrary initial (and final) point.

11. Evaluate the integral

$$\int_C z\,dz$$

directly from the definition

$$\int_C f(z)\,dz = \lim_{\lambda \to 0} \sum_{k=1}^{n} f(\zeta_k)\,\Delta z_k$$

(cf. Sec. 5.11).

12. Given a function continuous in the whole z-plane, let

$$M(R) = \max_{|z|=R} |f(z)|$$

and suppose $RM(R) \longrightarrow 0$ as $R \longrightarrow \infty$. Prove that

$$\lim_{R\to\infty} \int_{|z|=R} f(z)\, dz = 0.$$

13. Let $f(z)$ be continuous on a piecewise smooth curve C. Prove that

$$\left| \int_C f(z)\, dz \right| \le \int_C |f(z)|\, ds(t),$$

where $s(t)$ is the same as in Prob. 6. Deduce Theorem 5.23 from this sharper estimate.

14. Using Green's theorem (comment to Sec. 5.8), prove the following weaker version of Cauchy's integral theorem: If $f(z)$ has a continuous derivative $f'(z)$ at every point of a simply connected domain G, then

$$\int_C f(z)\, dz = 0$$

for every piecewise smooth closed Jordan curve C contained in G.

Comment. Of course, the assumption of continuous $f'(z)$ is quite unnecessary (see Theorem 5.41) and in fact completely redundant (see Sec. 5.72). There is also no need for C to be a Jordan curve.

15. Prove that

$$\int_C \frac{dz}{z^2+1} = 0$$

if C is any piecewise smooth closed Jordan curve contained in the annulus $1 < |z| < R$.

16. Evaluate the integral

$$\int_C \frac{dz}{z^2+1}$$

if C is

a) The circle $|z-i| = 1$;
b) The circle $|z+i| = 1$;
c) The circle $|z| = 2$

(all traversed in the counterclockwise direction).

17. Does the function

$$f(z) = \frac{1}{z(1-z^2)}$$

have an indefinite integral in the domain $0 < |z| < 1$?

18. Describe the behavior of the integral

$$\int_{|z-a|=R} \frac{z^4+z^2+1}{z(z^2+1)}\, dz$$

as a function of $R > 0$.

19. Find all possible values of the integral

$$\int_C \frac{dz}{z^2+1},$$

where C is a piecewise smooth curve with initial point 0 and final point 1. What restriction must be imposed on C?

20. Verify that the function $u = x^2 - y^2 + x$ is harmonic in the whole plane. Find its harmonic conjugate v and the corresponding analytic function $f(z) = u + iv$.

21. Verify that the function

$$v = y - \frac{y}{x^2+y^2}$$

is harmonic except at the origin. Find its harmonic conjugate and the corresponding analytic function $f(z) = u + iv$.

22. If u is harmonic, is the same true of u^2? More generally, if u is harmonic, which functions $f(u)$ are also harmonic?

23. Prove that if u_1 and u_2 are harmonic, then so is any linear combination $c_1u_1 + c_2u_2$ with arbitrary real coefficients c_1 and c_2.

24. Suppose u and v are conjugate harmonic functions in a domain G. Show that the same is true of the functions

$$U = au - bv, \qquad V = bu + av,$$

where a and b are arbitrary real constants.

25. Let $u = u(x, y)$, $v = v(x, y)$ be conjugate harmonic functions in a domain G, while $U = U(u, v)$, $V = V(u, v)$ are conjugate harmonic functions in a domain D, and suppose $u + iv \in D$ for every $x + iy \in G$. Prove that

$$U(u(x, y), v(x, y)), \qquad V(u(x, y), v(x, y))$$

are conjugate harmonic functions in G.

26. Prove that a harmonic function of an analytic function is harmonic. More exactly, prove that if the function $U(w)$ is harmonic in a domain D and if the function $w = f(z)$ is analytic in a domain G and takes its values in D, then the function $U(f(z))$ is harmonic in G.

27. Let $f(z)$ be an *integral of the Cauchy type*, i.e., let

$$f(z) = \frac{1}{2\pi i}\int_C \frac{\varphi(\zeta)}{\zeta - z}\,d\zeta,$$

where C is any piecewise smooth curve (not necessarily closed) and $\varphi(\zeta)$ is any function continuous (but not necessarily analytic) on C. Prove that $f(z)$ is analytic in any domain G containing no points of C, with derivatives given by formula (33).

28. Given any simply connected domain G in the z-plane and any piecewise smooth curve Γ in the ζ-plane, let $f(z, \zeta)$ be a function of two complex variables z

and ζ such that
(1) $f(z, \zeta)$ is analytic in G for every $\zeta \in \Gamma$;
(2) $f(z, \zeta)$ is continuous in both variables z and ζ for all $z \in G$, $\zeta \in \Gamma$.†
Prove that the function defined by the integral

$$F(z) = \int_{\Gamma} f(z, \zeta)\, d\zeta$$

is analytic in G.

†More exactly, given any $z_0 \in G$, $\zeta_0 \in \Gamma$, $\epsilon > 0$, there is a number $\delta = \delta(\epsilon) > 0$ such that $|f(z, \zeta) - f(z_0, \zeta_0)| < \epsilon$ for all $z \in G$, $\zeta \in \Gamma$ such that $|z - z_0| < \delta$, $|\zeta - \zeta_0| < \delta$.

CHAPTER SIX

Complex Series

6.1. Convergence vs. Divergence

6.11. Series and partial sums. Let

$$\sum_{n=1}^{\infty} z_n = z_1 + z_2 + \cdots + z_n + \cdots \qquad (1)$$

be an infinite series all of whose terms are complex numbers (briefly a *complex series*), and let

$$s_n = z_1 + z_2 + \cdots + z_n$$

be the *nth partial sum* of (1), i.e., the sum of the first n terms of (1). Then

$$s_1, s_2, \ldots, s_n, \ldots \qquad (2)$$

is an infinite sequence of complex numbers "generated" by (1). Conversely, starting from (2) we can easily deduce the original series (1) by merely observing that (1) can be written in the form

$$s_1 + (s_2 - s_1) + \cdots + (s_n - s_{n-1}) + \cdots$$

6.12. Definition. We say that the series (1) is *convergent* (or *converges*) if the corresponding sequence (2) converges to a finite limit s; the limit s is then called the *sum* of the series (1). More concisely, the series (1) is convergent with sum s if and only if

$$\lim_{n\to\infty} s_n = s.$$

If the sequence (2) fails to converge to a finite limit, we say that the series (1) is *divergent* (or *diverges*). A series may diverge either because s_n approaches infinity or because s_n has no limit at all; the series (1) is said to be *properly divergent* in the first case and *oscillatory* in the second.

6.13. It is clear from the foregoing that investigating the convergence of the series (1) is equivalent to investigating that of the sequence (2). For example, the *geometric series*

$$1 + q + q^2 + \cdots + q^n + \cdots \tag{3}$$

is convergent for $|q| < 1$ and divergent for $|q| > 1$. In fact, the nth partial sum of (3) is given by

$$s_n = 1 + q + q^2 + \cdots + q^{n-1} = \frac{1 - q^n}{1 - q} = \frac{1}{1 - q} - \frac{q^n}{1 - q}.$$

But $|q^n| = |q|^n$ and hence $q^n \to 0$ as $n \to \infty$ if $|q| < 1$, while $q^n \to \infty$ if $|q| > 1$. It follows that

$$\lim_{n\to\infty} s_n = \begin{cases} \dfrac{1}{1-q} & \text{if } |q| < 1, \\ \infty & \text{if } |q| > 1. \end{cases}$$

6.14. Next we prove a necessary condition for convergence of a series:

THEOREM. *If the series* (1) *is convergent, then*

$$\lim_{n\to\infty} z_n = 0. \tag{4}$$

Proof. If (1) is convergent, the sequence (2) converges to a finite limit s. But then

$$\lim_{n\to\infty} z_n = \lim_{n\to\infty} (s_n - s_{n-1}) = \lim_{n\to\infty} s_n - \lim_{n\to\infty} s_{n-1} = s - s = 0. \quad \blacksquare$$

Thus the series (1) diverges whenever (4) fails to hold.

6.15. If $|q| = 1$ then $|q^n| = 1$ for all n, while if $|q| > 1$ then $q^n \to \infty$ as $n \to \infty$. It follows from Theorem 6.14 that the series (3) is divergent for $|q| \geq 1$. In fact, (3) is properly divergent for $|q| > 1$ (see Sec. 6.13) and oscillatory for $|q| = 1$,† since the partial sum

$$s_n = \frac{1 - q^n}{1 - q} \qquad (q \neq 1)$$

does not approach infinity as $n \to \infty$.

6.16. Remark. The familiar harmonic series

$$1 + \frac{1}{2} + \frac{1}{3} + \cdots + \frac{1}{n} + \cdots \tag{5}$$

---

†Unless $q = 1$, in which case (3) is again properly divergent.

is an example of a *divergent* series satisfying (4).† Thus the condition (4) is necessary *but not sufficient* for convergence of a series.

6.2. Absolute vs. Conditional Convergence

6.21. Given a complex series

$$z_1 + z_2 + \cdots + z_n + \cdots, \tag{6}$$

consider the new series

$$|z_1| + |z_2| + \cdots + |z_n| + \cdots \tag{6'}$$

whose terms are the absolute values of those of (6). Investigating the convergence of (6′) is a much simpler matter than investigating that of (6). In fact, the partial sums

$$\sigma_n = |z_1| + |z_2| + \cdots + |z_n|$$

of (6′) form a positive nondecreasing sequence. Hence the sequence σ_n is either bounded for all n or else approaches infinity as $n \to \infty$. In the first case σ_n has a finite limit so that (6′) is convergent, while in the second case (6′) is properly divergent.

6.22. THEOREM. *If the series* (6′) *is convergent, then so is the series* (6).

Proof. It follows from

$$|s_{n+p} - s_n| = |z_{n+1} + \cdots + z_{n+p}| \leq |z_{n+1}| + \cdots + |z_{n+p}|$$

that

$$|s_{n+p} - s_n| \leq \sigma_{n+p} - \sigma_n. \tag{7}$$

If (6′) converges, then by the Cauchy convergence criterion (cf. Chap. 2, Prob. 14), given any $\epsilon > 0$, there exists an integer $N = N(\epsilon) > 0$ such that

$$\sigma_{n+p} - \sigma_n < \epsilon$$

for all $n > N$ and $p = 1, 2, \ldots$ Therefore by (7)

$$|s_{n+p} - s_n| < \epsilon$$

for the same n and p. But then (6) converges by another application of the Cauchy convergence criterion. ∎

If the series (6′) converges, we say that the original series (6) is *absolutely* or (*unconditionally*) *convergent*. Hence the theorem states that an absolutely

---

†A proof of the divergence of (5) is given, say, in G. M. Fichtenholz, *Infinite Series: Rudiments* (translated by R. A. Silverman), Gordon and Breach, Science Publishers, Inc., New York (1970), p. 10.

convergent series is automatically convergent. The importance of absolutely convergent series lies in the fact that operations on them obey the same rules as operations on finite sums (see below). The converse of Theorem 6.22 is false, i.e., there exist convergent series which are not absolutely convergent; such series are said to be *conditionally convergent.*

6.23. Examples

a. The series

$$1 - \frac{1}{2} + \frac{1}{3} - \frac{1}{4} + \cdots \tag{8}$$

is conditionally convergent. In fact, (8) converges by Leibniz's test† (see also Chap. 7, Prob. 12), but the harmonic series

$$1 + \frac{1}{2} + \frac{1}{3} + \frac{1}{4} + \cdots$$

made up of the absolute values of the terms of (8) is divergent, as already noted.

b. The series

$$1 - \frac{1}{2^2} + \frac{1}{3^2} - \frac{1}{4^2} + \cdots$$

is absolutely convergent.‡

6.24. THEOREM ***(Addition and subtraction of series).*** *If the two complex series*

$$z_1 + z_2 + \cdots + z_n + \cdots, \tag{9}$$

$$z_1' + z_2' + \cdots + z_n' + \cdots \tag{9'}$$

are convergent with sums s and s', respectively, then the series

$$(z_1 \pm z_1') + (z_2 \pm z_2') + \cdots + (z_n \pm z_n') + \cdots \tag{10}$$

is convergent with sum $s \pm s'$.

Proof. Let s_n and s_n' denote the nth partial sums of (9) and (9′), respectively, so that

$$\lim_{n\to\infty} s_n = s, \qquad \lim_{n\to\infty} s_n' = s'$$

by hypothesis. Then

$$\lim_{n\to\infty} [(z_1 \pm z_1') + \cdots + (z_n \pm z_n)] = \lim_{n\to\infty} (s_n \pm s_n')$$
$$= \lim_{n\to\infty} s_n \pm \lim_{n\to\infty} s_n' = s \pm s'. \quad ▮$$

---

†G. M. Fichtenholz, *op. cit.*, p. 73.

‡*Ibid.*, pp. 11, 73.

Thus two convergent series can be added or subtracted "term by term." Hence the operations of addition and subtraction can be carried over from the class of all finite sums to the class of all (conditionally or unconditionally) convergent series. Things are different for the operation of multiplication, which in general cannot be applied to conditionally convergent series (see Sec. 6.28).

6.25. Starting from a given series

$$z_1 + z_2 + \cdots + z_n + \cdots,$$

or more concisely

$$z_1 + z_2 + \cdots, \tag{11}$$

let

$$\begin{gathered} z_{\alpha_1} + z_{\alpha_2} + \cdots, \\ z_{\beta_1} + z_{\beta_2} + \cdots, \\ \cdots \end{gathered} \tag{12}$$

be an infinite set of new series made up of the terms of (11) such that each term of (11) appears in one and only one of the series (12). For example, one such set (there are infinitely many others) is

$$\begin{aligned} &z_1 + z_2 + z_4 + z_7 + z_{11} + \cdots, \\ &z_3 + z_5 + z_8 + z_{12} + \cdots, \\ &z_6 + z_9 + z_{13} + \cdots, \\ &z_{10} + z_{14} + \cdots, \\ &z_{15} + \cdots, \\ &\qquad \cdots \end{aligned}$$

(what is the law of formation of this set?).

LEMMA. *Let the series* (11) *be absolutely convergent with sum s. Then each of the series* (12) *is absolutely convergent. Moreover the series*

$$Z_1 + Z_2 + \cdots$$

where

$$\begin{gathered} Z_1 = z_{\alpha_1} + z_{\alpha_2} + \cdots, \\ Z_2 = z_{\beta_1} + z_{\beta_2} + \cdots, \\ \cdots \end{gathered}$$

is itself absolutely convergent with sum s.

Proof. Assuming that (11) is absolutely convergent, let

$$\sigma = |z_1| + |z_2| + \cdots.$$

Then obviously every partial sum of each of the series

$$|z_{\alpha_1}| + |z_{\alpha_2}| + \cdots,$$
$$|z_{\beta_1}| + |z_{\beta_2}| + \cdots,$$
$$\cdots$$

is less than σ. But then each of the series (12) is absolutely convergent, by the same argument as in Sec. 6.21. Moreover, adding the inequalities

$$|Z_1| \le |z_{\alpha_1}| + |z_{\alpha_2}| + \cdots,$$
$$|Z_2| \le |z_{\beta_1}| + |z_{\beta_2}| + \cdots,$$
$$\cdots$$
$$|Z_m| \le |z_{\mu_1}| + |z_{\mu_2}| + \cdots$$

(see Prob. 6), we get

$$|Z_1| + |Z_2| + \cdots + |Z_m| \le \sigma$$

for every $m = 1, 2, \ldots$ This proves the absolute convergence of the series $Z_1 + Z_2 + \cdots$.

To prove that

$$Z_1 + Z_2 + \cdots = s$$

as asserted, we need only show that the difference $s - (Z_1 + \cdots + Z_m)$ converges to zero as $m \to \infty$. Clearly

$$|s - (Z_1 + \cdots + Z_m)| \le |z_{\nu_1}| + |z_{\nu_2}| + \cdots,$$

where $\nu_1, \nu_2, \ldots$ are the indices of all the terms of (11) that do not appear in any of the first m series of the set (12). Given any positive integer n, let m be so large that the integers $\nu_1, \nu_2, \ldots$ all exceed n. Then obviously

$$|s - (Z_1 + \cdots + Z_m)| \le |z_{n+1}| + |z_{n+2}| + \cdots.$$

But the right-hand side is less than any given $\epsilon > 0$ for all sufficiently large n, by the absolute convergence of (11). It follows that

$$|s - (Z_1 + \cdots + Z_m)| < \epsilon$$

for all sufficiently large m. ▮

6.26. THEOREM ***(Rearrangement of series).*** *The terms of an absolutely convergent series can be rearranged arbitrarily without changing its sum.*

Proof. Given any absolutely convergent series $z_1 + z_2 + \cdots$ with sum s, let $\alpha_1, \alpha_2, \ldots$ be any rearrangement of the sequence $1, 2, \ldots$, and let

$$z_{\alpha_1} + z_{\alpha_2} + \cdots \tag{13}$$

be the corresponding rearrangement of the given series. Applying Lemma 6.25 with

$$Z_1 = z_{\alpha_1}, \qquad Z_2 = z_{\alpha_2}, \ldots,$$

we see that (13) is itself absolutely convergent with sum s. ▮

6.27. By the *product* (in the Cauchy sense) of two given series

$$z_1 + z_2 + \cdots + z_n + \cdots \tag{14}$$

and

$$z'_1 + z'_2 + \cdots + z'_n + \cdots, \tag{14'}$$

we mean the series

$$z_1z'_1 + (z_1z'_2 + z_2z'_1) + \cdots + (z_1z'_n + z_2z'_{n-1} + \cdots + z_nz'_1) + \cdots. \tag{15}$$

THEOREM *(Multiplication of series). If the two series* (14) *and* (14′) *are absolutely convergent with sums s and s', respectively, then the series* (15) *is absolutely convergent with sum ss'.*

Proof. Dropping parentheses in (15), we get the series

$$z_1z'_1 + z_1z'_2 + z_2z'_1 + z_1z'_3 + z_2z'_2 + \cdots. \tag{16}$$

This series is absolutely convergent. In fact, given any sum of the form

$$|z_1z'_1| + |z_1z'_2| + |z_2z'_1| + \cdots + |z_jz'_k|, \tag{17}$$

let n be the largest subscript of any term of the series (14) and (14′) appearing in (17). Then clearly (17) cannot exceed the product

$$(|z_1| + \cdots + |z_n|)(|z'_1| + \cdots + |z'_n|)$$

and hence cannot exceed $\sigma\sigma'$, where we write

$$\sigma = |z_1| + \cdots + |z_n| + \cdots, \qquad \sigma' = |z'_1| + \cdots + |z'_n| + \cdots,$$

assuming the absolute convergence of (14) and (14′). But then (16) is absolutely convergent (cf. Sec. 6.21). Applying Lemma 6.25 with

$$Z_1 = z_1z'_1, \qquad Z_2 = z_1z'_2 + z_2z'_1, \ldots,$$
$$Z_n = z_1z'_n + z_2z'_{n-1} + \cdots + z_nz'_1, \ldots,$$

we find that the original series (15) is also absolutely convergent. Let S be the sum of (15). To prove that $S = ss'$, we introduce the following set of new series made up of terms of (15):

$$z_1z'_1 + z_1z'_2 + z_1z'_3 + \cdots,$$
$$z_2z'_1 + z_2z'_2 + z_2z'_3 + \cdots,$$
$$\cdots$$

These series are absolutely convergent with sums $z_1s', z_2s', \ldots$ respectively (cf. Prob. 1). Hence by Lemma 6.25 again, this time with

$$Z_1 = z_1z'_1 + z_1z'_2 + z_1z'_3 + \cdots = z_1s',$$
$$Z_2 = z_2z'_1 + z_2z'_2 + z_2z'_3 + \cdots = z_2s',$$
$$\cdots$$

we have

$$S = Z_1 + Z_2 + \cdots = (z_1 + z_2 + \cdots)s' = ss'. \quad \blacksquare$$

6.28. Remark. We cannot assert that $S = ss'$ if the series (14) and (14′) are merely conditionally convergent. In fact, the product of two such series may well be divergent (see Prob. 10). However $S = ss'$ even for conditionally convergent series if it is known in advance that the product series is convergent (see Chap. 7, Prob. 13).

6.3. Uniform Convergence

6.31. Series of functions. Let

$$\sum_{n=1}^{\infty} f_n(z) = f_1(z) + \cdots + f_n(z) + \cdots \tag{18}$$

be a series each of whose terms is a single-valued complex function defined in the same set E. Suppose (18) converges in E, i.e., at every point of E. Then the sum of (18) is uniquely defined at every point of E and hence represents a single-valued function $s(z)$ defined in E. Suppose further that all the terms of (18) are continuous in E.† Then, unlike the case of finite sums, the sum function $s(z)$ need not be continuous in E. This is illustrated by the series

$$z + (z^2 - z) + \cdots + (z^n - z^{n-1}) + \cdots \tag{19}$$

whose nth partial sum $s_n(z)$ is just z^n. Clearly (19) is convergent in the region E consisting of the disk $|z| < 1$ plus the extra point $z = 1$ and divergent elsewhere (why?), with sum

$$s(z) = \lim_{n\to\infty} s_n(z) = \begin{cases} 0 & \text{if } |z| < 1, \\ 1 & \text{if } z = 1. \end{cases} \tag{20}$$

But the function (20) is not continuous in E, since it obviously fails to be continuous at the point $z = 1$.

Thus a further condition must be imposed on a convergent series of continuous functions if its sum is to be continuous. As will be shown below (see Theorem 6.35), the appropriate condition is that the series be *uniformly convergent*.

6.32. Definition. Let $s_n(z)$ be the nth partial sum of a series (18) which converges to a sum function $s(z)$ at every point of a set E. Then (18) is said to be *uniformly convergent in E* if given any $\epsilon > 0$, there exists an integer $N = N(\epsilon) > 0$ such that

$$|s_n(z) - s(z)| < \epsilon \tag{21}$$

for all $n > N$ and *all* z in E.

6.33. Remark. Thus, loosely speaking, the sum of a series which is uniformly convergent in E can be approximated everywhere in E to any

†Note that this tacitly requires E to be a region or a curve (see comment to Sec. 3.3).

desired accuracy by the sum of its first n terms, provided n is large enough. Here again (cf. Sec. 3.42) the key observation is that the number N must be *independent* of the point z in E. In the general case of a convergent series which may or may not be uniformly convergent, the inequality (21) holds for any given z provided n exceeds an integer $N = N(\epsilon, z) > 0$ which depends on z. If $N(\epsilon, z)$ becomes arbitrarily large for suitable z in E, then uniform convergence is clearly impossible. However if $N(\epsilon, z)$ does not exceed some fixed positive integer $N = N(\epsilon)$ for all z in E, then (21) holds for all z in E provided only that $n > N$ and the convergence is indeed uniform.

6.34. Example. The series (19) is not uniformly convergent in the open disk $|z| < 1$ and hence certainly not uniformly convergent in the region E consisting of the disk and the point $z = 1$. In fact, to satisfy the inequality

$$|s_n(z) - s(z)| = |z^n| < \epsilon \tag{22}$$

at points of the disk, we must require that $|z|^n < \epsilon$ or equivalently that

$$n > \frac{\ln \dfrac{1}{\epsilon}}{\ln \dfrac{1}{|z|}}.$$

Let $N(\epsilon, z)$ be the largest integer not exceeding

$$\frac{\ln \dfrac{1}{\epsilon}}{\ln \dfrac{1}{|z|}}.$$

Then (22) holds if and only if $n > N(\epsilon, z)$. But $N(\epsilon, z)$ approaches infinity as $|z| \longrightarrow 1$, and hence there is no positive integer N exceeding $N(\epsilon, z)$ for all z in the disk $|z| < 1$. Therefore the series (19) cannot be uniformly convergent in the disk $|z| < 1$.

On the other hand, the series (19) is uniformly convergent in every closed disk

$$|z| \leq r < 1. \tag{23}$$

In fact, we now have

$$\ln \frac{1}{|z|} \geq \ln \frac{1}{r},$$

which implies

$$\frac{\ln \dfrac{1}{\epsilon}}{\ln \dfrac{1}{|z|}} \leq \frac{\ln \dfrac{1}{\epsilon}}{\ln \dfrac{1}{r}}.$$

Hence choosing $N = N(\epsilon)$ to be the largest integer not exceeding

$$\frac{\ln \frac{1}{\epsilon}}{\ln \frac{1}{r}},$$

we find that $N(\epsilon, z) \leq N$ for all z in the set (23).

6.35. THEOREM. *If the series* (18) *is uniformly convergent in a set E and if every term of* (18) *is continuous at a point* z_0 *in E, then the sum* $s(z)$ *of the series is also continuous at* z_0.

Proof. Suppose $z_0 + h$ belongs to E. Then

$$s(z_0 + h) - s(z_0) = [s(z_0 + h) - s_N(z_0 + h)] + [s_N(z_0 + h) - s_N(z_0)] + [s_N(z_0) - s(z_0)],$$

where $s_N(z)$ is the Nth partial sum of (18), and hence

$$|s(z_0 + h) - s(z_0)| \leq |s(z_0 + h) - s_N(z_0 + h)| + |s_N(z_0 + h) - s_N(z_0)| + |s_N(z_0) - s(z_0)|. \tag{24}$$

Since (18) is uniformly convergent by hypothesis, there is a choice of the integer N such that

$$|s(z) - s_N(z)| < \frac{\epsilon}{3} \tag{25}$$

for all z in E. Choosing first $z = z_0$, and then $z = z_0 + h$ in (25), we get

$$|s(z_0) - s_N(z_0)| < \frac{\epsilon}{3}, \qquad |s(z_0 + h) - s_N(z_0 + h)| < \frac{\epsilon}{3}. \tag{26}$$

On the other hand, $s_N(z)$ is continuous at z_0, being the sum of a finite number of functions continuous at z_0 (see Sec. 3.33). Therefore

$$|s_N(z_0 + h) - s_N(z_0)| < \frac{\epsilon}{3} \tag{27}$$

for all h of modulus less than some number δ. Combining (24), (26) and (27), we find that

$$|s(z_0 + h) - s(z_0)| < \frac{\epsilon}{3} + \frac{\epsilon}{3} + \frac{\epsilon}{3} = \epsilon$$

for all $|h| < \delta$, i.e., $s(z)$ is continuous at z_0. ∎

It is an immediate consequence of the theorem that the sum of a uniformly convergent series of functions continuous in E is itself continuous in E (since continuity in E means continuity at every point of E).

6.36. Next we give a simple test for uniform convergence of a series of functions:

THEOREM. *Suppose the terms of the series* (18) *satisfy the inequalities*

$$|f_n(z)| \leq a_n \qquad (n = 1, 2, \ldots)$$

for all z in a set E, where the numerical series

$$a_1 + a_2 + \cdots + a_n + \cdots \tag{28}$$

converges. Then the series (18) *is uniformly (and absolutely) convergent in E.*

Proof. Suppose (28) converges. Then the series

$$|f_1(z)| + |f_2(z)| + \cdots + |f_n(z)| + \cdots$$

is convergent at every point z of E by the comparison test (see Prob. 3), i.e., the series (18) is absolutely convergent in E. Moreover, if $s(z)$ is the sum of (18) and $s_n(z)$ is its nth partial sum, then

$$\begin{aligned} |s_n(z) - s(z)| &\leq |f_{n+1}(z) + f_{n+2}(z) + \cdots| \\ &\leq |f_{n+1}(z)| + |f_{n+2}(z)| + \cdots \leq a_{n+1} + a_{n+2} + \cdots. \end{aligned}$$

But the remainder $a_{n+1} + a_{n+2} + \cdots$ of the convergent series (28) is less than any given $\epsilon > 0$ for sufficiently large N. It follows that $|s_n(z) - s(z)| < \epsilon$ for all $n > N$ and all z in E, i.e., (18) is uniformly convergent in E. ▮

6.37. Example. For the series (19) we have

$$|f_n(z)| = |z^n - z^{n-1}| = |z^{n-1}||z - 1|$$

and hence

$$|f_n(z)| \leq r^{n-1}(r + 1)$$

provided that $|z| \leq r$. But

$$\sum_{n=1}^{\infty} r^{n-1}(r + 1) = (r + 1)\sum_{n=1}^{\infty} r^{n-1}$$

converges if $r < 1$ (Sec. 6.13). It follows from Theorem 6.36 that the series (19) is uniformly (and absolutely) convergent in every closed disk $|z| \leq r < 1$, in keeping with the last part of Example 6.34.

6.38. Applying Theorem 5.22 repeatedly, we see that the sum of a finite number of continuous functions $f_1(z), \ldots, f_n(z)$, can be integrated "term by term" along any piecewise smooth curve C, i.e.,

$$\int_C [f_1(z) + \cdots + f_n(z)]\, dz = \int_C f_1(z)\, dz + \cdots + \int_C f_n(z)\, dz.$$

We now show that the same is true of an infinite series of continuous functions if the series is uniformly convergent on C.

THEOREM ***(Integration of series)***. *Let*

$$s(z) = f_1(z) + \cdots + f_n(z) + \cdots \tag{29}$$

be an infinite series with sum $s(z)$, *each of whose terms is a function continuous on a piecewise smooth curve* C. *Then*

$$\int_C s(z)\,dz = \int_C f_1(z)\,dz + \cdots + \int_C f_n(z)\,dz + \cdots, \qquad (30)$$

provided that (29) *is uniformly convergent on* C.

Proof. Since the functions $f_1(z), \ldots, f_n(z), \ldots$ are all continuous on C, so is the function $s(z)$, by Theorem 6.35. But then $s(z)$ is integrable along C, by Theorem 5.12. Let

$$s_n(z) = f_1(z) + \cdots + f_n(z)$$

be the nth partial sum of the series (29). By the uniform convergence of (29) on C, given any $\epsilon > 0$, there is an integer $N = N(\epsilon) > 0$ such that

$$|s(z) - s_n(z)| < \epsilon$$

for all $n > N$ and all $z \in C$ (see Sec. 6.32). But then, by Theorem 5.23,

$$\left|\int_C [s(z) - s_n(z)]\,dz\right| \leq \epsilon l,$$

where l is the length of C. Since ϵ is arbitrarily small, it follows that

$$\lim_{n\to\infty}\left|\int_C [s(z) - s_n(z)]\,dz\right| = 0$$

and hence

$$\lim_{n\to\infty}\int_C [s(z) - s_n(z)]\,dz = 0.$$

Therefore

$$\int_C s(z)\,dz = \lim_{n\to\infty}\int_C s_n(z)\,dz = \lim_{n\to\infty}\left\{\int_C f_1(z)\,dz + \cdots + \int_C f_n(z)\,dz\right\},$$

which is equivalent to (30). ∎

6.39. The following theorem involving series of *analytic* functions is an important tool of complex analysis:

THEOREM ***(Weierstrass).*** *Let* (29) *be an infinite series with sum* $s(z)$, *each of whose terms is a function analytic in a domain* G, *and suppose* (29) *is uniformly convergent in every bounded closed domain* $\bar{D}$ *contained in* G. *Then the sum* $s(z)$ *of the series is analytic in* G. *Moreover, the series* (29) *can be differentiated term by term any number of times, i.e.,*

$$s^{(k)}(z) = f_1^{(k)}(z) + \cdots + f_n^{(k)}(z) + \cdots \qquad (k = 1, 2, \ldots) \qquad (31)$$

for all $z \in G$, *and each differentiated series is uniformly convergent in every bounded closed domain* $\bar{D}$ *contained in* G.

Proof. Let z_0 be an arbitrary point of G, and let γ_R be a circle of radius R centered at z_0 which is so small that G contains both γ_R and its interior I.

Being uniformly convergent in $\bar{I}$ by hypothesis, the series (29) is uniformly convergent on γ_R, and hence the same is true of each of the series

$$\frac{k!}{2\pi i}\frac{s(z)}{(z-z_0)^{k+1}} = \frac{k!}{2\pi i}\frac{f_1(z)}{(z-z_0)^{k+1}} + \cdots + \frac{k!}{2\pi i}\frac{f_n(z)}{(z-z_0)^{k+1}} + \cdots \qquad (k = 0, 1, 2, \ldots), \tag{32}$$

since

$$\left|\frac{k!}{2\pi i}\frac{1}{(z-z_0)^{k+1}}\right| = \frac{k!}{2\pi R^{k+1}}$$

for all $z \in \gamma_R$ (see Prob. 12). Therefore, by Theorem 6.38, we can integrate (32) term by term along γ_R, obtaining

$$\frac{k!}{2\pi i}\int_{\gamma_R}\frac{s(z)}{(z-z_0)^{k+1}}\,dz = \frac{k!}{2\pi i}\int_{\gamma_R}\frac{f_1(z)}{(z-z_0)^{k+1}}\,dz + \cdots + \frac{k!}{2\pi i}\int_{\gamma_R}\frac{f_n(z)}{(z-z_0)^{k+1}}\,dz + \cdots \qquad (k = 0, 1, 2, \ldots) \tag{33}$$

For $k = 0$, (33) reduces to

$$\frac{1}{2\pi i}\int_{\gamma_R}\frac{s(z)}{z-z_0}\,dz = \frac{1}{2\pi i}\int_{\gamma_R}\frac{f_1(z)}{z-z_0}\,dz + \cdots + \frac{1}{2\pi i}\int_{\gamma_R}\frac{f_n(z)}{z-z_0}\,dz + \cdots.$$

But the sum on the right equals

$$f_1(z_0) + \cdots + f_n(z_0) + \cdots = s(z_0)$$

by Cauchy's integral formula, since every term of (29) is analytic inside and on γ_R. It follows that

$$s(z_0) = \frac{1}{2\pi i}\int_{\gamma_R}\frac{s(z)}{z-z_0}\,dz \qquad (z_0 \in G), \tag{34}$$

i.e., $s(z)$ also satisfies Cauchy's integral formula. But in proving Theorem 5.71, we used the analyticity of the given function only to establish formula (33), p. 65, for $n = 0$. Therefore if the function $s(z)$ satisfies (34), it must be infinitely differentiable at every point $z_0 \in G$ and hence certainly analytic in G. For $k > 0$, (33) reduces to (31) with z replaced by z_0, since, by Theorem 5.71,

$$s^{(k)}(z_0) = \frac{k!}{2\pi i}\int_{\gamma_R}\frac{s(z)}{(z-z_0)^{k+1}}\,dz \qquad (z_0 \in G,\ k = 0, 1, 2, \ldots),$$

as well as

$$f_n^{(k)}(z_0) = \frac{k!}{2\pi i}\int_{\gamma_R}\frac{f_n(z)}{(z-z_0)^{k+1}}\,dz \qquad (z_0 \in G,\ k = 0, 1, 2, \ldots),$$

now that $s(z)$, as well as every $f_n(z)$, is known to be analytic in G.

We must still show that each differentiated series

$$f_1^{(k)}(z) + \cdots + f_n^{(k)}(z) + \cdots \qquad (k = 0, 1, 2, \ldots) \tag{35}$$

is uniformly convergent in every bounded closed domain $\bar{D}$ contained in G. Let γ_{z_0} be the circle of radius R centered at z_0, previously denoted by γ_R, and let K_{z_0} be the open disk of radius $\frac{1}{2}R$ centered at z_0. Since the series (29) is uniformly convergent on γ_{z_0}, as already shown, there is an integer $N_{z_0} = N_{z_0}(\epsilon) > 0$ such that

$$|s_n(\zeta) - s(\zeta)| < \epsilon \tag{36}$$

for all $n > N_{z_0}$ and all $\zeta \in \gamma_{z_0}$, where $s_n(z)$ is the nth partial sum of (29). It follows from (36) and Theorem 5.23 that

$$\begin{aligned} &\left| \frac{k!}{2\pi i} \int_{\gamma_{z_0}} \frac{f_1(\zeta)}{(\zeta - z)^{k+1}} \, d\zeta + \cdots + \frac{k!}{2\pi i} \int_{\gamma_{z_0}} \frac{f_n(\zeta)}{(\zeta - z)^{k+1}} \, d\zeta \right. \\ &\qquad\qquad \left. - \frac{k!}{2\pi i} \int_{\gamma_{z_0}} \frac{s(\zeta)}{(\zeta - z)^{k+1}} \, d\zeta \right| \\ &\qquad = \left| \frac{k!}{2\pi i} \int_{\gamma_{z_0}} \frac{s_n(\zeta) - s(\zeta)}{(\zeta - z)^{k+1}} \, d\zeta \right| \le \frac{k!}{2\pi} \frac{\epsilon}{(R/2)^{k+1}} 2\pi R \end{aligned} \tag{37}$$

for all $n > N_{z_0}$ and all $z \in K_{z_0}$, where the right-hand side of (37) obviously approaches zero as $\epsilon \to 0$. Therefore the series (35) is uniformly convergent in K_{z_0}. But by the Heine–Borel theorem, any bounded closed domain $\bar{D}$ contained in G can be covered by a finite number of the disks K_{z_0} ($z_0 \in G$), say $K_{z_1}, \ldots, K_{z_n}$. Choosing $N = \max\{N_{z_1}, \ldots, N_{z_n}\}$, we see that the inequality holds for all $n > N$ and all $z \in \bar{D}$, i.e., each differentiated series (35) is uniformly convergent in $\bar{D}$. ∎

COMMENTS

6.1. The definitions and theorems of Secs. 6.1. and 6.2 are the exact analogues of those for real series. The reader already familiar with this material can refer back to it only as needed.

6.2. According to a theorem of Riemann which will not be proved here,† a conditionally convergent real series can be rearranged to have any real number (including $\pm\infty$) as its sum. As for conditionally convergent complex series, it can be shown that the set of all such series can be divided into two classes C_1 and C_2 such that

1) For every series in C_1 we can find a straight line L in the complex plane such that there is a rearrangement of the series with any given point of L as its sum, while no rearrangement of the series has a sum which is not a point of L;

---

†See e.g., G. E. Shilov, *Elementary Real and Complex Analysis* (translated by R. A. Silverman), The MIT Press, Cambridge, Mass. (1973), Theorem 6.37.

2) For every series in C_2 we can find a rearrangement of the series with any given point of the (extended) complex plane as its sum.†

For example,

$$1 - \frac{1}{2} + \frac{1}{3} - \frac{1}{4} + \cdots$$

is a series in C_1, while

$$1 + i - \frac{1}{2} - \frac{i}{2} + \frac{1}{3} + \frac{i}{3} - \frac{1}{4} - \frac{i}{4} + \cdots$$

is a series in C_2.

6.3. Note the crucial distinction between *uniform continuity*, involving the behavior (in regard to closeness of values at neighboring points) of a single function $f(z)$ in a set E, and *uniform convergence*, involving the behavior (in regard to rate of convergence) of a whole series $f_1(z) + f_2(z) + \cdots$ or sequence $f_1(z), f_2(z), \ldots$ of functions (cf. Probs. 18, 20, 23) in a set E.

PROBLEMS

1. Prove that if the series $z_1 + z_2 + \cdots$ is convergent with sum s, then the series $\alpha z_1 + \alpha z_2 + \cdots$ (α any complex number) is convergent with sum αs.

2. Prove that a complex series $z_1 + z_2 + \cdots$ with general term $z_n = x_n + iy_n$ converges if and only if both real series $x_1 + x_2 + \cdots$ and $y_1 + y_2 + \cdots$ converge.

3. Prove the following *comparison test* for a complex series $z_1 + z_2 + \cdots$. If $|z_n| \leq a_n$ for all sufficiently large n, where $a_1 + a_2 + \cdots$ is a convergent real series, then $z_1 + z_2 + \cdots$ is absolutely convergent.

4. Prove that if the series $z_1 + z_2 + \cdots$ converges absolutely, then so does the series $z_1^2 + z_2^2 + \cdots$.

5. Is the converse of the preceding problem true?

6. Prove that if the series $z_1 + z_2 + \cdots$ is absolutely convergent with sum s, then $|s| \leq |z_1| + |z_2| + \cdots$.

7. Give an example showing the breakdown of Lemma 6.25 for conditionally convergent series.

8. State and prove appropriate analogues for complex series of the familiar ratio and root tests for real series.‡

†This is the two-dimensional version of *Steinitz's theorem* (G. E. Shilov, *op. cit.*, Chap. 6, Probs. 17–22).

‡G. M. Fichtenholz, *op. cit.*, Sec. 5.

9. Investigate the convergence of the following series:

a) $\sum_{n=0}^{\infty} \frac{1}{(1+i)^n}$; b) $\sum_{n=0}^{\infty} \left(\cos\frac{n\pi}{4} + i\sin\frac{n\pi}{4}\right)$;

c) $\sum_{n=0}^{\infty} \frac{1}{(1+i)^n}\left(\cos\frac{n\pi}{4} + i\sin\frac{n\pi}{4}\right)$.

10. Prove that the series

$$1 - \frac{1}{\sqrt{2}} + \frac{1}{\sqrt{3}} - \frac{1}{\sqrt{4}} + \cdots$$

is conditionally convergent. Prove that the product of this series with itself is divergent.

11. Give an example of a series which fails to be uniformly convergent in a region $\bar{G}$ but has a continuous sum in $\bar{G}$.

12. Prove that if the series

$$\sum_{n=1}^{\infty} f_n(z) \tag{38}$$

is uniformly convergent in a set E with sum $s(z)$ and if $|\varphi(z)| \leq$ constant for all z in E, then the series

$$\sum_{n=1}^{\infty} \varphi(z) f_n(z)$$

is uniformly convergent in E, with sum $\varphi(z)s(z)$.

13. State and prove the analogue of the Cauchy convergence criterion (Theorem 2.33 and Chap. 2, Prob. 15) for

a) Convergence of a numerical series;

b) "Pointwise" convergence of a series of functions;

c) Uniform convergence of a series of functions.

14. Use the Cauchy convergence criterion to prove Theorem 6.36. Prove that the theorem remains true if $|f_n(z)| \leq a_n$ for all z in E and all $n = N, N+1, \ldots$ (i.e., for all sufficiently large n).

15. Prove that if the series

$$\sum_{n=1}^{\infty} |g_n(z)|$$

is uniformly convergent in a set E and if $|f_n(z)| \leq |g_n(z)|$ for all z in E and all (sufficiently large) n, then the series (38) is uniformly convergent in E.

16. Show that the geometric series

$$1 + z + z^2 + \cdots + z^n + \cdots$$

is uniformly convergent in every closed disk $|z| \leq r < 1$ but not in the open disk $|z| < 1$.

17. Prove that the series

$$\sum_{n=1}^{\infty} \frac{z^n}{n^2}$$

is uniformly convergent in the closed disk $|z| \leq 1$ and divergent elsewhere.

How about the series

$$\sum_{n=1}^{\infty} \frac{1}{n^2}\left(z^n + \frac{1}{z^n}\right)?$$

18. Define and discuss uniform convergence for *sequences* of functions

$$f_1(z), f_2(z), \ldots, f_n(z), \ldots$$

19. Give an example showing how (30) can break down if the series (29) fails to be uniformly convergent.

20. State and prove the analogue of Theorem 6.38 for uniformly convergent sequences of functions.

21. Give an example showing that even if the series

$$s(z) = f_1(z) + \cdots + f_n(z) + \cdots$$

in Weierstrass' theorem is uniformly convergent in the whole domain G, the same need not be true of the differentiated series

$$s'(z) = f'_1(z) + \cdots + f'_n(z) + \cdots.$$

22. Give an example showing that term-by-term differentiation of a uniformly convergent series of analytic functions may not be permissible if the series is defined in a set other than a domain.

23. State and prove the analogue of Weierstrass' theorem for uniformly convergent sequences of functions.

24. Give another proof of the analyticity of $s(z)$ in Weierstrass' theorem, starting from Theorem 5.73.

CHAPTER SEVEN

Power Series

7.1. Basic Theory

7.11. Definition. We now consider a class of series of key importance in complex analysis. Suppose that in a general series of functions of the form

$$f_0(z) + f_1(z) + f_2(z) + \cdots + f_n(z) + \cdots$$

we make the special choice

$$f_n(z) = c_n(z - a)^n$$

involving the complex variable z and arbitrary complex numbers a, c_0, c_1, $c_2, \ldots, c_n, \ldots$ Then the resulting series

$$c_0 + c_1(z - a) + c_2(z - a)^2 + \cdots + c_n(z - a)^n + \cdots$$

is called a *power series*. Choosing $a = 0$, which clearly entails no loss of generality (see Prob. 1), we get the somewhat simpler power series

$$c_0 + c_1 z + c_2 z^2 + \cdots + c_n z^n + \cdots. \tag{1}$$

7.12. By the *region of convergence* of the power series (1) we mean the set of all points for which (1) converges.† The series obviously converges for $z = 0$, and hence the region of convergence of (1) always includes the point $z = 0$. Moreover, it is easy to see that there exist power series for which the region of convergence consists of the single point $z = 0$. For example, consider the series

$$1 + z + 2^2 z^2 + \cdots + n^n z^n + \cdots. \tag{2}$$

†As we will see below (Sec. 7.16), the region of convergence of (1) is indeed a region in the sense of Sec. 3.23b.

If n is sufficiently large, we have $n|z| > 2$ and hence

$$|n^n z^n| = (n|z|)^n > 2^n$$

for any given $z \neq 0$. But then the general term $n^n z^n$ of (2) approaches infinity as $n \longrightarrow \infty$. It follows from Theorem 6.14 that the series (2) is divergent for all $z \neq 0$. Such "degenerate" power series are clearly of only academic interest. Hence we now turn our attention to power series for which the region of convergence contains at least one nonzero point.

7.13. Lemma. *Suppose the power series* (1) *is convergent at a point* $z_0 \neq 0$. *Then* (1) *is convergent (in fact absolutely convergent) for all* z *such that* $|z| < |z_0|$.†

Proof. If the series

$$c_0 + c_1 z_0 + \cdots + c_n z_0^n + \cdots$$

converges, then

$$\lim_{n \to \infty} c_n z_0^n = 0 \tag{3}$$

by Theorem 6.14. But (3) implies that the points $c_0 z_0^n$ all lie in some neighborhood of the origin, i.e., that

$$|c_n z_0^n| < M$$

for all $n = 1, 2, \ldots$ and some sufficiently large positive number M. Let $|z| < |z_0|$. Then

$$|c_n z^n| = |c_n z_0^n| \left|\frac{z}{z_0}\right|^n < M \left|\frac{z}{z_0}\right|^n = Mk^n,$$

where

$$k = \left|\frac{z}{z_0}\right| < 1,$$

so that the modulus of every term of the series (1) is less than the corresponding term of the convergent geometric series

$$M + Mk + \cdots + Mk^n + \cdots \qquad (k < 1).$$

But then (1) is absolutely convergent, by the comparison test (Chap. 6, Prob. 3), and hence convergent. ∎

Geometrically this means that if the series (1) is convergent at the point z_0, then it is absolutely convergent at every point z inside the circle $|z| = |z_0|$ with center at the origin going through the point z_0.

7.14. As noted in Sec. 7.12, there exist power series for which the region of convergence consists of the single point $z = 0$. At the other extreme there

†It follows that if (1) is divergent at z_0, then (1) is also divergent for all z such that $|z| > |z_0|$, since otherwise (1) would be convergent at z_0.

exist power series which converge at every finite point, i.e., for which the region of convergence is the whole (finite) plane. For example, consider the series

$$1 + z + \frac{z^2}{2^2} + \cdots + \frac{z^n}{n^n} + \cdots. \tag{4}$$

If n is sufficiently large, we have

$$\left|\frac{z}{n}\right| < \frac{1}{2}$$

and hence

$$\left|\frac{z^n}{n^n}\right| = \left(\frac{|z|}{n}\right)^n < \left(\frac{1}{2}\right)^n$$

for any given z, i.e., all but a finite number of terms of the series (4) are of modulus less than the corresponding terms of the convergent geometric series

$$1 + \frac{1}{2} + \frac{1}{2^2} + \cdots.$$

Hence the series (4) is (absolutely) convergent for all finite z.

Apart from the two types of power series just discussed (one type converging only for $z = 0$, the other for all z), there is a third type, namely power series which converge at some but not all nonzero points. We now investigate the region of convergence for series of this type.

7.15. THEOREM. *Suppose the power series* (1) *is convergent for some but not all nonzero values of* z. *Then there exists a number* $R > 0$ *such that* (1) *is convergent (in fact absolutely convergent) for all* $|z| < R$ *and divergent for all* $|z| > R$.†

Proof. Any ray σ drawn from the origin O contains a nonzero point at which the series (1) converges and a point at which it diverges (the existence of such points follows from Lemma 7.13 and the fact that the series is of the third type). Let E be the set of all points $P \in \sigma$ such that the series (1) converges at P, and let R be the least upper bound of the set Ω of all numbers OP with $P \in E$,‡ where the existence of R follows from the boundedness of Ω (the series diverges at some point of σ) and the completeness of the real number system. Clearly $R > 0$ (why?). Let P_0 be

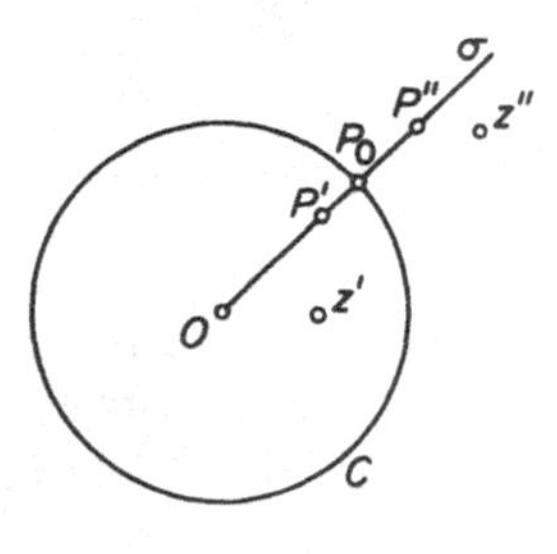

Figure 20

†For simplicity, we say "for all $|z| < R$" rather than "for all z such that $|z| < R$," and so on.

‡Here and elsewhere we use the same notation (like OP) to denote both a line segment and its length.

the point of σ such that $OP_0 = R$, and let C be the circle with center O going through P_0 (see Figure 20). Given any point $P \in \sigma$, it follows from Lemma 7.13 and the definition of R that the series (1) converges at P if $OP < R$ and diverges at P if $OP > R$. But then by Lemma 7.13 again, the series is (absolutely) convergent at every point z' inside C (choose $P' \in \sigma$ such that $|z'| < OP' < R$) and divergent at every point z'' outside C(choose $P'' \in \sigma$ such that $R < OP'' < |z''|$). ∎

7.16. Radius of convergence and circle of convergence. The number R figuring in the above theorem is called the *radius of convergence* of the power series (1), and the circle C, i.e., the circle $|z| = R$, is called the *circle of convergence* of (1). The theorem says nothing at all about the behavior of the series on C itself, and in fact (1) may or may not converge at a given point of C, depending on the precise nature of the series. Thus the region of convergence of the power series (1) is the open circular disk $|z| < R$ and some (possibly all or none) of its boundary points (see Sec. 7.25).

Finally we set $R = 0$ in the case of a series converging only at the point $z = 0$, and $R = +\infty$ in the case of a series converging in the whole plane. We then see that *every* power series has a well-defined radius of convergence R in the interval $0 \leq R \leq \infty$.

7.17. We know from the example of the geometric series

$$1 + z + z^2 + \cdots + z^n + \cdots$$

(see Chap. 6, Prob. 16) that a power series with radius of convergence R need not be *uniformly* convergent in the open disk $|z| < R$. However, the series is uniformly convergent in any smaller disk, as shown by the following

THEOREM. *If the power series* (1) *has radius of convergence* R, *then the series is uniformly convergent in every closed disk* $|z| \leq r < R$.

Proof. We need only apply Theorem 6.36, noting that $|z| \leq r$ implies

$$|c_n z^n| = |c_n||z|^n \leq |c_n| r^n,$$

where the numerical series

$$\sum_{n=0}^{\infty} |c_n| r^n$$

converges for all $r < R$. ∎

7.18. THEOREM. *The sum* $s(z)$ *of a power series* (1) *with radius of convergence* R *is a continuous function in the disk* $|z| < R$.

Proof. It follows from the preceding theorem and Theorem 6.35 that $s(z)$ is continuous in every closed disk $|z| \leq r < R$. Hence $s(z)$ is continuous in the open disk $|z| < R$. ∎

7.19. As noted in Sec. 4.15, every polynomial is an analytic function in the whole complex plane. We now prove the analogous result for power series, which can be regarded as "polynomials of infinitely high degree":†

THEOREM. *Let*

$$s(z) = c_0 + c_1 z + c_2 z^2 + \cdots + c_n z^n + \cdots \tag{5}$$

be a power series with radius of convergence R. Then the sum $s(z)$ is analytic in the disk $|z| < R$. Moreover, the series (5) can be differentiated term by term, i.e.,

$$s'(z) = c_1 + 2c_2 z + \cdots + nc_n z^{n-1} + \cdots, \tag{6}$$

where the differentiated series (6) has the same radius of convergence R as the original series (5).

Proof. By Theorem 7.17, the series (5) is uniformly convergent in every closed disk $|z| \leq r < R$ and hence in every (bounded) closed domain $\bar{D}$ contained in the disk $|z| < R$ (why?). Moreover, every term of (5) is obviously analytic in the disk $|z| < R$ (for that matter, in the whole plane). Hence by Theorem 6.39, $s(z)$ is analytic in the disk $|z| < R$ and the series (5) can be differentiated term by term, giving formula (6) valid for all z in the disk $|z| < R$. Let R' be the radius of convergence of the series (6). Then obviously $R' \geq R$. Suppose $R' > R$, and let C be any piecewise smooth curve in the disk $|z| < R'$ joining the points 0 and z. The series (6) is uniformly convergent on C, being uniformly convergent in some closed domain $\bar{D}$ containing C (this follows from either Theorem 7.17 or Theorem 6.39 itself). Therefore we can use Theorem 6.38 to integrate (6) term by term along C, finding with the help of Theorems 5.51 and 5.53 that the original series (5) converges in the disk $|z| < R'$ and hence has a radius of convergence larger than R, contrary to hypothesis‡. This contradiction shows that $R' = R$, i.e., the differentiated series (6) has the same radius of convergence as the original series (5).§ ∎

Since (6) is itself a power series with radius of convergence R, we can differentiate (6) term by term, getting a new power series with the same radius of convergence R, and so on repeatedly. Thus the series (5) can be differenti-

---

†Conversely, polynomials are power series of a very special kind (with only finitely many nonzero coefficients). For such series the radius of convergence is automatically infinite (why?).

‡Note that

$$n\int_0^z \zeta^{n-1}\,d\zeta = z^n$$

(cf. Sec. 5.15), so that term-by-term integration of (6) along C gives back (5) except for the constant term c_0.

§For an alternative proof of this fact, see Prob. 10.

ated term by term any number of times, i.e.,

$$s^{(k)}(z) = 2\cdot 3\cdots kc_k + 2\cdot 3\cdots k(k+1)c_{k+1}z + \cdots \qquad (k = 1, 2, \ldots), \tag{6'}$$

where each differentiated series has the same radius of convergence R as (5) itself.

7.2. Determination of the Radius of Convergence

7.21. We now consider the problem of determining the radius of convergence R of an arbitrary power series

$$c_0 + c_1 z + c_2 z^2 + \cdots + c_n z^n + \cdots,$$

starting from a knowledge of the coefficients c_n. The complete solution of this problem (Theorem 7.24) involves the notion of the "upper limit" of a real sequence, which we introduce for the special case of a nonnegative sequence (sufficient for our purposes).

7.22. Definition. Let

$$a_1, a_2, \ldots, a_n, \ldots \tag{7}$$

be a sequence of nonnegative real numbers. Then by the *upper limit* of (7), denoted by

$$l = \overline{\lim_{n\to\infty}}\, a_n,$$

we mean the largest limit point of (7) if the sequence is bounded† and $+\infty$ otherwise.

7.23. Remark. Obviously

$$\overline{\lim_{n\to\infty}}\, a_n = \lim_{n\to\infty} a_n$$

in the case where a_n is convergent (cf. Chap. 2, Prob. 4b).

7.24. THEOREM (***Cauchy–Hadamard***). *Given a power series*

$$\sum_{n=0}^{\infty} c_n z^n = c_0 + c_1 z + c_2 z^2 + \cdots + c_n z^n + \cdots \tag{8}$$

with coefficients c_n, let

$$l = \overline{\lim_{n\to\infty}}\, \sqrt[n]{|c_n|}.$$

---

†A bounded sequence necessarily has at least one limit point (Theorem 2.25). The neighborhoods figuring in the definition of a limit point α of the real sequence (7) are open intervals of the form $|x - \alpha| < \epsilon$ rather than disks $|z - \alpha| < \epsilon$ as in Sec. 2.22.

Then the radius of convergence of (8) *is given by*

$$R = \frac{1}{l}, \tag{9}$$

with the understanding that

$$R = \begin{cases} +\infty & \text{if } l = 0, \\ 0 & \text{if } l = +\infty. \end{cases}$$

Proof. We distinguish three cases.

Case 1. Let $l = +\infty$, so that the sequence of nonnegative real numbers

$$|c_1|, \quad \sqrt{|c_2|}, \quad \sqrt[3]{|c_3|}, \ldots, \quad \sqrt[n]{|c_n|}, \ldots \tag{10}$$

is unbounded. We must show that $R = 0$, i.e., that the power series (8) diverges at every point $z_0 \neq 0$. Suppose to the contrary that (8) converges at some point $z_0 \neq 0$. Then

$$\lim_{n \to \infty} c_n z_0^n = 0$$

by Theorem 6.14, and hence there exists a number $M > 0$ such that

$$|c_n z_0^n| < M \tag{11}$$

for all $n = 0, 1, \ldots$ There is obviously no loss of generality in assuming that $M > 1$. The inequality (11) then implies

$$\sqrt[n]{|c_n|}\,|z_0| < \sqrt[n]{M}$$

or†

$$\sqrt[n]{|c_n|} < \frac{M}{|z_0|},$$

which is impossible since the sequence (10) is unbounded. This contradiction shows that (8) actually diverges at every point $z_0 \neq 0$.

Case 2. Let $l = 0$. Then we must show that $R = +\infty$, i.e., that the series (8) converges at every point $z_0 \neq 0$. Since every limit point of a nonnegative sequence is necessarily nonnegative, $l = 0$ implies $\sqrt[n]{|c_n|} \to 0$ as $n \to \infty$. Therefore

$$\sqrt[n]{|c_n|} < \epsilon$$

for arbitrary $\epsilon > 0$ and sufficiently large n, say

$$\sqrt[n]{|c_n|} < \frac{1}{2|z_0|}.$$

Therefore

$$\sqrt[n]{|c_n|}\,|z_0| < \frac{1}{2}$$

†Note that $\sqrt[n]{M} < M$ if $M > 1$.

or equivalently

$$|c_n|\,|z_0|^n = |c_n z_0^n| < \frac{1}{2^n}.$$

Since the series with general term $1/2^n$ is convergent, this implies the absolute convergence of the series (8) at the point z_0.

Case 3. Finally suppose $l \neq 0$, $l \neq \infty$. Then to prove (9) we need only show that the series (8) is absolutely convergent at every point z_1 such that $|z_1| < 1/l$ and divergent at every point z_2 such that $|z_2| > 1/l$. Since l is the largest limit point of the sequence (10), we have

$$\sqrt[n]{|c_n|} < l + \epsilon \tag{12}$$

for arbitrary $\epsilon > 0$ and all sufficiently large n. Noting that $l|z_1| < 1$, let

$$\epsilon = \frac{1 - l|z_1|}{2|z_1|}.$$

Then (12) takes the form

$$\sqrt[n]{|c_n|} < l + \frac{1 - l|z_1|}{2|z_1|} = \frac{1 + l|z_1|}{2|z_1|}$$

or

$$\sqrt[n]{|c_n|}\,|z_1| < \frac{1 + l|z_1|}{2} = q < 1.$$

Raising both sides to the nth power, we get

$$|c_n|\,|z_1|^n < q^n$$

or

$$|c_n z_1^n| < q^n. \tag{13}$$

Since the series with general term q^n $(q < 1)$ converges, (13) implies the absolute convergence of the series (8) at z_1. On the other hand, again by the very meaning of l, we have

$$\sqrt[n]{|c_n|} > l - \epsilon \tag{14}$$

for arbitrary $\epsilon > 0$ and infinitely many values of n. Noting that $l|z_2| > 1$, let

$$\epsilon = \frac{l|z_2| - 1}{|z_2|}.$$

Then (14) takes the form

$$\sqrt[n]{|c_n|} > \frac{1}{|z_2|}$$

or

$$\sqrt[n]{|c_n|}\,|z_2| > 1,$$

which implies

$$|c_n|\,|z_2|^n > 1$$

or

$$|c_n z_2^n| > 1. \tag{15}$$

Since (15) holds for infinitely many values of n, $c_n z^n$ cannot approach zero as $n \to \infty$. It follows from Theorem 6.14 that the series (8) diverges at z_2. ∎

7.25. Examples

a. The radius of convergence of the series

$$1 + z + z^4 + z^9 + \cdots$$

equals 1. In fact, here $c_n = 1$ if n is the square of an integer m and $c_n = 0$ otherwise. Therefore $\sqrt[n]{|c_n|} = 1$ or 0 depending on whether or not $n = m^2$, so that the sequence (10) has two limit points 0 and 1. But then $l = 1$, $R = 1$.

b. The radius of convergence of the series

$$1 + \frac{z}{1^s} + \frac{z^2}{2^s} + \cdots + \frac{z^n}{n^s} + \cdots \qquad (s \geq 0)$$

equals 1. In fact, here

$$\sqrt[n]{|c_n|} = \frac{1}{n^{s/n}} = \frac{1}{e^{(s \ln n)/n}}.$$

But $(\ln n)/n \to 0$ as $n \to \infty$ and hence $\sqrt[n]{|c_n|} \to 1$, so that $l = 1$, $R = 1$.

c. The series

$$1 + z + \frac{z^2}{2!} + \cdots + \frac{z^n}{n!} + \cdots$$

converges in the whole complex plane. To see this, we first observe that

$$(n!)^2 = (1 \cdot n)(2(n-1)) \cdots (n \cdot 1),$$

where each of the n terms in parentheses is no smaller than n since

$$k(n-k+1) - n = (k-1)(n-k) \geq 0 \qquad (k = 1, 2, \ldots, n).$$

Therefore

$$(n!)^2 \geq n^n$$

or

$$n! \geq (\sqrt{n})^n,$$

which implies

$$\sqrt[n]{n!} \geq \sqrt{n},$$

and hence

$$\sqrt[n]{\frac{1}{n!}} \leq \frac{1}{\sqrt{n}}.$$

It follows that in this case,

$$\lim_{n\to\infty} \sqrt[n]{|c_n|} = 0,$$

so that $l = 0$, $R = +\infty$.

d. In the same way, we see at once that both series

$$1 - \frac{z^2}{2!} + \frac{z^4}{4!} - \cdots,$$

$$z - \frac{z^3}{3!} + \frac{z^5}{5!} - \cdots$$

converge in the whole complex plane ($R = +\infty$), while the series

$$1 + z + 2!z^2 + \cdots + n!z^n + \cdots$$

converges only at the point $z = 0$ ($R = 0$).

7.26. It has already been noted in Sec. 7.16 that a power series may behave in various ways on its circle of convergence. For example, setting $s = 0, 1, 2$ in turn in Example 7.25b, we get three different series with the same radius of convergence $R = 1$, namely

$$1 + z + z^2 + \cdots + z^n + \cdots,$$

$$1 + z + \frac{z^2}{2} + \cdots + \frac{z^n}{n} + \cdots,$$

$$1 + z + \frac{z^2}{2^2} + \cdots + \frac{z^n}{n^2} + \cdots.$$

The first series diverges at every point of the circle of convergence C (the unit circle $|z| = 1$), while the third series converges at every point of C. As for the second series, it converges at some points of C (e.g., at $z = -1$) and diverges at others (e.g., at $z = 1$).†

COMMENTS

7.1. Concerning the importance of power series in complex analysis, we note that not only does every power series represent an analytic function inside its circle of convergence (Theorem 7.19), but conversely every function $f(z)$ analytic in a disk K can be represented as the sum of a power series convergent in K (Theorem 10.13). Thus, remarkably enough, the class of functions analytic in K coincides with the class of functions which can be represented by power series in K!

7.2. Since a bounded infinite set need not have a largest member (con-

†Actually, it can be shown that this series diverges only at $z = 1$ and converges at every other point of C (cf. A. I. Markushevich, *op. cit.*, Volume I, Example 1, p. 408).

sider the set $\frac{1}{2}, \frac{2}{3}, \frac{3}{4}, \frac{4}{5}, \ldots$), Definition 7.22 relies implicitly on the fact that the set E of all limit points of a bounded sequence of nonnegative numbers has a largest member. To see this, we need only note that the least upper bound of E is itself a member of E (why?), and hence obviously the largest member of E.

PROBLEMS

1. Find the analogues of the various results of this chapter for a general power series

$$c_0 + c_1(z - a) + c_2(z - a)^2 + \cdots + c_n(z - a)^n + \cdots$$

"centered" at the point $a \neq 0$.

2. Suppose a power series

$$\sum_{n=0}^{\infty} c_n z^n \tag{16}$$

with radius of convergence R is absolutely convergent at some point on its circle of convergence. Prove that the series is absolutely and uniformly convergent for all $|z| \leq R$.

3. Prove that if

$$\lim_{n\to\infty} \left|\frac{c_n}{c_{n+1}}\right| = R,$$

then the series (16) has radius of convergence R.

4. Find the radius of convergence of each of the following power series:

a) $\sum_{n=1}^{\infty} n^{\ln n} z^n$; b) $\sum_{n=1}^{\infty} n^n z^n$; c) $\sum_{n=1}^{\infty} z^{n!}$; d) $\sum_{n=0}^{\infty} [3 + (-1)^n]^n z^n$;

e) $\sum_{n=0}^{\infty} (n + a^n) z^n$.

5. Prove that the radius of convergence of the series

$$z + \frac{2^2}{2!} z^2 + \cdots + \frac{n^n}{n!} z^n + \cdots$$

equals $1/e$.

6. Find the radius of convergence of the series

$$z + \frac{z^{2^2}}{2!} + \cdots + \frac{z^{n^2}}{n!} + \cdots.$$

7. Given that the radius of convergence of the series (16) is R ($0 < R < \infty$), find the radius of convergence of each of the following series:

a) $\sum_{n=0}^{\infty} n^k c_n z^n$; b) $\sum_{n=0}^{\infty} \frac{c_n}{n!} z^n$; c) $\sum_{n=0}^{\infty} c_n^k z^n$; d) $\sum_{n=0}^{\infty} c_n z^{kn}$

($k = 1, 2, \ldots$).

8. Suppose the two power series

$$\sum_{n=1}^{\infty} c_n z^n, \qquad \sum_{n=1}^{\infty} c_n' z^n \tag{17}$$

have radii of convergence r, r' respectively. What can be said about the radius of convergence R of each of the following series:

a) $\sum_{n=0}^{\infty} (c_n \pm c_n')z^n$; b) $\sum_{n=0}^{\infty} c_n c_n' z^n$; c) $\sum_{n=0}^{\infty} \frac{c_n}{c_n'} z^n \quad (c_n' \neq 0)$?

9. Give an example of two power series (17) with the same finite radius of convergence such that the radius of convergence of the series

$$\sum_{n=0}^{\infty} (c_n + c_n')z^n$$

is infinite.

10. Use the Cauchy–Hadamard theorem to show that the series obtained by term-by-term differentiation of a power series has the same radius of convergence as the original series.

11. Prove the following proposition known as *Abel's theorem*: If the power series

$$s(z) = \sum_{n=0}^{\infty} c_n z^n \tag{18}$$

converges at a point z_0 of its circle of convergence $|z| = R$, then $s(z)$ approaches $s(z_0)$ as $z \longrightarrow z_0$ along the radius Oz_0.

12. Use the real version of Abel's theorem to show that the series

$$1 - \tfrac{1}{2} + \tfrac{1}{3} - \tfrac{1}{4} + \cdots$$

is convergent with sum ln 2.

13. Use Abel's theorem to prove the following generalization of Theorem 6.27 on multiplication of series: If two (numerical) series

$$\sum_{n=1}^{\infty} z_n, \qquad \sum_{n=1}^{\infty} z_n' \tag{19}$$

are convergent (in general only conditionally) with sums s and s' respectively, and if their formal product

$$\sum_{n=1}^{\infty} (z_1 z_n' + z_2 z_{n-1}' + \cdots + z_n z_1')$$

is convergent with sum S, then $S = ss'$.

14. Give an example showing that the converse of Abel's theorem is false.

Comment. However, the next problem shows that the converse of Abel's theorem does in fact hold if the coefficients c_n are suitably restricted.

15. Prove the following proposition known as *Tauber's theorem*: If the coefficients of the power series (18) satisfy the condition

$$\lim_{n\to\infty} nc_n = 0$$

and if

$$\lim_{z \to 1} s(z) = A \qquad (0 < z < 1),$$

then the series

$$\sum_{n=0}^{\infty} c_n$$

converges and has the sum A.

CHAPTER EIGHT

Some Special Mappings

8.1. The Exponential and Related Functions

8.11. A complex function $f(z)$ is said to be *entire* if it is analytic at every point of the finite z-plane. It has already been noted in Sec. 7.25 that the radius of convergence of each of the power series

$$1 + z + \frac{z^2}{2!} + \cdots, \qquad 1 - \frac{z^2}{2!} + \frac{z^4}{4!} - \cdots, \qquad z - \frac{z^3}{3!} + \frac{z^5}{5!} - \cdots \tag{1}$$

is infinite. Hence, by Theorem 7.19, the sum of each series is an entire function. But for real $z = x$ the sums of the series (1) are just the functions e^x, $\cos x$, $\sin x$ familiar from calculus. Guided by these considerations, we now set

$$e^z = 1 + z + \frac{z^2}{2!} + \cdots + \frac{z^n}{n!} + \cdots, \tag{2}$$

$$\cos z = 1 - \frac{z^2}{2!} + \frac{z^4}{4!} - \cdots + (-1)^k \frac{z^{2k}}{(2k)!} + \cdots, \tag{3}$$

$$\sin z = z - \frac{z^3}{3!} + \frac{z^5}{5!} - \cdots + (-1)^k \frac{z^{2k+1}}{(2k+1)!} + \cdots, \tag{4}$$

thereby *defining* the exponential, cosine and sine *for arbitrary complex z.*

8.12. The key formula

$$e^{z_1+z_2} = e^{z_1}e^{z_2} \tag{5}$$

valid for real z_1 and z_2 continues to hold for arbitrary complex z_1 and z_2. To see this, we need only use Theorem 6.27 to multiply the (absolutely convergent) power series

$$e^{z_1} = 1 + z_1 + \frac{z_1^2}{2!} + \cdots + \frac{z_1^n}{n!} + \cdots,$$

$$e^{z_2} = 1 + z_2 + \frac{z_2^2}{2!} + \cdots + \frac{z_2^n}{n!} + \cdots,$$

obtaining

$$1 + (z_1 + z_2) + \frac{(z_1 + z_2)^2}{2!} + \cdots + \frac{(z_1 + z_2)^n}{n!} + \cdots. \tag{6}$$

Formula (5) follows at once, since (6) can be immediately recognized as the power series for $e^{z_1+z_2}$.

Setting $z_1 = z$, $z_2 = -z$ in (5), we get

$$e^z e^{-z} = e^0 = 1,$$

so that

$$e^{-z} = \frac{1}{e^z}. \tag{7}$$

It follows from (7) that

$$\frac{e^{z_1}}{e^{z_2}} = e^{z_1}e^{-z_2} = e^{z_1 - z_2}.$$

8.13. a. Replacing z by iz in (2) and separating real and imaginary parts of the resulting series, we get

$$e^{iz} = \left(1 - \frac{z^2}{2!} + \frac{z^4}{4!} - \cdots\right) + i\left(z - \frac{z^3}{3!} + \frac{z^5}{5!} - \cdots\right).$$

Recognizing the two series in parentheses as $\cos z$ and $\sin z$, respectively, we find at once that

$$e^{iz} = \cos z + i \sin z. \tag{8}$$

This remarkable result, known as *Euler's formula*, exhibits an intimate relation between the exponential and trigonometric functions (when regarded as functions of a complex variable).

b. Clearly

$$\cos(-z) = \cos z, \qquad \sin(-z) = -\sin z, \tag{9}$$

since the series (3) for $\cos z$ contains only even powers of z, while the series (4) for $\sin z$ contains only odd powers of z. Hence replacing z by $-z$ in (8), we get

$$e^{-iz} = \cos z - i \sin z. \tag{8'}$$

Adding (8) and (8′), we find that

$$\cos z = \frac{e^{iz} + e^{-iz}}{2}. \tag{10}$$

Similarly subtracting (8′) from (8), we get

$$\sin z = \frac{e^{iz} - e^{-iz}}{2i}. \qquad (10')$$

c. The function e^z is periodic with period $2\pi i$, i.e.,

$$e^{z+2\pi i} = e^z$$

for all z. In fact,

$$e^{z+2\pi i} = e^z e^{2\pi i}$$

because of (5), while

$$e^{2\pi i} = \cos 2\pi + i \sin 2\pi = 1$$

because of (8).

d. Any complex number z can be written in the trigonometric form

$$z = r(\cos\theta + i \sin\theta) \qquad (11)$$

(see Sec. 1.33). Using (8), we can write (11) in the equivalent *exponential form*

$$z = re^{i\theta}. \qquad (11')$$

e. The function e^z does not vanish for any complex z. In fact, if $z = x + iy$ then

$$e^z = e^{x+iy} = e^x e^{iy} = e^x(\cos y + i \sin y)$$

because of (5) and (8), so that

$$|e^z| = e^x.$$

But e^x is nonvanishing for all real x. Therefore $|e^z|$ (and hence e^z itself) is nonvanishing for all complex z.

8.14. a. The trigonometric addition formulas

$$\begin{aligned} \cos(z_1 + z_2) &= \cos z_1 \cos z_2 - \sin z_1 \sin z_2, \\ \sin(z_1 + z_2) &= \sin z_1 \cos z_2 + \cos z_1 \sin z_2, \end{aligned} \qquad (12)$$

familiar for real z_1 and z_2, continue to hold for complex z_1 and z_2. In fact,

$$e^{i(z_1+z_2)} = e^{iz_1} e^{iz_2}$$

and hence, by Euler's formula,

$$\begin{aligned} \cos(z_1 + z_2) &+ i \sin(z_1 + z_2) \\ &= (\cos z_1 + i \sin z_1)(\cos z_2 + i \sin z_2) \\ &= (\cos z_1 \cos z_2 - \sin z_1 \sin z_2) + i(\sin z_1 \cos z_2 + \cos z_1 \sin z_2), \end{aligned}$$

from which (12) follows at once by taking real and imaginary parts. Changing z_2 to $-z_2$ in (12) and then using (9), we get the corresponding subtraction formulas

$$\cos(z_1 - z_2) = \cos z_1 \cos z_2 + \sin z_1 \sin z_2,$$
$$\sin(z_1 - z_2) = \sin z_1 \cos z_2 - \cos z_1 \sin z_2. \qquad (12')$$

Furthermore, setting $z_1 = z$, $z_2 = 2\pi$ in (12), we find that

$$\cos(z + 2\pi) = \cos z \cos 2\pi - \sin z \sin 2\pi = \cos z,$$
$$\sin(z + 2\pi) = \sin z \cos 2\pi + \cos z \sin 2\pi = \sin z,$$

i.e., the cosine and sine functions are periodic with period 2π in the whole complex plane as well as on the real line. Finally, setting $z_1 = z_2 = z$ in the first of the formulas (12′), we get

$$\cos 0 = \cos^2 z + \sin^2 z,$$

i.e.,

$$\cos^2 z + \sin^2 z = 1 \qquad (13)$$

just as in the real case. However, unlike the real case, it can no longer be concluded from (13) that $|\cos z| \leq 1$, $|\sin z| \leq 1$. In fact,

$$\cos i = \cosh 1 = 1.543, \qquad \sin i = i \sinh 1 = 1.175i$$

(see Sec. 8.15).

b. Next we find the *zeros* of $\sin z$ and $\cos z$, i.e., the points of the complex plane at which $\sin z$ and $\cos z$ vanish. Because of (10′), the equation $\sin z = 0$ reduces to $e^{iz} = e^{-iz}$ or $e^{2iz} = 1$, which takes the form

$$e^{2i(x+iy)} = 1$$

or

$$e^{2ix}e^{-2y} = 1 \qquad (14)$$

after writing $z = x + iy$. The left-hand side of (14) is a complex number of argument $2x$ and modulus e^{-2y}, while the right-hand side is just the number 1. Hence we must have

$$2x = 2\pi k, \qquad e^{-2y} = 1,$$

where k is an integer, so that $x = \pi k$, $y = 0$. Therefore $\sin z$ vanishes if and only if

$$z = \pi k \qquad (k = 0, \pm 1, \pm 2, \ldots).$$

In the same way, by solving the equation $e^{iz} = -e^{-iz}$ or $e^{2iz} = -1$, we find that $\cos z$ vanishes if and only if

$$z = \frac{\pi}{2} + \pi k \qquad (k = 0, \pm 1, \pm 2, \ldots).$$

8.15. Hyperbolic functions. In calculus the hyperbolic cosine and hyperbolic sine are defined by the formulas

$$\cosh z = \frac{e^z + e^{-z}}{2}, \tag{15}$$

$$\sinh z = \frac{e^z - e^{-z}}{2}, \tag{15'}$$

where $z = x$ is real. We now use the same formulas to define $\cosh z$ and $\sinh z$ for arbitrary complex z. Substituting the power series for e^z and e^{-z} into (15) and (15′), we find that

$$\cosh z = 1 + \frac{z^2}{2!} + \frac{z^4}{4!} + \cdots + \frac{z^{2k}}{(2k)!} + \cdots,$$

$$\sinh z = z + \frac{z^3}{3!} + \frac{z^5}{5!} + \cdots + \frac{z^{2k+1}}{(2k+1)!} + \cdots,$$

where both series are absolutely convergent in the whole plane. Comparing (15) and (15′) with (10) and (10′), we see at once that

$$\cosh z = \cos iz,$$

$$\sinh z = -i \sin iz,$$

$$\cos iz = \cosh z,$$

$$\sin iz = i \sinh z.$$

(In particular, note that the cosine and sine cannot be bounded in the complex plane, since $|\cos ix|$ and $|\sin ix|$ become arbitrarily large for sufficiently large real x.) Using these relations, we can deduce a formula involving the hyperbolic functions $\cosh z$ and $\sinh z$ from any formula involving the trigonometric functions $\cos z$ and $\sin z$. For example,

$$\cos^2 iz + \sin^2 iz = 1$$

by (13) and hence

$$\cosh^2 z - \sinh^2 z = 1,$$

while

$$\cos i(z_1 + z_2) = \cos iz_1 \cos iz_2 - \sin iz_1 \sin iz_2,$$

$$\sin i(z_1 + z_2) = \sin iz_1 \cos iz_2 + \cos iz_1 \sin iz_2$$

by (12) and hence

$$\cosh (z_1 + z_2) = \cosh z_1 \cosh z_2 + \sinh z_1 \sinh z_2,$$

$$\sinh (z_1 + z_2) = \sinh z_1 \cosh z_2 + \cosh z_1 \sinh z_2.$$

8.16. To differentiate the functions e^z, $\cos z$, $\sin z$, $\cosh z$ and $\sinh z$, we need only invoke Theorem 7.19 on term-by-term differentiation of power series. In this way, we find at once that

$$\frac{de^z}{dz} = 1 + z + \frac{z^2}{2!} + \cdots = e^z,$$

$$\frac{d\cos z}{dz} = -z + \frac{z^3}{3!} - \frac{z^5}{5!} + \cdots = -\sin z,$$

$$\frac{d\sin z}{dz} = 1 - \frac{z^2}{2!} + \frac{z^4}{4!} - \cdots = \cos z,$$

$$\frac{d\cosh z}{dz} = z + \frac{z^3}{3!} + \frac{z^5}{5!} + \cdots = \sinh z,$$

$$\frac{d\sinh z}{dz} = 1 + \frac{z^2}{2!} + \frac{z^4}{4!} + \cdots = \cosh z,$$

just as in the case of real z.

8.17. According to Sec. 4.33, the mapping $w = e^z$ is conformal at every point of the finite z-plane, since its derivative $w' = e^z$ is nonzero for all z. Suppose z describes a straight line

$$z = \alpha + it \qquad (\alpha \text{ real}, -\infty < t < \infty) \tag{16}$$

parallel to the imaginary axis (see Figure 21a). Then the image point $w = e^z$ describes the curve

$$w = e^{\alpha + it} = e^{\alpha}(\cos t + i \sin t), \tag{16'}$$

i.e., w traces out a circle of radius e^{α} with its center at the origin (see Figure 21b). In fact, as z describes the line (16) once in the upward direction (the direction of increasing t), w describes the circle (16′) an infinite number of times in the counterclockwise direction.

Similarly, suppose z describes a straight line

$$z = t + i\beta \qquad (\beta \text{ real}, -\infty < t < \infty) \tag{17}$$

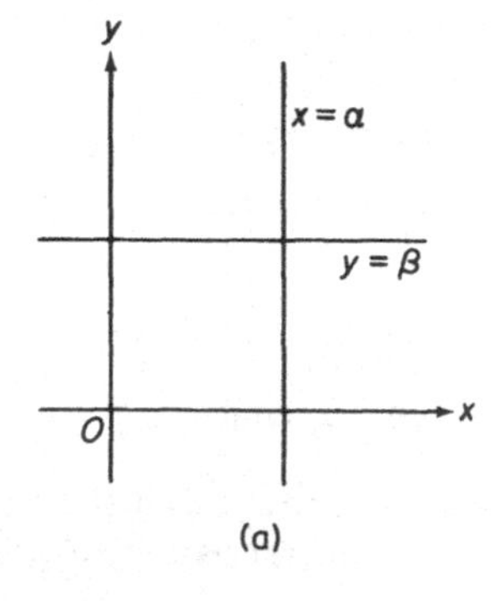

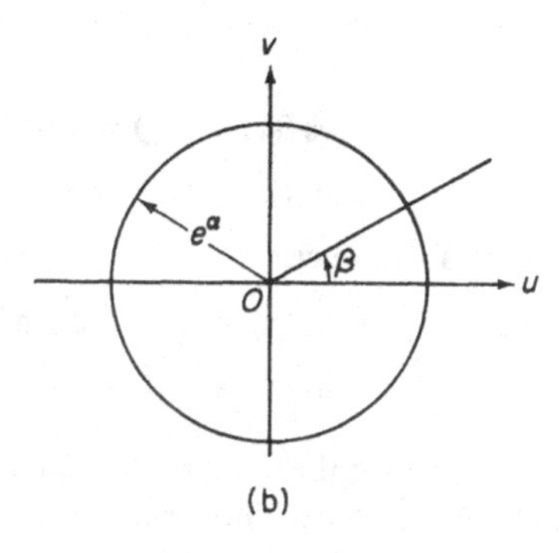

Figure 21

parallel to the real axis (again see Figure 21a). Then the point $w = e^z$ describes the curve

$$w = e^{t+i\beta} = e^t(\cos\beta + i\sin\beta), \tag{17'}$$

i.e., w traces out a ray (why not a whole line?) of slope $\tan\beta$ emanating from (but not including!) the origin, as shown in Figure 21b. Note that since the lines (16) and (17) are orthogonal, the conformality of the mapping guarantees the orthogonality of the circle (16′) and the ray (17′), a fact which is geometrically obvious.

8.2. Fractional Linear Transformations

8.21. The *fractional linear transformation*

$$w = f(z) = \frac{az+b}{cz+d} \qquad (ad - bc \neq 0) \tag{18}$$

has already been encountered in Chap. 4, Probs. 22–28, where it was shown that (18) is a one-to-one mapping of the extended z-plane Π onto itself, which is conformal at every point of Π.† We now continue our study of this important mapping. First we prove that the fractional linear transformation is circle-preserving, i.e., that it carries every circle or straight line in the z-plane into a circle or straight line in the w-plane.‡

8.22. LEMMA. *The transformation*

$$w = \rho z \qquad (\rho > 0)$$

is circle-preserving.

Proof. The general equation of a circle or straight line in the z-plane is

$$Az\bar{z} + \bar{E}z + E\bar{z} + D = 0 \qquad (A, D \text{ real}), \tag{19}$$

where we have a circle if $A \neq 0$, $E\bar{E} - AD > 0$ or a straight line if $A = 0$, $E \neq 0$ (see Chap. 1, Prob. 25). Substituting $z = w/\rho$ into (19), we get

$$\frac{A}{\rho^2}w\bar{w} + \frac{\bar{E}}{\rho}w + \frac{E}{\rho}\bar{w} + D = 0,$$

which is obviously the equation of a circle or straight line in the w-plane. ∎

8.23. LEMMA. *The transformation*

$$w = \frac{1}{z}$$

is circle-preserving.

---

†If $c = 0$, (18) reduces to an entire linear transformation (Chap. 4, Probs. 16–21).

‡A straight line is regarded as a limiting case of a circle, corresponding to infinite radius.

Proof. This time we substitute $z = 1/w$ in (19), obtaining

$$\frac{A}{w\bar{w}} + \frac{\bar{E}}{w} + \frac{E}{\bar{w}} + D = 0$$

or equivalently

$$Dw\bar{w} + \bar{E}\bar{w} + Ew + A = 0.$$

But this is the equation of a circle in the w-plane (if $D \neq 0$) or a straight line (if $D = 0$).† ∎

8.24. THEOREM. *The fractional linear transformation* (18) *is circle-preserving.*

Proof. If $c = 0$, (18) reduces to the entire linear transformation

$$w = \alpha z + \beta \qquad (\alpha \neq 0) \tag{20}$$

where $\alpha = a/d$, $\beta = b/d$. If $\alpha = 1$, this is a pure shift which is obviously circle-preserving. If $\alpha \neq 1$, let

$$\alpha = \rho e^{i\theta} \qquad (\rho > 0).$$

Then (20) is the result of the three consecutive transformations

$$z^* = ze^{i\theta}, \qquad w^* = \rho z^*, \qquad w = w^* + \beta.$$

But all three transformations are circle-preserving, the first because it is a rotation, the second because of Lemma 8.22, and the third because it is a shift. Therefore the combined transformation (20) is circle-preserving.

Now suppose $c \neq 0$. Then (18) can be written in the form

$$w = \frac{a}{c} + \frac{bc - ad}{c(cz + d)} \qquad (bc - ad \neq 0), \tag{18'}$$

which is the result of the three consecutive transformations

$$z^* = cz + d, \qquad w^* = \frac{1}{z^*}, \qquad w = \frac{a}{c} + \frac{bc - ad}{c} w^*.$$

But all three transformations are circle-preserving, the first and third because they are entire linear transformations, the second because of Lemma 8.23. Therefore the combined transformation (18′) is circle-preserving. ∎

8.25. THEOREM. *There exists a unique fractional linear transformation carrying three given points z_1, z_2, z_3 of the z-plane into three given points w_1, w_2, w_3 of the w-plane.*

Proof. We assume that the given six points are all finite (otherwise see Prob. 14). It follows from (18) and

$$w_j = \frac{az_j + b}{cz_j + d} \qquad (j = 1, 2, 3) \tag{21}$$

†Verify the validity of the extra conditions $E\bar{E} - AD > 0$ if $D \neq 0$ and $E \neq 0$ if $D = 0$.

that

$$w - w_1 = \frac{(ad - bc)(z - z_1)}{(cz + d)(cz_1 + d)},$$

$$w - w_2 = \frac{(ad - bc)(z - z_2)}{(cz + d)(cz_2 + d)},$$

$$w_3 - w_1 = \frac{(ad - bc)(z_3 - z_1)}{(cz_3 + d)(cz_1 + d)},$$

$$w_3 - w_2 = \frac{(ad - bc)(z_3 - z_2)}{(cz_3 + d)(cz_2 + d)}.$$

Dividing the first of these equations by the second and the third by the fourth, and then dividing the resulting equations by each other, we get the required fractional linear transformation

$$\frac{w - w_1}{w - w_2} : \frac{w_3 - w_1}{w_3 - w_2} = \frac{z - z_1}{z - z_2} : \frac{z_3 - z_1}{z_3 - z_2} \tag{22}$$

carrying z_1, z_2, z_3 into w_1, w_2, w_3. The transformation is obviously unique, being derived under the sole assumption that the *general* transformation (18) satisfies the conditions (21).† ∎

8.26. **a.** COROLLARY. *The quantity*

$$\frac{z - z_1}{z - z_2} : \frac{z_3 - z_1}{z_3 - z_2},$$

called the ***cross ratio*** *of the four points z_1, z_2, z, z_3 (in that order), is invariant under every fractional linear transformation.*

Proof. We need only note that the right-hand side of (22) is the cross ratio of the points z_1, z_2, z, z_3, while the left-hand side of (22) is the cross ratio of their images under the given fractional linear transformation. ∎

b. COROLLARY. *Let C be the circle determined by three given points z_1, z_2, z_3 in the z-plane, and let Γ be the circle determined by three given points w_1, w_2, w_3 in the w-plane. Then there exists a unique fractional linear transformation carrying C into Γ, with the points z_1, z_2, z_3 going into the points w_1, w_2, w_3.*‡

Proof. According to Theorem 8.25 there is a unique fractional linear transformation carrying z_1, z_2, z_3 into w_1, w_2, w_3, while according to Theorem

†The general fractional linear transformation (18) involves only three parameters, namely the ratios of any three of the numbers a, b, c, d, to the fourth number. These parameters can be determined from (21) and then eliminated from (18). This is in effect what we have done (more elegantly) in going from (18) and (21) to (22).

‡If C (or Γ) is a straight line, one of the points z_j (or w_j) is the point at infinity.

8.24 this transformation must carry C into a circle. But the circle in question can only be the circle Γ. ∎

c. Suppose C has interior G, while Γ has interior I and exterior E. Then the fractional linear transformation $w = f(z)$ figuring in Corollary 8.26b maps G into either I or E. In fact, let z_1, z_2 be any two points of G, with images w_1, w_2 under $w = f(z)$, and let γ be any curve contained in G joining the points z_1 and z_2. Then the image of γ under the transformation $w = f(z)$ joins w_1 and w_2 but cannot intersect the circle Γ since γ does not intersect C. It follows that w_1 and w_2 both belong to the same domain I or E. Denoting this domain by D, we see that every other point $w \in D$ is also the image of a point in G, by the same argument applied to the inverse of the transformation $w = f(z)$. In other words, the "image of G under $w = f(z)$," i.e., the set of all values $w = f(z)$ taken at points $z \in G$, is precisely the domain D.

d. Apart from the fact that D is obviously just the domain containing the image of any point of G, we can use the following argument based on the conformality of the transformation $w = f(z)$ to determine whether $D = I$ or $D = E$ from the behavior of the transformation on the circle Γ. Let z describe C in the counterclockwise direction, so that G lies to the left of an observer moving with z. Then the image of G under $w = f(z)$ is precisely that domain which an observer moving along Γ with the image point $w = f(z)$ finds on his left, be it the interior domain I or the exterior domain E (think this through). The same considerations apply to the case where G is the *exterior* of C, except that now z must describe C in the clockwise direction in order to make the exterior of C the domain on the left of an observer moving with z. The case where C is a straight line and G one of the half-planes with C as its boundary is handled similarly (give the details).

8.27. It will be recalled from Sec. 1.42 that two points P and P' are said to be *symmetric with respect to a given circle* C of radius R centered at O if they lie on the same ray emanating from O and if the product of the distances OP and OP' equals R^2. We now establish an important characterization of such symmetric points.

THEOREM. *Two points P and P' are symmetric with respect to a circle C if and only if every circle or straight line γ passing through P and P' is orthogonal to C.*

Proof. Suppose P and P' are symmetric with respect to the circle C, let γ be any circle passing through P and P', and let OA be the tangent to γ drawn from O, the center of C (see Figure 22). Then

$$(OA)^2 = OP \cdot OP' \tag{23}$$

by elementary geometry ("the tangent to a circle is the mean proportional between the secant and the external segment"). But $OP \cdot OP' = R^2$ and hence $OA = R$. Therefore OA is a radius of C and hence is orthogonal to C. Since OA is also tangent to γ, it follows that γ is orthogonal to C.†

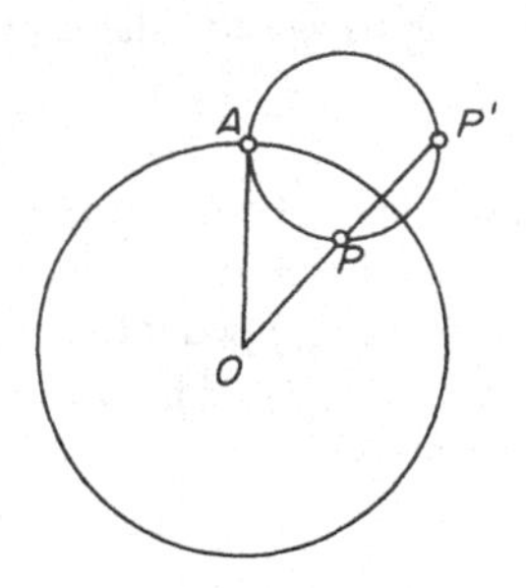

Figure 22

Conversely, suppose every circle through P and P' is orthogonal to C. Then the straight line through P and P' (a special case of a circle!) must be orthogonal to C and hence must go through the point O. Let γ be any circle through P and P', and draw the tangent to γ intersecting γ in a point A. Since γ is orthogonal to C, OA must be a radius of C. It follows from (23) that $OP \cdot OP' = R^2$, i.e., that P and P' are symmetric with respect to C. ∎

8.28. Next we prove that fractional linear transformations "preserve symmetric points" in the sense of the following

THEOREM. *Given two points z_1 and z_2 which are symmetric with respect to a circle C, let w_1, w_2 and Γ be the images of z_1, z_2 and C under the fractional linear transformation* (18). *Then the points w_1 and w_2 are symmetric with respect to the circle Γ.*

Proof. According to Theorem 8.27, we need only show that every circle (or straight line) L passing through w_1 and w_2 is orthogonal to Γ. Let $z = \varphi(w)$ be the inverse of the transformation $w = f(z)$, itself a fractional linear transformation (see Chap. 4, Prob. 24), and let γ be the image of L under $\varphi(w)$. Then, by Theorem 8.24, γ is a circle or straight line passing through z_1 and z_2. Since z_1 and z_2 are symmetric with respect to C, by hypothesis, it follows from Theorem 8.27 again that γ is orthogonal to C. But then L is orthogonal to Γ, by the conformality of (18). ∎

8.29. Examples

a. Let D be the upper half-plane $\operatorname{Im} z > 0$, and let z_0 be any point of D. Find the fractional linear transformation

$$w = f(z) = \frac{az + b}{cz + d} \tag{24}$$

†If γ reduces to a straight line, then γ is obviously orthogonal to C, since P and P' lie on a ray emanating from O.

which maps D onto the unit disk $|w| < 1$ and satisfies the conditions†

$$f(z_0) = 0, \qquad f'(z_0) > 0. \tag{25}$$

Solution. The function (24) must vanish for $z = z_0$. Moreover, by Theorem 8.28, the point $\bar{z}_0$ symmetric to z_0 with respect to the real axis (see Prob. 18) must be mapped into the point symmetric to the point $w = 0$ with respect to the circle $|w| = 1$, i.e., into the point at infinity. It follows that (24) is of the form

$$w = \frac{a}{c}\frac{z - z_0}{z - \bar{z}_0}. \tag{26}$$

Since $|w| = 1$ if $z = 0$, (26) implies $|a/c| = 1$ or $a/c = e^{i\theta}$, and hence

$$w = f(z) = e^{i\theta}\frac{z - z_0}{z - \bar{z}_0}, \tag{27}$$

where the parameter θ is still unspecified. To determine $e^{i\theta}$ (there being no need to determine θ separately), we invoke the second of the conditions (25), which in this case becomes

$$f'(z_0) = \frac{e^{i\theta}}{z_0 - \bar{z}_0} = \frac{e^{i\theta}}{2i \operatorname{Im} z_0} > 0.$$

Since $\operatorname{Im} z_0 > 0$, it follows that $e^{i\theta} = i$, so that the required transformation (24) is just

$$w = i\frac{z - z_0}{z - \bar{z}_0}.$$

b. Let K be the unit disk $|z| < 1$, and let z_0 be any point of K. Find the fractional linear transformation (24) which maps K onto the unit disk $|w| < 1$ and satisfies the conditions (25).

Solution. If $z_0 = 0$, the required transformation is obviously just the trivial transformation $w = z$. Thus let $0 < |z_0| < 1$. The function (24) must again vanish for $z = z_0$, but this time the point $1/\bar{z}_0$ symmetric to z_0 with respect to the circle $|z| = 1$ (see Sec. 1.41) must be mapped into the point at infinity. Therefore (24) must now be of the form

$$w = \frac{a}{c}\frac{z - z_0}{z - \dfrac{1}{\bar{z}_0}},$$

or equivalently

$$w = k\frac{z - z_0}{1 - \bar{z}_0 z} \tag{28}$$

$(k = -a\bar{z}_0/c)$. Since $|w| = 1$ if $z = 1$, (28) implies $|k| = 1$ or $k = e^{i\theta}$, and

†Note that $f'(z_0) > 0$ implies $\arg f'(z_0) = 0$ (what does this mean geometrically?).

hence

$$w = f(z) = e^{i\theta}\frac{z - z_0}{1 - \bar{z}_0 z}.$$

To determine $e^{i\theta}$, we again use the condition $f'(z_0) > 0$, which now takes the form

$$f'(z_0) = \frac{e^{i\theta}}{1 - |z_0|^2} > 0.$$

Since $0 < |z_0| < 1$, it follows that $e^{i\theta} = 1$, so that the required fractional linear transformation is this time just

$$w = \frac{z - z_0}{1 - \bar{z}_0 z}. \tag{29}$$

c. Let D be any half-plane, and let z_0 be any point of D. Find the fractional linear transformation (24) which maps D onto the unit disk $|w| < 1$ and satisfies the conditions (25).

Solution. Clearly D can be mapped onto the upper half-plane $\operatorname{Im} z > 0$ by making a suitable preliminary entire linear transformation $\varphi(z) = \alpha z + \beta$, where $|\alpha| = 1$ and hence $\alpha = e^{i\tau}$ (τ fixed), since the transformation corresponds to a rotation and a shift (without expansion). We then use (27) to map the half-plane $\operatorname{Im} z > 0$ onto the disk $|w| < 1$. The combined transformation

$$w = f(z) = e^{i(\theta+\tau)}\frac{z - z_0}{\varphi(z) - \overline{\varphi(z_0)}} = e^{i\lambda}\frac{z - z_0}{\varphi(z) - \overline{\varphi(z_0)}} \tag{30}$$

then maps D onto the disk $|w| < 1$. Clearly (30) satisfies the condition $f(z_0) = 0$, but to satisfy the condition $f'(z_0) > 0$ we must require that

$$f'(z_0) = \frac{e^{i\lambda}}{\varphi(z_0) - \overline{\varphi(z_0)}} = \frac{e^{i\lambda}}{2i \operatorname{Im} \varphi(z_0)} > 0.$$

Hence $e^{i\lambda} = i$, so that (30) reduces to

$$w = i\frac{z - z_0}{\varphi(z) - \overline{\varphi(z_0)}}.$$

d. Let K be any disk, and let z_0 be any point of K. Find the fractional linear transformation (24) which maps K onto the unit disk $|w| < 1$ and satisfies the conditions (25).

Solution. Clearly K can be mapped onto the unit disk $|z| < 1$ by making a suitable preliminary entire linear transformation, $\varphi(z) = \alpha z + \beta$, where this time $\alpha > 0$ since the transformation corresponds to an expansion and a shift (without rotation). Hence we can now use (29) directly, obtaining

$$w = \frac{\varphi(z) - \varphi(z_0)}{1 - \overline{\varphi(z_0)}\varphi(z)} = \alpha\frac{z - z_0}{1 - \overline{\varphi(z_0)}\varphi(z)}$$

for the required fractional linear transformation.

COMMENTS

8.1. Setting $z = \pi$ in (8), we get

$$e^{i\pi} = -1.$$

With tongue in cheek, we might say that this remarkable formula reveals the intimate relationship between analysis (e), algebra (i), geometry (π), and arithmetic (-1).

8.2. The subject of fractional linear transformations constitutes a digression from the central theme of the book (analytic functions and their properties), and can be postponed until just before Sec. 13.24, after which fractional linear transformations become indispensable.

PROBLEMS

1. Express the complex numbers $\pm 1, \pm i, 1 \pm i, -1 \pm i$ in exponential form.
2. Calculate the following complex numbers:
   a) e^{2+i}; b) $\cos(5 - i)$; c) $\sin(1 - 5i)$.
3. Verify directly that the real and imaginary parts of the functions $\cos z$, $\sin z$, $\cosh z$ and $\sinh z$ satisfy the Cauchy–Riemann equations.
4. Prove that
   $$\tfrac{1}{4}|z| < |e^z - 1| < \tfrac{7}{4}|z|$$
   if $0 < |z| < 1$, while
   $$|e^z - 1| \le e^{|z|} - 1 \le |z|e^{|z|}$$
   for all z.
5. Describe the limiting behavior of e^z as $z \to \infty$ along the ray $\arg z = \alpha$.
6. Prove that the function
   $$f(z) = \begin{cases} e^{-1/z^4} & \text{if } z \neq 0, \\ 0 & \text{if } z = 0 \end{cases}$$
   satisfies the Cauchy–Riemann equations at every point of the plane without being analytic in the whole plane.
7. Find all the zeros of the functions $\cosh z$ and $\sinh z$.
8. Prove that
   a) $|\cos z| = \sqrt{\cosh^2 y - \sin^2 x}$; b) $|\sin z| = \sqrt{\cosh^2 y - \cos^2 x}$;
   c) $|\cosh z| = \sqrt{\cosh^2 x - \sin^2 y}$; d) $|\sinh z| = \sqrt{\cosh^2 x - \cos^2 y}$;
   e) $|\sinh y| \le |\cos z| \le \cosh y$; f) $|\sinh y| \le |\sin z| \le \cosh y$.
9. Find all the solutions of the equation $\cosh z = \frac{1}{2}$.
10. Find the image Γ under the mapping $w = e^z$ of the straight line
    $$z = (1 + i\alpha)t + i\beta \qquad (-\infty < t < \infty)$$
    with slope $\alpha \neq 0$ and y-intercept β.

11. Why does the curve Γ in the preceding problem intersect all rays emanating from the origin at the same angle arc tan α?

12. The tangent, cotangent, hyperbolic tangent and hyperbolic cotangent are defined for arbitrary complex z by the same formulas

$$\tan z = \frac{\sin z}{\cos z}, \qquad \cot z = \frac{\cos z}{\sin z}, \qquad \tanh z = \frac{\sinh z}{\cosh z}, \qquad \coth z = \frac{\cosh z}{\sinh z}$$

as for the case of real z. Where is each of these functions analytic and what is its derivative?

13. Find the fractional linear transformation of the upper half-plane onto the unit disk carrying the points $-1, 0, 1$ of the real axis into the points $1, i, -1$ of the circle.

14. Show that the right-hand side of formula (22) should be replaced by

$$\frac{1}{z - z_2} : \frac{1}{z_3 - z_2}$$

if $z_1 = \infty$, by

$$(z - z_1) : (z_3 - z_1)$$

if $z_2 = \infty$, and by

$$\frac{z - z_1}{z - z_2}$$

if $z_3 = \infty$. Similarly, show that the left-hand side of (22) should be replaced by

$$\frac{1}{w - w_2} : \frac{1}{w_3 - w_2}$$

if $w_1 = \infty$, by

$$(w - w_1) : (w_3 - w_1)$$

if $w_2 = \infty$, and by

$$\frac{w - w_1}{w - w_2}$$

if $w_3 = \infty$.

15. Find the fractional linear transformation carrying the points $-1, \infty, i$ into the points
a) $i, 1, 1 + i$; b) $\infty, i, 1$; c) $0, \infty, 1$.

16. Prove that the cross ratio

$$\frac{z - z_1}{z - z_2} : \frac{z_3 - z_1}{z_3 - z_2}$$

is real if the points z_1, z_2, z, z_3 lie on the same circle or straight line.

17. Find the point symmetric to the point $2 + i$ with respect to
a) The circle $|z| = 1$; b) The circle $|z - i| = 3$.

18. As usual, two points P and P' are said to be *symmetric with respect to a given straight line* L if L is the perpendicular bisector of the line segment joining P and P' (or equivalently if each of the points P and P' is the reflection of the other in the line L). Prove that two points P and P' are symmetric with respect to a straight line L if and only if every circle or straight line γ passing through P and P' is orthogonal to L (cf. Theorem 8.27).

19. Prove that the mapping (27) is uniquely determined if we require it to carry a given point $z = \xi$ of the real axis into the point $w = 1$ of the unit circle. Write the mapping in this case.

20. Map the half-planes $\operatorname{Re} z > 0$ and $\operatorname{Im} z < 1$ onto the unit disk $|w| < 1$. Do the same for the disks $|z + 1| < 1$ and $|z - i| < 2$.

21. Describe the mapping (29) for $|z_0| > 1$.

22. Find the image of each of the following domains under the indicated fractional linear transformations:

a) The quadrant $x > 0, y > 0$ under $w = \dfrac{z - i}{z + i}$;

b) The sector $0 < \arg z < \dfrac{\pi}{4}$ under $w = \dfrac{z}{z - 1}$;

c) The strip $0 < x < 1$ under $w = \dfrac{z - 1}{z}$.

23. By a *fixed point* of a mapping $w = f(z)$ we mean any solution of the equation $z = f(z)$, i.e., any point which is carried into itself by the mapping (cf. Chap. 4, Prob. 17). Prove that the fractional linear transformation

$$w = f(z) = \frac{az + b}{cz + d}$$

has two distinct fixed points if $c \neq 0$, $(a - d)^2 + 4bc \neq 0$. What happens if $(a - d)^2 + 4bc = 0$? If $c = 0$?

24. Let $w = f(z)$ be a fractional linear transformation with two distinct finite fixed points z_1 and z_2. Prove that

$$\frac{w - z_1}{w - z_2} = k\frac{z - z_1}{z - z_2} \tag{31}$$

for a suitable complex constant k.

25. Find the fractional linear transformation such that

a) The points 1 and i are fixed, and the point 0 goes into the point -1;

b) The points $\frac{1}{2}$ and 2 are fixed, and the point $\frac{5}{4} + \frac{3}{4}i$ goes into ∞;

c) The point $z = i$ is the only fixed point, and the point 1 goes into ∞.

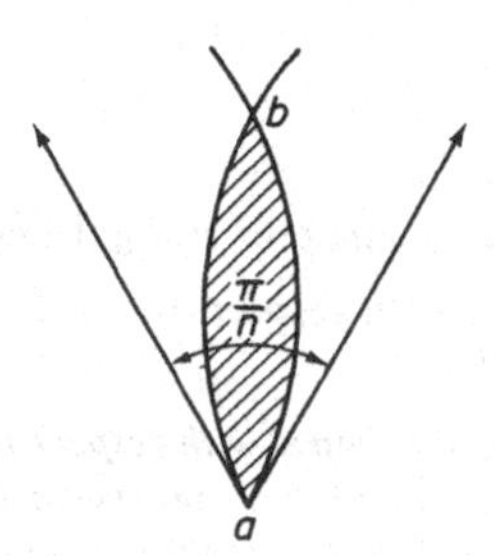

Figure 23

26. Map each of the following domains onto the upper half-plane:

a) The half-disk $|z| < 1$, $0 < \arg z < \pi$;

b) The angular sector $|z| < 1$, $0 < \arg z < \pi/n$ ($n = 1, 2, \ldots$);

c) The circular lune shown in Figure 23;

d) The half-strip $\operatorname{Re} z > 0$, $0 < \operatorname{Im} z < \pi$.

27. By the *product* (or *composition*) of two

mappings or transformations $w = f_1(z)$ and $w = f_2(z)$ we mean the result of applying first one transformation and then the other. There are two such transformations, namely

$$w = f_1 \circ f_2(z) = f_1(f_2(z))$$

and

$$w = f_2 \circ f_1(z) = f_2(f_1(z)),$$

depending on which transformation is carried out first. Prove that multiplication of transformations is *associative*, i.e., that

$$(f_1 \circ f_2) \circ f_3 = f_1 \circ (f_2 \circ f_3)$$

for any three transformations f_1, f_2, f_3.† Prove that multiplication of transformations is in general *noncommutative*, i.e., that in general

$$f_1 \circ f_2 \neq f_2 \circ f_1.$$

28. A set G of elements $a, b, c, \ldots$ equipped with a product (denoted by $\circ$, say) is said to be a *group* if

a) G is *closed* under multiplication, i.e., $a \circ b \in G$ whenever $a, b \in G$;

b) Multiplication is associative, i.e., $(a \circ b) \circ c = a \circ (b \circ c)$ for all $a, b, c \in G$;

c) G contains an element e (the *unit element*) such that $a \circ e = e \circ a = a$ for all $a \in G$;

d) Given any $a \in G$, there is an element $a^{-1} \in G$ (the *inverse* of a) such that $a \circ a^{-1} = a^{-1} \circ a = e$.

Prove that the set of all fractional linear transformations

$$w = f(z) = \frac{az + b}{cz + d} \qquad (ad - bc \neq 0) \tag{32}$$

forms a group (with $\circ$ interpreted as in the preceding problem). What is the unit element of this group? The inverse of a given element?

29. Prove that the set of six fractional linear transformations of the special form

$$f_1(z) = z, \qquad f_2(z) = \frac{1}{z}, \qquad f_3(z) = 1 - z,$$
$$f_4(z) = \frac{1}{1 - z}, \qquad f_3(z) = \frac{z - 1}{z}, \qquad f_4(z) = \frac{z}{z - 1} \tag{33}$$

forms a group.

Comment. This fact is summarized by saying that the transformations (33) form a *subgroup* of the group of all fractional linear transformations (32).

30. Prove that the set of all fractional linear transformations

$$w = f(z) = \frac{az + b}{cz + d} \qquad (ad - bc = 1) \tag{32'}$$

is a subgroup of the group of all fractional linear transformations (32).

---

†For simplicity, we write f_j instead of $f_j(z)$. In writing the expression $f_j \circ f_k$ it is tacitly assumed that the range of f_k is contained in the domain of definition of f_j.

CHAPTER NINE

Multiple-Valued Functions

9.1. Domains of Univalence

9.11. A function $f(z)$ is said to be *univalent* in a domain G if it is one-to-one and analytic in G; by the same token, G is then called a *domain of univalence* for $f(z)$. As will be shown later (see Theorem 12.24), if $f(z)$ is univalent in a domain G, then the derivative $f'(z)$ is nonvanishing in G.

9.12. THEOREM. *Let*

$$w = f(z) = u(x, y) + iv(x, y) \qquad (1)$$

be univalent in a domain G, and let E be the image of G under the mapping (1). *Then E is also a domain* (*in the w-plane*).

Proof. We must prove that E is both open and connected (see Sec. 3.22b). The connectedness of E is almost obvious. Let w_1, w_2 be any two points of E, and let z_1, z_2 be the corresponding points of G, so that $w_1 = f(z_1)$, $w_2 = f(z_2)$. Suppose we join z_1 and z_2 by a curve C lying entirely in G. Then as the point z goes along C from z_1 to z_2, the point $w = f(z)$ traces out a curve Γ joining w_1 and w_2 without leaving the set E, by the very definition of E. Hence E is connected.

To prove that E is open, we must show that every point of E is an interior point of E. To this end, let $w_0 = u_0 + iv_0$ be any point of E, and let $z_0 = x_0 + iy_0$ be the corresponding point of G, so that $w_0 = f(z_0)$. Then we can apply the implicit function theorem† to the system of equations

$$u - u(x, y) = 0, \qquad v - v(x, y) = 0, \qquad (2)$$

†See e.g., V. I. Smirnov, *Linear Algebra and Group Theory* (translated by R. A. Silverman), Dover Publications, Inc., New York (1970), Sec. 19.

since the left-hand sides vanish for $x = x_0, y = y_0, u = u_0, v = v_0$, are continuous in all four variables, and have continuous partial derivatives (cf. Sec. 5.72) with the nonvanishing Jacobian determinant

$$\begin{vmatrix} -\frac{\partial u}{\partial x} & -\frac{\partial u}{\partial y} \\ -\frac{\partial v}{\partial x} & -\frac{\partial v}{\partial y} \end{vmatrix} = \frac{\partial u}{\partial x}\frac{\partial v}{\partial y} - \frac{\partial u}{\partial y}\frac{\partial v}{\partial x} = \left(\frac{\partial u}{\partial x}\right)^2 + \left(\frac{\partial u}{\partial y}\right)^2 = |f'(z)|^2$$

(note that the Cauchy–Riemann equations have been used in the last step!). Hence for (u, v) sufficiently close to (u_0, v_0) there exists a unique pair of functions $x = x(u, v), y = y(u, v)$ which are continuous and satisfy the system (2) subject to the conditions $x_0 = x(u_0, v_0)$, $y_0 = y(u_0, v_0)$. Therefore the point $(x(u, v), y(u, v))$ will lie in any given neighborhood of (x_0, y_0) and in particular will lie in G, provided only that the point (u, v) lies in a sufficiently small neighborhood of (u_0, v_0). But this means that some neighborhood of the point $w_0 = u_0 + iv_0$ consists entirely of points which are images of points of G under the mapping (1), i.e., that w_0 is an interior point of the set E. ▮

9.13. THEOREM. *Let $w = f(z)$, G and E be the same as in the preceding theorem, and let $z = \varphi(w)$ be the inverse of the function $w = f(z)$. Then $\varphi(w)$ is univalent in E, with derivative*

$$\varphi'(w) = \frac{1}{f'(z)}.$$

Proof. The inverse function $z = \varphi(w)$ as defined in Sec. 3.14 is obviously single-valued and one-to-one in E, since $w = f(z)$ is one-to-one in G. To prove the analyticity of $\varphi(w)$, let w_0, w be any two points of E, and let z_0, z be the corresponding points of G. The function $\varphi(w)$ is continuous in E, because of the continuity of the functions $\text{Re}\,\varphi(w) = x(u, v)$, $\text{Im}\,\varphi(w) = y(u, v)$ figuring in the proof of Theorem 9.12. Therefore $z \to z_0$ as $w \to w_0$ and hence

$$\varphi'(w_0) = \lim_{w \to w_0} \frac{z - z_0}{w - w_0} = \lim_{z \to z_0} \frac{1}{\dfrac{w - w_0}{z - z_0}} = \frac{1}{f'(z_0)},$$

i.e., the derivative of $\varphi(w)$ exists at every point $w_0 \in E$ and equals†

$$\varphi'(w_0) = \frac{1}{f'(z_0)}. \quad ▮$$

9.14. Examples

a. If

$$w = z^n,$$

then given any point

$$w = re^{i\theta} \qquad (r > 0)$$

†As noted in Sec. 9.11, $f'(z_0) \neq 0$ for all $z_0 \in G$.

in the w-plane, there are precisely n distinct points of the z-plane with w as their image, namely the points

$$z = \sqrt[n]{r}\, e^{i(\theta+2k\pi)/n} \qquad (k = 0, 1, \ldots, n-1)$$

(recall Sec. 1.36). But these points lie at the vertices of a regular n-gon inscribed in the circle of radius $\sqrt[n]{r}$ with its center at the origin. Therefore z^n is univalent in any "wedge" of the form

$$c < \arg z < c + \frac{2\pi}{n} \qquad (c \text{ real}) \tag{3}$$

with vertex angle $2\pi/n$. In other words, every such wedge is a domain of univalence for z^n, which is "maximal" in the sense that it cannot be enlarged (while staying a domain) without destroying the univalence of z^n. Choosing $c = 0$ in (3), we get the wedge

$$0 < \arg z < \frac{2\pi}{n} \tag{3'}$$

shown in Figure 24a. The function $w = z^n$ maps this wedge into the domain E shown in Figure 24b, namely the w-plane "cut along the positive real axis" (the w-plane minus all the points corresponding to nonnegative real numbers)†, since (3′) implies

$$0 < \arg w = n \arg z < 2\pi$$

(cf. Sec. 1.35).

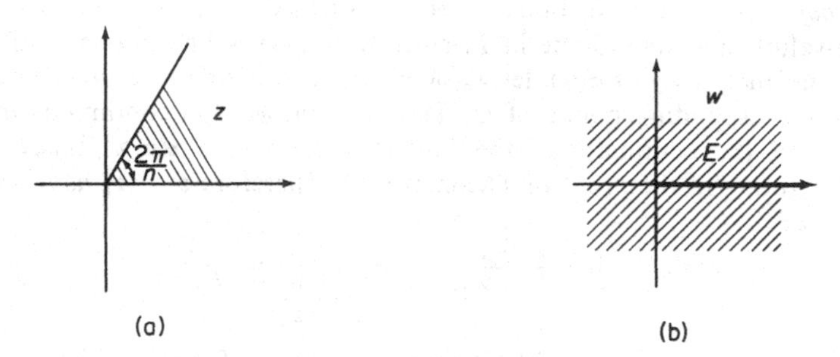

Figure 24

The inverse of the function $w = z^n$ is just

$$z = \sqrt[n]{w}, \tag{4}$$

i.e., the nth root of w (cf. Sec. 1.36). It follows from Theorem 9.13 that (4) is univalent in E, with derivative

†By convention, the "positive real axis" contains the origin O, and so does the "negative real axis." The more exact terms "nonnegative real axis" and "nonpositive real axis" are a bit clumsy.

$$\frac{d\sqrt[n]{w}}{dw} = \frac{1}{\dfrac{dz^n}{dz}} = \frac{1}{nz^{n-1}} = \frac{\sqrt[n]{w}}{nw}.$$

b. Next let

$$w = e^z.$$

If $z_1 = x_1 + iy_1$, $z_2 = x_2 + iy_2$, then $|e^{z_1}| = e^{x_1}$, $|e^{z_2}| = e^{x_2}$, and hence e^{z_1} cannot equal e^{z_2} unless $x_1 = x_2$. If $x_1 = x_2 = x$, $y_1 \neq y_2$, then

$$\begin{aligned} e^{z_1} - e^{z_2} &= e^x(e^{iy_1} - e^{iy_2}) = e^x[(\cos y_1 + i \sin y_1) - (\cos y_2 + i \sin y_2)] \\ &= e^x\left[-2 \sin \frac{y_1 + y_2}{2} \sin \frac{y_1 - y_2}{2} + 2i \cos \frac{y_1 + y_2}{2} \sin \frac{y_1 - y_2}{2}\right] \\ &= 2i \sin \frac{y_1 - y_2}{2} e^x e^{i(y_1+y_2)/2}. \end{aligned}$$

But the expression on the right vanishes if and only if

$$\sin \frac{y_1 - y_2}{2} = 0,$$

i.e., if and only if

$$y_1 - y_2 = 2k\pi \qquad (k = 0, \pm 1, \pm 2, \ldots).$$

Thus e^z is univalent in any strip of the form

$$c < \operatorname{Im} z < c + 2\pi \qquad (c \text{ real}) \tag{5}$$

with sides parallel to the real axis. In other words, every such strip is a (maximal) domain of univalence for e^z. Choosing $c = 0$ in (5), we get the strip

$$0 < \operatorname{Im} z < 2\pi \tag{5'}$$

shown in Figure 25a. The function $w = e^z$ maps this strip into the same domain E as in the preceding example, namely the w-plane cut along the positive real axis (see Figure 25b). In fact, since

$$w = e^z = e^{x+iy} = e^x e^{iy}$$

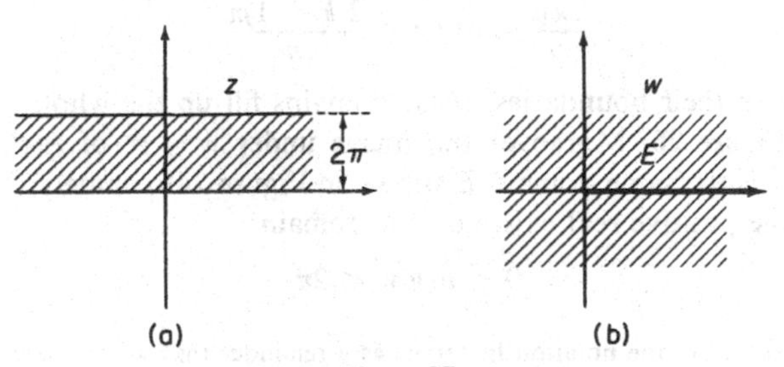

Figure 25

we have

$$|w| = e^x, \qquad \arg w = y = \operatorname{Im} z, \tag{6}$$

and hence (5′) again implies $0 < \arg w < 2\pi$.

The inverse of the function $w = e^z$ is denoted by

$$z = \ln w \tag{7}$$

and is called the *logarithm* of w (just as in the real case).† It follows from Theorem 9.13 that (7) is univalent in E, with derivative

$$\frac{d(\ln w)}{dw} = \frac{1}{\dfrac{de^z}{dz}} = \frac{1}{e^z} = \frac{1}{w}.$$

According to (6),

$$x = \ln |w|, \qquad y = \arg w$$

and hence

$$z = \ln w = \ln |w| + i \arg w. \tag{8}$$

Formula (8) can be used to calculate logarithms of negative numbers and complex numbers. For example,

$$\ln(-1) = \ln 1 + i \arg(-1) = (2k + 1)\pi i \qquad (k = 0, \pm 1, \pm 2, \ldots),$$

while

$$\ln i = \ln 1 + i \arg i = (2k + \tfrac{1}{2})\pi i \qquad (k = 0, \pm 1, \pm 2, \ldots).$$

9.2. Branches and Branch Points

9.21. Suppose we choose the constant c in (3) to have the values

$$0, \frac{2\pi}{n}, \frac{4\pi}{n}, \ldots, \frac{2(n-1)\pi}{n} \tag{9}$$

in turn. This gives n overlapping domains of univalence for the function $w = z^n$, each a wedge of vertex angle $2\pi/n$, which we denote by G_0, G_1, G_2, $\ldots$, G_{n-1} respectively, where G_k is the domain

$$\frac{2k\pi}{n} < \arg z < \frac{2(k+1)\pi}{n} \tag{10}$$

Together with their boundaries, these domains fill up the whole z-plane, as shown in Figure 26. Moreover the image under $w = z^n$ of each of the n domains G_k is the same domain E shown in Figure 24b, namely the w-plane cut along the positive real axis, i.e., the domain

$$0 < \arg w < 2\pi. \tag{11}$$

---

†The letter "n" in the notation ln serves as a reminder that we are dealing with the *natural* logarithm, i.e., the logarithm to the base e.

To see this, we need only note that (10) implies

$$2k\pi < \arg w = n \arg z < 2(k+1)\pi,$$

which is equivalent to (11) for every $k = 0, 1, \ldots, n-1$.

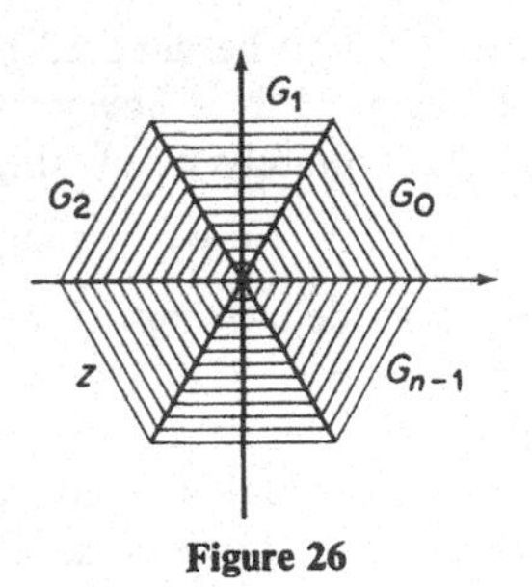

Figure 26

The function $w = z^n$ with domain of definition G_k and range E has a single-valued (in fact univalent) inverse with domain of definition E and range G_k. Let this inverse function be denoted by $z = (\sqrt[n]{w})_k$, so that the function considered in Example 9.14a is now denoted by $z = (\sqrt[n]{w})_0$ rather than by $z = \sqrt[n]{w}$. Then the n single-valued functions

$$(\sqrt[n]{w})_0, (\sqrt[n]{w})_1, \ldots, (\sqrt[n]{w})_{n-1} \tag{12}$$

can be regarded as "parts" of one "master" multiple-valued function $z = \sqrt[n]{w}$, called the *nth root* of w (cf. Sec. 1.36), namely the inverse of the function $w = z^n$, the latter being defined this time in the whole z-plane with the whole w-plane as its range. The separate functions (12) are called (single-valued) *branches* of the multiple-valued function $z = \sqrt[n]{w}$.

9.22. Similarly, suppose we choose the constant c in (5) to have the values

$$0, 2\pi, -2\pi, 4\pi, -4\pi, \ldots$$

in turn. This gives infinitely many nonoverlapping domains of univalence for the function $w = e^z$, each a strip of width 2π, which we denote by G_0, G_1, G_{-1}, G_2, G_{-2}, ... respectively, where G_k is the domain

$$2k\pi < \operatorname{Im} z < 2(k+1)\pi. \tag{13}$$

Together with their boundaries, these domains fill up the whole z-plane, as shown in Figure 27. Moreover the image under $w = e^z$ of each of the infinitely many domains G_k is the same domain E shown in Figure 25b, again the w-plane cut along the positive real axis. To see this, we need only note that (13) implies

$$2k\pi < \arg w = \operatorname{Im} z < 2(k+1)\pi,$$

which is equivalent to (11) for every $k = 0, \pm 1, \pm 2, \ldots$

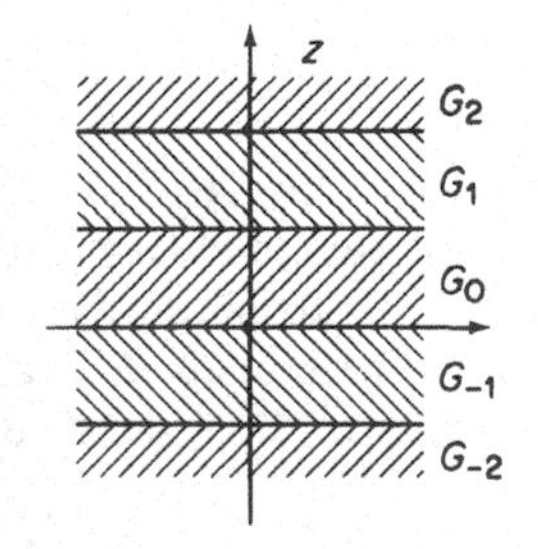

Figure 27

The function $w = e^z$ with domain of definition G_k and range E has a single-valued inverse with domain of definition E and range G_k. Let this inverse function be denoted by $z = (\ln w)_k$, so that the function considered in Example 9.14b is now denoted by $z = (\ln w)_0$ rather than by $z = \ln w$. Then the infinitely many single-valued functions

$$(\ln w)_0, (\ln w)_1, (\ln w)_{-1}, (\ln w)_2, (\ln w)_{-2}, \ldots \tag{14}$$

can be regarded as "parts" of one multiple-valued function $z = \ln w$, called the *logarithm* of w, namely the inverse of the function $w = e^z$, the latter being defined this time in the whole z-plane with the whole w-plane as its range. The separate single-valued functions (14) are again called *branches* of the multiple-valued function $z = \ln w$.†

9.23. It should be kept in mind that the notion of a branch of a multiple-valued function is intimately related to the choice of the corresponding domain of univalence, and hence inevitably contains an element of arbitrariness. For example, suppose that for the function $w = z^n$ we choose the domains of univalence $D_0, D_1, D_2, \ldots, D_{n-1}$ shown in Figure 28, obtained

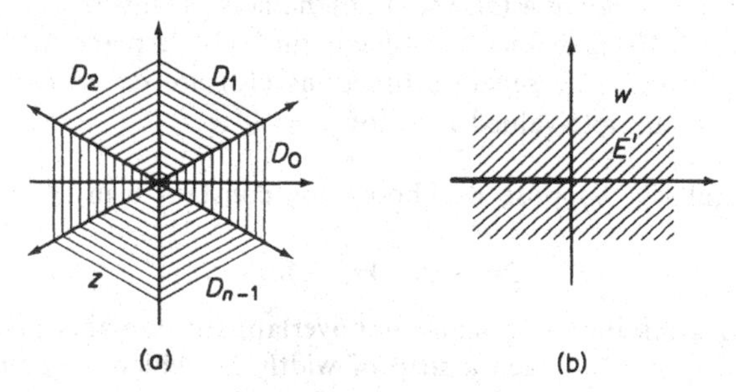

Figure 28

by assigning the constant c in (3) the values

$$-\frac{\pi}{n}, \frac{\pi}{n}, \frac{3\pi}{n}, \ldots, \frac{(2n-3)\pi}{n}, \tag{9'}$$

π/n less than the values (9). Then D_k is the domain

$$\frac{(2k-1)\pi}{n} < \arg z < \frac{(2k+1)\pi}{n}, \tag{10'}$$

made up of the "top half" of G_{k-1} (as seen by an observer moving around the origin in the counterclockwise direction) and the "bottom half" of G_k, where $G_{-1} = G_{n-1}$ by definition. The image under $w = z^n$ of each of the domains D_k is the same domain E', namely the w-plane cut along the *negative* real axis (see Figure 28b), i.e., the domain

$$-\pi < \arg w < \pi. \tag{11'}$$

†The symbols $\sqrt[n]{w}$ and $\ln w$ will henceforth always denote multiple-valued functions, rather than the single-valued branches of these functions considered in Sec. 9.14.

To see this, we need only note that (10′) implies

$$(2k - 1)\pi < \arg w = n \arg z < (2k + 1)\pi,$$

which is equivalent to (11′) for every $k = 0, 1, \ldots, n - 1$.

The function $w = z^n$ with domain of definition D_k and range E' has a single-valued inverse with domain of definition E' and range D_k. Let this function be denoted by $z = \{\sqrt[n]{w}\}_k$, where we use braces instead of the parentheses used in Sec. 9.21. Then the n single-valued functions

$$\{\sqrt[n]{w}\}_0, \{\sqrt[n]{w}\}_1, \ldots, \{\sqrt[n]{w}\}_{n-1} \tag{12'}$$

can be regarded as new branches of the multiple-valued function $z = \sqrt[n]{w}$, which differ from the old branches (12). Nevertheless, the branch $\{\sqrt[n]{w}\}_k$ coincides with the branch $(\sqrt[n]{w})_{k-1}$ in the lower half-plane $-\pi < \arg w < 0$ and with the branch $(\sqrt[n]{w})_k$ in the upper half-plane $0 < \arg w < \pi$, where $(\sqrt[n]{w})_{-1} = (\sqrt[n]{w})_{n-1}$ by definition. This is hardly surprising since the various branches (12) and (12′) are all "generated" by the same underlying multiple-valued function $z = \sqrt[n]{w}$.

9.24. Next suppose that at any given point $w_0 = |w_0|e^{i\theta_0} \in E$ we choose a value of the multiple-valued function $\sqrt[n]{w}$ corresponding to the branch $(\sqrt[n]{w})_k$ and represented by the point

$$z_0 = \sqrt[n]{|w_0|}\left(\cos\frac{\theta_0}{n} + i\sin\frac{\theta_0}{n}\right)$$

belonging to the domain G_k of Sec. 9.21. Let the variable point $w = |w|e^{i\theta}$ describe a closed Jordan curve Γ in the w-plane with initial and final point w_0. Then the corresponding point

$$z = \sqrt[n]{|w|}\left(\cos\frac{\theta}{n} + i\sin\frac{\theta}{n}\right) \tag{15}$$

describes a curve C in the z-plane. There are now just two possibilities:

a. If the point $w = 0$ (the origin of the w-plane) lies outside the curve Γ, then θ returns to its initial value θ_0 after w makes one circuit around Γ in either direction. Hence the point z describes a closed Jordan curve C with initial and final point z_0. In the course of describing Γ, w may well cross the ray $\arg w = 0$ excluded from the domain E. The point z will then "move onto other branches" of $\sqrt[n]{w}$ before eventually returning to z_0.†

b. If the point $w = 0$ lies inside Γ, then one circuit around Γ in the counterclockwise direction causes θ to increase from θ_0 to $\theta_0 + 2\pi$ ("arg w

†The ray $\arg w = 0$ is called a *branch cut* of the function $\sqrt[n]{w}$, since it separates the branches of $\sqrt[n]{w}$ in this way. Note that θ represents that branch of the multiple-valued function arg w which equals θ_0 when $w = w_0$ (possibly together with certain "adjacent branches"). One customarily (but rather loosely) says "arg w," meaning "θ," in locutions like "arg w does not change," "arg w increases by 2π," and so on.

increases by 2π"). Correspondingly, the point (15) describes an arc going from z_0 to the point

$$z_1 = \sqrt[n]{|w_0|}\left(\cos\frac{\theta_0 + 2\pi}{n} + i\sin\frac{\theta_0 + 2\pi}{n}\right)$$

obtained by rotating z_0 through the angle $2\pi/n$ in the counterclockwise direction. But this is just the value of $\sqrt[n]{w_0}$ corresponding to the branch $(\sqrt[n]{w})_{k+1}$. Since the point w_0 is arbitrary, we see that one circuit around Γ in the counterclockwise direction carries the branch $(\sqrt[n]{w})_k$ into the branch $(\sqrt[n]{w})_{k+1}$, in the sense that every value of $\sqrt[n]{w}$ on the branch $(\sqrt[n]{w})_k$ changes continuously into the corresponding value of $\sqrt[n]{w}$ on the branch $(\sqrt[n]{w})_{k+1}$. Similarly, one circuit around Γ in the *clockwise* direction carries the branch $(\sqrt[n]{w})_k$ into the branch $(\sqrt[n]{w})_{k-1}$. Moreover, an appropriate number of circuits around Γ in one direction or the other carries any branch $(\sqrt[n]{w})_k$ into any other branch $(\sqrt[n]{w})_l$, provided only that the point $w = 0$ lies inside Γ. In particular, a total of n circuits around Γ in the same direction carries the point z_0 back into itself, and hence carries every branch $(\sqrt[n]{w})_k$ into itself.

9.25. A point η with the property that a circuit around any closed Jordan curve with η in its interior (briefly, a "circuit around η") carries every branch of a given multiple-valued function into another branch in the way just described is called a *branch point* of the function. If a finite number of circuits around η in the same direction carries every branch of the function into itself, then η is called a branch point *of finite order*, specifically *of order $n - 1$*.† Thus we have just proved that $w = 0$ is a branch point of order $n - 1$ of the function $\sqrt[n]{w}$. Moreover, since a complete circuit in either direction around the point $w = 0$ is also a complete circuit around the point $w = \infty$ (think of things on the Riemann sphere), the point at infinity is also a branch point of $\sqrt[n]{w}$, of order $n - 1$. Clearly $\sqrt[n]{w}$ has no other branch points, since, as already noted, if the point $w = 0$ does not lie inside Γ, the value of $(\sqrt[n]{w})_k$ remains unchanged after making a circuit around Γ.

9.26. Turning our attention to the logarithm, suppose that at a given point $w_0 = |w_0|e^{i\theta_0} \in E$ we choose a value of $\ln w$ corresponding to the branch $(\ln w)_k$ and represented by the point

$$z_0 = \ln|w_0| + i\theta_0$$

belonging to the domain G_k of Sec. 9.22. Again let $w = |w|e^{i\theta}$ describe a closed Jordan curve Γ in the w-plane with initial and final point w_0. Then the corresponding point

$$z = \ln|w| + i\theta$$

†Such a branch point is also said to be *algebraic*, provided the function in question has a limit (finite or infinite) at η.

describes a curve C in the z-plane. The analysis of the case where the point $w = 0$ lies outside Γ is exactly the same as for the function $\sqrt[n]{w}$. If the point $w = 0$ lies inside Γ, then a circuit around Γ in the counterclockwise direction carries the branch $(\ln w)_k$ into the branch $(\ln w)_{k+1}$, while a circuit around Γ in the clockwise direction carries the branch $(\ln w)_k$ into the branch $(\ln w)_{k-1}$, so that the point $w = 0$ is a branch point of $\ln w$. Moreover $w = \infty$ is the only other branch point of $\ln w$, for the same reason as in the case of the function $\sqrt[n]{w}$. However, unlike the case of $\sqrt[n]{w}$, no matter how many circuits we make around Γ in a given direction, the branch $(\ln w)_k$ is never carried back into itself,† and instead we "keep generating new branches" of $\ln w$, as described schematically by the infinite series of transformations

$$(\ln w)_k \longrightarrow (\ln w)_{k+1} \longrightarrow (\ln w)_{k+2} \longrightarrow \cdots$$

or

$$(\ln w)_k \longrightarrow (\ln w)_{k-1} \longrightarrow (\ln w)_{k-2} \longrightarrow \cdots.$$

For this reason, the branch points $w = 0$ and $w = \infty$ are said to be *of infinite order* (or *logarithmic*).

9.3. Riemann Surfaces

9.31. a. As we now show, a multiple-valued function can actually be regarded as single-valued if we suitably generalize its domain of definition. Consider first the multiple-valued function $z = \sqrt[n]{w}$. Suppose we take n "copies" $E_0, E_1, \ldots, E_{n-1}$ of the domain E of Figure 24b (the w-plane cut along the positive real axis), regarding the point $w = 0$ as the same for all n copies. Let δ_k^+ and δ_k^- denote the upper and lower edges of the positive real axis, regarded as the boundary of E_k. We now "paste" δ_0^- to δ_1^+, δ_1^- to $\delta_2^+, \ldots,$ δ_{n-2}^- to δ_{n-1}^+, and finally δ_{n-1}^- to δ_0^+ (think of each E_k as a cut sheet of paper).‡ This has the effect of joining E_0 to E_1, E_1 to $E_2, \ldots, E_{n-2}$ to E_{n-1}, and finally E_{n-1} back to E_0. As a result, we get an "n-sheeted structure" (shown schematically in Figure 29 for the case $n = 4$), called the *Riemann surface* of the function $z = \sqrt[n]{w}$. The fact that $n - 1$ "sheets" get in the way of the final past-

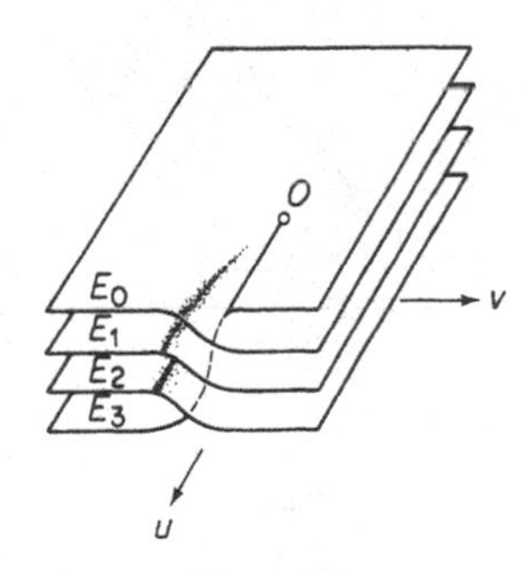

Figure 29

†In particular, no matter how many times w winds around the origin in the same direction, the point z never describes a closed curve.

‡Mathematically, "pasting" δ_k^- to δ_{k+1}^+ merely means identifying opposite points of δ_k^- and δ_{k+1}^+. Note that δ_k^- is the ray $\arg w = 2(k+1)\pi$ while δ_k^+ is the ray $\arg w = 2k\pi$, so that this identification is perfectly natural.

ing of δ_{n-1}^- to δ_0^+ if we try to make a paper model of the Riemann surface does not matter, since the intervening sheets do not in the least prevent us from identifying opposite points of δ_{n-1}^- and δ_0^+.

b. Let S be the Riemann surface just constructed. To define $z = \sqrt[n]{w}$ on the surface S (rather than in the domain E), we choose the value of $\sqrt[n]{w}$ on the kth sheet E_k of S to be the value corresponding to the kth branch $(\sqrt[n]{w})_k$ introduced in Sec. 9.21, i.e., the unique value of $\sqrt[n]{w}$ lying in the domain of univalence G_k.† The function $z = \sqrt[n]{w}$ is then one-to-one (and single-valued!) on the surface S, which it maps onto the whole z-plane. By the same token, the function $w = z^n$ is a one-to-one mapping of the whole z-plane onto S.

c. Suppose the point z makes a circuit around the origin $z = 0$ of the z-plane. Then the image point $w = z^n$ makes a circuit around the origin of the Riemann surface S, moving successively onto all n sheets. Conversely, suppose the point w makes a circuit around the origin $w = 0$ of S. Then the image point $z = \sqrt[n]{w}$ moves continuously from one branch $(\sqrt[n]{w})_k$, corresponding to the initial position of w, to every other branch $(\sqrt[n]{w})_l$ in turn.

9.32. Next we construct the Riemann surface of the logarithm $z = \ln w$, starting this time with infinitely many sheets E_k $(k = 0, \pm 1, \pm 2, \ldots)$, all copies of the domain E of Figure 25b (once again the w-plane cut along the positive real axis). The sheets are pasted together in the same way as in Sec. 9.31a, i.e., they are all joined together at the point $w = 0$ and for each k the lower edge of the cut on the sheet E_k is pasted to the upper edge of the cut on the sheet E_{k+1}. However, we no longer have a "first sheet" and a "last sheet" to be pasted together at the end of the construction. Instead we get the "infinite-sheeted" Riemann surface shown schematically in Figure 30. The function $z = \ln w$ can be regarded as one-to-one on this surface, which it maps onto the whole z-plane (give the details).

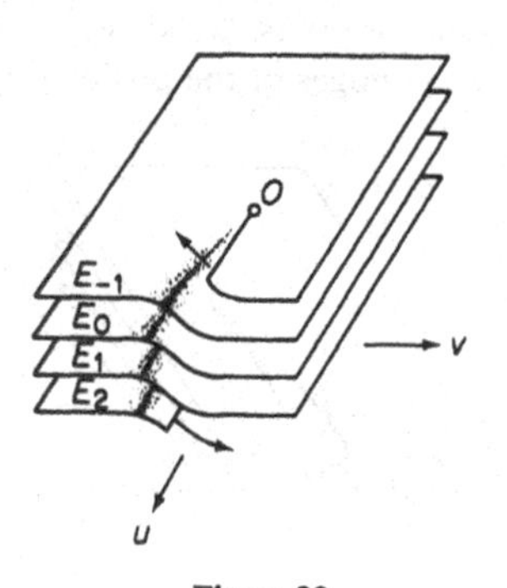

Figure 30

COMMENTS

9.1. It can be shown that Theorem 9.12 remains true if $f(z)$ is one-to-one and merely continuous in G (A. I. Markushevich, *op. cit.*, Volume I, Theorem

†If w lies on the common boundary of E_k and E_{k+1} (an admissible part of S), we choose z to be the unique value of $\sqrt[n]{w}$ lying on the common boundary of G_k and G_{k+1}.

6.1), or if $f(z)$ is analytic but not necessarily univalent in G (Chap. 12, Probs. 10 and 11).

9.2. Problem 12 is particularly germane in connection with the arbitrariness of the branches of a multiple-valued function.

9.3. It must not be thought that the notion of a Riemann surface is a mere "mathematical gimmick." There is in fact a vast literature devoted to the important subject of Riemann surfaces, to which Springer's notable text is recommended as an introduction.†

PROBLEMS

1. Prove that the function
$$f(z) = (1 - iz)^2$$
is univalent in the upper half-plane but not in the whole plane.

2. Prove that the function
$$f(z) = (1 - iz)^3$$
fails to be univalent in the upper half-plane.

3. Prove that the interior of the unit circle $|z| = 1$ and its exterior are both (maximal) domains of univalence for the function
$$w = \frac{1}{2}\left(z + \frac{1}{z}\right). \tag{16}$$

4. What is the image of the domain $|z| < 1$ under the mapping (16)? How about the domain $|z| > 1$?

5. Find all values of
a) $\ln e$; b) $\ln \dfrac{1+i}{\sqrt{2}}$; c) $\ln (2 - 3i)$.

6. If a and b are arbitrary complex numbers, we set
$$a^b = e^{b \ln a}$$
by definition. Find all values of
a) $(-2)^{\sqrt{2}}$; b) i^i; c) 1^{-i}.

7. When is the function $w = z^\alpha$ multiple-valued? What are its branch points?

8. Prove that
$$\frac{d(\sqrt[n]{w})_k}{dw} = \frac{\sqrt[n]{w}}{nw} \qquad (k = 0, 1, \ldots, n-1),$$
$$\frac{d(\ln w)_k}{dw} = \frac{1}{w} \qquad (k = 0, \pm 1, \pm 2, \ldots).$$
More generally, prove that the branches of a multiple-valued function all have the same derivative (at the same point).

---

†G. Springer, *Introduction to Riemann Surfaces*, Addison-Wesley Publishing Company, Inc., Reading, Mass. (1957). See also A. I. Markushevich, *op. cit.*, Volume III, Chap. 7.

9. Is the point at infinity a branch point of the function

a) $w = \sqrt{z(z-1)}$; b) $w = \sqrt{z^2(z-1)}$; c) $w = \sqrt{\dfrac{z-1}{z+1}}$?

10. Suppose z makes a circuit around the circle $|z| = 2$ in the counterclockwise direction, starting and ending at the point $z = 2$. Assuming that the initial value of arg $f(z)$ is zero and that arg $f(z)$ varies continuously with z, find the final value of arg $f(z)$ if

a) $f(z) = \sqrt[3]{z-1}$; b) $f(z) = \sqrt{z^2-1}$; c) $f(z) = \sqrt{z^2+2z-3}$.

11. Again suppose z makes a circuit around the circle $|z| = 2$ in the counterclockwise direction, starting and ending at the point $z = 2$. Assuming that the value of Im $f(z)$ is zero at $z = 2$ and that $f(z)$ varies continuously with z, find the final value of $f(z)$ if

a) $f(z) = \ln \dfrac{1}{z}$; b) $f(z) = \ln z - \ln(z+1)$; c) $f(z) = \ln z + \ln(z+1)$.

12. Let D be any domain on the Riemann surface S of Sec. 9.31 such that no point $z_1 \in D$ "lies under" another point $z_2 \in D$ (with z_1 and z_2 on different sheets of S). Let G be the image of D under the mapping $z = \sqrt[n]{w}$. Prove that G is a domain of univalence for $w = z^n$. Define a corresponding single-valued branch of $z = \sqrt[n]{w}$.

13. Prove that every strip of the form

$$(k - \tfrac{1}{2})\pi < \text{Re}\, z < (k + \tfrac{1}{2})\pi \qquad (k = 0, \pm 1, \pm 2, \ldots) \tag{17}$$

is a (maximal) domain of univalence for the function $w = \sin z$. Prove that the same is true of every half-strip of the form

$$c < \text{Re}\, z < c + 2\pi, \qquad \text{Im}\, z > 0$$

or

$$c < \text{Re}\, z < c + 2\pi, \qquad \text{Im}\, z < 0$$

(c real). Why can't (17) be replaced by $c < \text{Re}\, z < c + \pi$?

14. Let G_k be the strip (17). Prove that the function $w = \sin z$ maps all the strips $G_0, G_1, G_{-1}, G_2, G_{-2}, \ldots$ onto the same domain E, namely the w-plane cut along the two infinite intervals $(-\infty, -1]$ and $[1, \infty)$ of the real axis. The function $w = \sin z$ with domain of definition G_k and range E has a single-valued inverse with domain of definition E and range G_k. Let this function be denoted by

$$z = (\text{arc sin}\, w)_k \qquad (k = 0, \pm 1, \pm 2, \ldots). \tag{18}$$

Prove that (18) is univalent in E, with derivative

$$\frac{d(\text{arc sin}\, w)_k}{dw} = \frac{1}{\sqrt{1-w^2}} \qquad (k = 0, \pm 1, \pm 2, \ldots).$$

15. Let $z = \text{arc sin}\, w$ denote the multiple-valued inverse of the function $w = \sin z$, the latter being defined this time in the whole z-plane with the whole w-plane as its range. Show that

$$\text{arc sin}\, w = \frac{1}{i} \ln i(w + \sqrt{w^2-1}) = \frac{1}{i} \ln (iw + \sqrt{1-w^2}).$$

Show that arc sin w has precisely three branch points, a logarithmic branch point at $w = \infty$ and algebraic branch points of order 1 at $w = -1$ and $w = 1$.

16. Describe all points mapped into a given point w_0 by the function $w = \sin z$. Use the result to solve Prob. 13.

17. Construct the Riemann surface of the function $z = \text{arc sin } w$.

18. The inverse trigonometric functions arc cos w and the inverse hyperbolic functions arc cosh w, arc sinh w, arc tanh w are defined as the inverses of the corresponding trigonometric and hyperbolic functions, just as $z = \text{arc sin } w$ is defined as the inverse of the function $w = \sin z$. Verify the following formulas:

a) $\text{arc cos } w = \frac{1}{i} \ln (w + \sqrt{w^2 - 1})$;

b) $\text{arc tan } w = \frac{i}{2} \ln \frac{i + z}{i - z} = \frac{1}{2i} \ln \frac{1 + iz}{1 - iz}$;

c) $\text{arc cosh } w = \ln (w + \sqrt{w^2 - 1})$;

d) $\text{arc sinh } w = \ln (w + \sqrt{w^2 + 1})$;

e) $\text{arc tanh } w = \frac{1}{2} \ln \frac{1 + w}{1 - w}$.

19. Find all values of the following quantities:
a) arc cos $\frac{1}{2}$; b) arc sin i; c) arc cosh $2i$.

20. Show that the function defined by the integral

$$L(z) = \int_1^z \frac{d\zeta}{\zeta}$$

is identical with the logarithm, i.e., with the multiple-valued function

$$\ln z = \ln |z| + i \arg z.$$

What restriction must be imposed on the curve joining the points 1 and z?

21. Given that $f(z)$ is analytic and nonvanishing in a domain G, which of the functions $|f(z)|$, $\arg f(z)$ and $\ln |f(z)|$ are harmonic in G?

CHAPTER TEN

Taylor Series

10.1. The Taylor Expansion of an Analytic Function

10.11. Let

$$f(z) = c_0 + c_1(z-a) + c_2(z-a)^2 + \cdots \tag{1}$$

be a power series with sum $f(z)$ and radius of convergence R. Differentiating (1) repeatedly as in Sec. 7.19, we obtain the new power series

$$f'(z) = c_1 + 2c_2(z-a) + 3c_3(z-a)^2 + \cdots, \tag{2}$$

$$f''(z) = 2c_2 + 2\cdot 3c_3(z-a) + \cdots, \tag{3}$$

$$\cdots$$

$$f^{(n)}(z) = 2\cdot 3 \cdots nc_n + 2\cdot 3 \cdots n(n+1)c_{n+1}(z-a) + \cdots, \tag{4}$$

$$\cdots$$

all with the same radius of convergence R as the original series (1). Setting $z = a$ in the series (1)–(4), we get

$$c_0 = f(a), \quad c_1 = f'(a), \quad c_2 = \frac{f''(a)}{2!}, \ldots, \quad c_n = \frac{f^{(n)}(a)}{n!}, \ldots \tag{5}$$

A power series with coefficients related to a given analytic function $f(z)$ by the formulas (5) is called a *Taylor series* (more exactly the Taylor series of $f(z)$ *at the point* a), and the numbers (5) themselves are called the *Taylor coefficients* of $f(z)$. Thus we see that every power series with a nonzero radius of convergence is the Taylor series of its own sum. The Taylor series of a function $f(z)$ is also called the *Taylor expansion* of $f(z)$.

10.12. Let C be a circle of radius $\rho < R$ centered at a. Then it follows from Cauchy's integral formula that

$$f(a) = \frac{1}{2\pi i}\int_C \frac{f(z)}{z-a}\,dz,$$

and more generally from Theorem 5.71 that

$$f^{(n)}(a) = \frac{n!}{2\pi i}\int_C \frac{f(z)}{(z-a)^{n+1}}\,dz. \tag{6}$$

Comparing (5) and (6), we find that

$$c_n = \frac{1}{2\pi i}\int_C \frac{f(z)}{(z-a)^{n+1}}\,dz \qquad (n = 0, 1, 2, \ldots). \tag{7}$$

Suppose $|f(z)| \leq M$ for all $|z| < R$. Then, applying Theorem 5.23 to (7), we get *Cauchy's inequalities*

$$|c_n| \leq \frac{M}{2\pi}\frac{2\pi\rho}{\rho^{n+1}} = \frac{M}{\rho^n} \qquad (n = 0, 1, 2, \ldots), \tag{8}$$

satisfied by the coefficients of the power series (1). Taking the limit of (8) as $\rho \to R$, we find that

$$|c_n| \leq \frac{M}{R^n} \qquad (n = 0, 1, 2, \ldots), \tag{8'}$$

where R is the radius of convergence of the series (1).

10.13. According to Theorem 7.19, a power series (1) with radius of convergence R represents an analytic function in the disk $|z-a| < R$. We now prove the converse proposition:

THEOREM. *Let K be the disk $|z-a| < R$, and suppose $f(z)$ is analytic in K. Then $f(z)$ has a Taylor expansion in K, i.e., the power series*

$$f(z) = \sum_{n=0}^{\infty} c_n(z-a)^n \tag{9}$$

with coefficients (5) *converges to $f(z)$ at every point of K.*

Proof. Given any point $\zeta \in K$, let C be a circle of radius $r < R$ centered at a and enclosing ζ. Then

$$f(\zeta) = \frac{1}{2\pi i}\int_C \frac{f(z)}{z-\zeta}\,dz \tag{10}$$

by Cauchy's integral formula, since $f(z)$ is analytic inside and on C (being analytic in K). To convert (10) into a power series, we first write

$$\frac{1}{z-\zeta} = \frac{1}{(z-a)-(\zeta-a)} = \frac{1}{(z-a)\left(1-\dfrac{\zeta-a}{z-a}\right)},$$

where $z \in C$ and the expression

$$\frac{1}{1-\dfrac{\zeta-a}{z-a}}$$

can be recognized as the sum of a convergent geometric series. In fact,

$$\left|\frac{\zeta - a}{z - a}\right| = \frac{|\zeta - a|}{r} < 1,$$

since $z \in C$ and ζ lies inside C, and hence

$$\frac{1}{1 - \dfrac{\zeta - a}{z - a}} = \sum_{n=0}^{\infty} \left(\frac{\zeta - a}{z - a}\right)^n \tag{11}$$

(see Sec. 6.13). It follows from (11) that

$$\frac{1}{z - \zeta} = \frac{1}{(z - a)\left(1 - \dfrac{\zeta - a}{z - a}\right)} = \sum_{n=0}^{\infty} \frac{(\zeta - a)^n}{(z - a)^{n+1}},$$

and hence

$$\frac{1}{2\pi i}\frac{f(z)}{z - \zeta} = \frac{1}{2\pi i}\sum_{n=0}^{\infty} \frac{f(z)}{(z - a)^{n+1}}(\zeta - a)^n. \tag{12}$$

If $M = \max_{z \in C} |f(z)|$, then

$$\left|\frac{1}{2\pi i}\frac{f(z)}{(z - a)^{n+1}}(\zeta - a)^n\right| \leq \frac{M}{2\pi r}\left(\frac{|\zeta - a|}{r}\right)^n,$$

where the right-hand side is the general term of a convergent geometric series. It follows from Theorem 6.36 that the series (12) is uniformly convergent on C, and hence can be integrated term by term along C, by Theorem 6.38. This gives

$$\frac{1}{2\pi i}\int_C \frac{f(z)}{z - \zeta}\,dz = \sum_{n=0}^{\infty} c_n(\zeta - a)^n, \tag{13}$$

where

$$c_n = \frac{1}{2\pi i}\int_C \frac{f(z)}{(z - a)^{n+1}}\,dz = \frac{f^{(n)}(a)}{n!} \tag{14}$$

because of (6). But (10) and (13) together imply (9) after changing ζ to z, and (9) is the Taylor series of $f(z)$ because of (14). ▮

10.14. Definition. If $f(z)$ is analytic at a point z_0, i.e., in some neighborhood of z_0, we call z_0 a *regular point* of $f(z)$; otherwise z_0 is called a *singular point* of $f(z)$. For example, the point $z = 1$ is a singular point and every point $z \neq 1$ is a regular point of the function $f(z) = 1/(1 - z)$. Moreover, according to Theorem 7.19, every point inside the circle of convergence of a power series is a regular point of the sum of the series.

10.15. THEOREM. *Suppose $f(z)$ has the Taylor expansion*

$$f(z) = c_0 + c_1(z - a) + c_2(z - a)^2 + \cdots + c_n(z - a)^n + \cdots \tag{15}$$

at the point a, with radius of convergence R. Then $f(z)$ has at least one singular point on the circle of convergence $|z - a| = R$.†

Proof. Let C be the circle of convergence $|z - a| = R$, and suppose $f(z)$ has no singular points on C. Then every point of C is a regular point of $f(z)$, i.e., every point $z \in C$ is the center of an open disk K_z in which $f(z)$ is analytic. But C can be covered by a finite number of these disks, say $K_{z_1}, \ldots, K_{z_n}$, by the Heine–Borel theorem (cf. Chap. 3, Prob. 11). The set of all points belonging to at least one of the disks $K_{z_1}, \ldots, K_{z_n}$, being obviously open and connected, is a domain G containing C. Let ρ be the distance between C and the boundary of G, as in Chap. 3, Prob. 17. Then $f(z)$ is analytic inside the circle C^* of radius $R + \rho$ concentric with C. It follows from Theorem 10.13 that $f(z)$ has a convergent Taylor series inside C^*, coinciding with (15) (why?), i.e., the radius of convergence of (15) cannot be less than $R + \rho > R$. This contradiction shows that $f(z)$ has at least one singular point on C. ∎

10.16. It is an immediate consequence of Theorems 10.13 and 10.15 that if a is a regular point of $f(z)$, then $f(z)$ has a power series expansion of the form (15), whose circle of convergence passes through the singular point of $f(z)$ nearest the point a. Thus there is an intimate connection between the radius of convergence of a power series and the behavior of the sum of the series. Hence the theory of power series achieves full clarity only in the complex case. For example, if we consider only real values of x, the divergence of the series

$$\frac{1}{1 + x^2} = 1 - x^2 + x^4 - x^6 + \cdots \tag{16}$$

for $x \leq -1$ and $x \geq 1$ is not accounted for by the nature of the function $1/(1 + x^2)$, which is defined for all real x and exhibits no exceptional behavior at the points $x = \pm 1$. The reason for the indicated divergence emerges when we replace x by a complex variable z. The left-hand side of (16) then becomes $1/(1 + z^2)$, a function with singular points at $z = \pm i$, so that the radius of convergence of the series must equal 1 (the distance between 0 and $\pm i$).

10.17. THEOREM ***(Liouville).*** *Every bounded entire function is a constant, i.e., if $f(z)$ is entire and if $|f(z)| \leq M$ for all finite z, then $f(z) \equiv$ constant.*‡

Proof. By Theorem 10.13, $f(z)$ has a Taylor expansion

$$f(z) = c_0 + c_1 z + c_2 z^2 + \cdots + c_n z^n + \cdots$$

†Here we tacitly assume that the domain of definition of the (analytic) function $f(z)$ is larger than the disk $|z-a| < R$. However, see Sec. 13.42c.

‡As in Sec. 8.11, a complex function $f(z)$ is said to be *entire* if it is analytic in the whole finite plane. The symbol $\equiv$ means "is identically equal to."

valid for every finite z, i.e., valid in every disk $|z| < R$. But then taking the limit as $R \to \infty$ in Cauchy's inequalities

$$|c_n| \leq \frac{M}{R^n} \qquad (n = 0, 1, 2, \ldots)$$

(Sec. 10.12), we get $c_n = 0$ if $n \geq 1$, and hence $f(z) \equiv c_0$. ∎

10.2. Uniqueness Theorems

10.21. Suppose two power series

$$a_0 + a_1(z - z_0) + \cdots + a_n(z - z_0)^n + \cdots, \tag{17}$$

$$b_0 + b_1(z - z_0) + \cdots + b_n(z - z_0)^n + \cdots, \tag{18}$$

with radii of convergence R_1 and R_2, respectively, have the same sum in a neighborhood of z_0, i.e., suppose

$$a_0 + a_1(z - z_0) + \cdots + a_n(z - z_0)^n + \cdots$$
$$= b_0 + b_1(z - z_0) + \cdots + b_n(z - z_0)^n + \cdots$$

for all z in some disk $|z - z_0| < r \leq \min\{R_1, R_2\}$. Then, according to (5),

$$a_0 = b_0 = f(z_0), \quad a_1 = b_1 = f'(z_0), \ldots, a_n = b_n = \frac{f^{(n)}(z_0)}{n!}, \ldots,$$

where $f(z)$ denotes the common sum of the two series. In particular, it follows from the Cauchy–Hadamard theorem that $R_1 = R_2$. Thus we have proved the following result, typical of a class of propositions known as *uniqueness theorems*: *If the sums of two power series in the variable $z - z_0$ coincide in a neighborhood of the point z_0, then identical powers of $z - z_0$ have identical coefficients, i.e., there is a unique power series in the variable $z - z_0$ which has a given sum in a neighborhood of z_0.*

10.22. The requirement that the series (17) and (18) coincide in a whole neighborhood of z_0 is actually much more than is required to guarantee the identity of (17) and (18), as shown by the following

THEOREM ***(Uniqueness theorem for power series).*** *If the sums of two power series in the variable $z - z_0$ coincide at every point of a set E with z_0 as a limit point,† then identical powers of $z - z_0$ have identical coefficients, i.e., there is a unique power series in the variable $z - z_0$ which has a given sum at the points of E.*

Proof. Let (17) and (18) be the given series, and let z_n ($z_n \neq z_0$) be a sequence of distinct points of E converging to z_0 (why does z_n exist?). Then,

†It will be recalled from Chap. 2, Prob. 2 that a point z_0 is said to be a *limit point* of a set E if every neighborhood of z_0 contains infinitely many (distinct) points of E. Note that the set E is tacitly assumed to contain infinitely many points.

since

$$a_0 + a_1(z_n - z_0) + a_2(z_n - z_0)^2 + \cdots$$
$$= b_0 + b_1(z_n - z_0) + b_2(z_n - z_0)^2 + \cdots \qquad (19)$$

for all $n = 1, 2, \ldots$ and since the sum of a power series is continuous inside its circle of convergence, we have

$$a_0 = \lim_{n\to\infty} [a_0 + a_1(z_n - z_0) + a_2(z_n - z_0)^2 + \cdots]$$
$$= \lim_{n\to\infty} [b_0 + b_1(z_n - z_0) + b_2(z_n - z_0)^2 + \cdots] = b_0,$$

and hence $a_0 = b_0$. Suppose it is known that

$$a_0 = b_0, a_1 = b_1, \ldots, a_k = b_k.$$

Then (19) implies

$$a_{k+1}(z_n - z_0)^{k+1} + a_{k+2}(z_n - z_0)^{k+2} + \cdots$$
$$= b_{k+1}(z_n - z_0)^{k+1} + b_{k+2}(z_n - z_0)^{k+2} + \cdots. \qquad (20)$$

Dividing (20) by $(z_n - z_0)^{k+1} \neq 0$ and taking the limit as $n \to \infty$, we get $a_{k+1} = b_{k+1}$. The proof now follows at once by induction. ▮

10.23. We are now in a position to prove one of the most important propositions of complex analysis, showing that the "uniqueness property" of Theorem 10.22 carries over to the case of analytic functions:

THEOREM ***(Uniqueness theorem for analytic functions).*** *Let $f(z)$ and $g(z)$ be two functions analytic in the same domain G, and suppose $f(z)$ and $g(z)$ coincide at all points of a subset E of G with a limit point $z_0 \in G$. Then $f(z)$ and $g(z)$ coincide in the whole domain G.*

Proof. Given any point $Z \neq z_0$ in G, we join z_0 to Z by a continuous curve C contained in G. Let ρ be the distance between C and the boundary of G (see Chap. 3, Prob. 17) unless G is the whole plane, in which case let ρ be any positive number, and let $z_0, z_1, z_2, \ldots, z_j, z_{j+1}, \ldots, z_{n-1}, z_n = Z$ be consecutive points of C such that†

$$|z_{j+1} - z_j| < \rho \qquad (j = 0, 1, \ldots, n-1). \qquad (21)$$

Then construct the "chain" of disks $K_0, K_1, K_2, \ldots, K_j, K_{j+1}, \ldots, K_{n-1}$, where K_j is the disk $|z - z_j| < \rho$ (see Figure 31). Clearly each disk K_j contains the center z_{j+1} of the "next" disk K_{j+1}. According to Theorem 10.13, both $f(z)$ and $g(z)$ have convergent power series expansions in every K_j $(j = 0, 1, \ldots, n-1)$. Since the functions $f(z)$ and $g(z)$ coincide at every point of the set $E \cap K_0$,‡ with limit point z_0, it follows from Theorem 10.22

†Let C have parametric equation $z = z(t)$, $a \leq t \leq b$. Then the existence of points z_j satisfying (21) follows from the uniform continuity of $z(t)$ in the interval $a \leq t \leq b$, by the "one-dimensional" analogue of Theorem 3.44.

‡Given two sets A and B, by the *intersection* of A and B, denoted by $A \cap B$, we mean the set of all points belonging to both A and B.

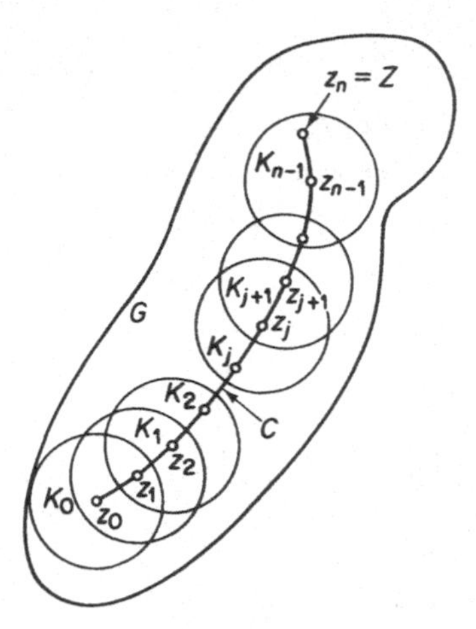

Figure 31

that they coincide in the whole disk K_0. But then they coincide at every point of the set $K_0 \cap K_1$, which has z_1 as a limit point (why?), and hence, by Theorem 10.22 again, they coincide in K_1. Repeating this argument $n - 2$ more times, we eventually find that $f(z)$ and $g(z)$ coincide in K_{n-1} and hence coincide at the point $z_n = Z$. But then $f(z)$ and $g(z)$ coincide in the whole domain G, since Z is an arbitrary point of G. ∎

10.24. Remark. According to Theorem 10.23, two functions $f(z)$ and $g(z)$ analytic in the same domain G actually coincide in G if they are known to coincide in an arbitrarily small neighborhood of a point of G or even on an arbitrarily small arc of a curve in G. In particular, if $f(z)$ is identically equal to a constant c in an arbitrarily small neighborhood of a point of G or on an arbitrarily small arc of a curve in G, then $f(z) \equiv c$ everywhere in G (since the function $g(z)$ identically equal to c throughout the whole domain G is obviously analytic in G and equal to $f(z)$ at all points of the indicated sets). In particular, choosing $c = 0$, we see that if $f(z)$ vanishes in an arbitrarily small neighborhood of a point of G or on an arbitrarily small arc of a curve in G, then $f(z)$ vanishes everywhere in G. It is indeed remarkable that the mere fact of being differentiable in G imposes such heavy restrictions on the possible behavior of complex functions defined in G.

10.25. By a *zero* of a function $f(z)$ we mean any root of the equation $f(z) = 0$. Suppose $f(z) \not\equiv 0$† is analytic in a domain G and has a zero at the point z_0. Then, by Theorem 10.13, $f(z)$ has a power series expansion of the form

$$f(z) = c_1(z - z_0) + c_2(z - z_0)^2 + \cdots + c_n(z - z_0)^n + \cdots \qquad (22)$$

in some neighborhood of the point z_0, since $c_0 = f(z_0) = 0$. At least one of the coefficients $c_1, c_2, \ldots, c_n, \ldots$ must be nonzero, since otherwise $f(z)$ would vanish everywhere in a neighborhood of z_0 and hence would vanish everywhere in G, by the uniqueness theorem for analytic functions. Suppose c_m is the first nonzero coefficient in (22), i.e., suppose

$$c_1 = c_2 = \cdots = c_{m-1} = 0, \quad c_m \neq 0,$$

† $f(z) \not\equiv 0$ is shorthand for a function not identically zero (in G).

so that (22) becomes

$$f(z) = c_m(z - z_0)^m + c_{m+1}(z - z_0)^{m+1} + \cdots \qquad (c_m \neq 0).$$

The point z_0 is then called a *zero of order m* of $f(z)$. In particular, z_0 is said to be a *simple zero* if $m = 1$ and a *multiple zero* if $m > 1$.

10.26. THEOREM. *Every zero z_0 of a function $f(z) \not\equiv 0$ analytic in a domain G is isolated, i.e., there is a neighborhood of z_0 in which $f(z)$ has no zeros other than z_0 itself.*

Proof. Suppose every neighborhood K of z_0 contains a zero of $f(z)$ other than z_0. Then z_0 is a limit point of zeros of $f(z)$. It follows that $f(z) \equiv 0$ (cf. Sec. 10.24), contrary to hypothesis. Therefore some neighborhood of z_0 contains no zeros of $f(z)$ other than z_0 itself. ∎

10.27. The following proposition, which smacks of a uniqueness theorem, will be needed in the next section:

THEOREM. *An analytic function of constant modulus is itself constant. More exactly, if $f(z)$ is analytic in a domain G and if $|f(z)|$ is constant in G, then $f(z)$ is constant in G.*

Proof. If $|f(z)| \equiv 0$ in G, then obviously $f(z) \equiv 0$ in G. Thus suppose $|f(z)| \equiv M > 0$ in G, so that

$$|f(z)|^2 = u^2 + v^2 = M^2, \tag{23}$$

where $f(z) = u + iv$. Differentiating (23) with respect to x and y, we obtain

$$\begin{aligned} u\frac{\partial u}{\partial x} + v\frac{\partial v}{\partial x} &= 0, \\ u\frac{\partial u}{\partial y} + v\frac{\partial v}{\partial y} &= 0 \end{aligned} \tag{24}$$

everywhere in G. The functions u and v cannot vanish simultaneously in G, since $M \neq 0$. It follows that

$$\frac{\partial u}{\partial x}\frac{\partial v}{\partial y} - \frac{\partial u}{\partial y}\frac{\partial v}{\partial x} = 0, \tag{25}$$

since the determinant of (24) regarded as a system of linear equations in u and v must vanish. But $f(z)$ satisfies the Cauchy–Riemann equation

$$\frac{\partial u}{\partial x} = \frac{\partial v}{\partial y}, \qquad \frac{\partial u}{\partial y} = -\frac{\partial v}{\partial x}$$

everywhere in G, being analytic in G. Hence (25) can be written in the form

$$\left(\frac{\partial u}{\partial x}\right)^2 + \left(\frac{\partial u}{\partial y}\right)^2 = 0,$$

which implies

$$\frac{\partial u}{\partial x} = \frac{\partial u}{\partial y} = 0$$

and hence

$$\frac{\partial v}{\partial x} = \frac{\partial v}{\partial y} = 0$$

everywhere in G, where we have again used the Cauchy–Riemann equations. But then

$$u_1 \equiv c_1, \qquad u_2 \equiv c_2, \qquad f(z) \equiv c_1 + ic_2$$

in G, where c_1 and c_2 are real constants. ∎

10.3. The Maximum Modulus Principle and Its Implications

10.31. The following theorem expresses a key property of analytic functions:

THEOREM ***(Maximum modulus principle).*** *If $f(z)$ is analytic and nonconstant in a domain G,† then $|f(z)|$ cannot have a maximum in G, i.e., there is no point $z_0 \in G$ such that $|f(z_0)| \geq |f(z)|$ for all $z \in G$.*

Proof. Suppose to the contrary that there is a point $z_0 \in G$ such that $|f(z_0)| \geq |f(z)|$ for all $z \in G$, and let $|f(z_0)| = M$. We can assume that $M > 0$, since $f(z) \equiv 0$ if $M = 0$. Let γ_R be a circle of radius R centered at z_0 so small that G contains both γ_R and its interior. Then, by Cauchy's integral formula,

$$f(z_0) = \frac{1}{2\pi i}\int_{\gamma_R} \frac{f(z)}{z - z_0}\, dz = \frac{1}{2\pi i}\int_{\gamma_R} \frac{f(z_0 + Re^{i\theta})}{Re^{i\theta}} iRe^{i\theta}\, d\theta$$
$$= \frac{1}{2\pi}\int_0^{2\pi} f(z_0 + Re^{i\theta})\, d\theta, \tag{26}$$

and hence

$$|f(z_0)| = M = \left|\frac{1}{2\pi}\int_0^{2\pi} f(z_0 + Re^{i\theta})\, d\theta\right| \leq \frac{1}{2\pi}\int_0^{2\pi} |f(z_0 + Re^{i\theta})|\, d\theta, \tag{26'}$$

where the last step is an immediate consequence of the definition of the integral as a limiting sum. By hypothesis, $|f(z_0 + Re^{i\theta})| \leq M$ for all θ $(0 \leq \theta \leq 2\pi)$, but the strict inequality $|f(z_0 + Re^{i\theta})| < M$ cannot hold for any θ. In fact, if $|f(z_0 + Re^{i\theta_0})| < M$ for some θ_0, then, because of the continuity of $|f(z)|$ on γ_R, the interval $[0, 2\pi]$ has a subinterval I of length $\delta > 0$, say, containing θ_0 such that $|f(z_0 + Re^{i\theta})| \leq M - \epsilon$ for some $\epsilon > 0$ and all $\theta \in I$. Let I' be the part or parts of the interval $[0, 2\pi]$ remaining

†Note that G is open, so that $f(z)$ need not be defined on the boundary of G. The case where $f(z)$ is also defined on the boundary of G is treated in Corollaries 10.32b, c, d.

after deletion of I. Then (26′) implies

$$\begin{aligned} M &\le \frac{1}{2\pi}\int_0^{2\pi} |f(z_0 + Re^{i\theta})|\,d\theta \\ &= \frac{1}{2\pi}\int_{I'} |f(z_0 + Re^{i\theta})|\,d\theta + \frac{1}{2\pi}\int_{I} |f(z_0 + Re^{i\theta})|\,d\theta \\ &\le \frac{1}{2\pi}M(2\pi - \delta) + \frac{1}{2\pi}(M - \epsilon)\delta < M, \end{aligned}$$

which is impossible. It follows that

$$|f(z_0 + Re^{i\theta})| = M \qquad (0 \le \theta \le 2\pi)$$

for all R such that G contains the circle γ_R and its interior. But then $|f(z)| \equiv M$ in any disk K centered at z_0 and contained in G, since every point of K lies on such a circle γ_R. Hence $f(z)$ is constant in K, by Theorem 10.27. It follows from Sec. 10.24 that $f(z)$ is constant in the whole domain G, contrary to hypothesis. This contradiction shows that there can be no point $z_0 \in G$ such that $|f(z_0)| \ge |f(z)|$ for all $z \in G$. ∎

10.32. Next we prove a number of interesting implications of the maximum modulus principle:

a. COROLLARY ***(Minimum modulus principle).*** *If $f(z)$ is analytic, nonconstant and nonvanishing in a domain G, then $|f(z)|$ cannot have a minimum in G, i.e., there is no point $z_0 \in G$ such that $|f(z_0)| \le |f(z)|$ for all $z \in G$.*

Proof. The function $\varphi(z) = 1/f(z)$ is analytic and nonconstant in G, by the assumed properties of $f(z)$. Hence $|\varphi(z)|$ cannot have a maximum in G, by the maximum modulus principle. But a minimum of $|f(z)|$ is automatically a maximum of $|\varphi(z)|$. Hence $|f(z)|$ cannot have a minimum in G. ∎

b. COROLLARY. *Given a bounded domain G, let $f(z)$ be analytic in G and continuous in $\bar{G}$. Then the boundary of G contains a point ζ such that*

$$|f(\zeta)| = \max_{z \in \bar{G}} |f(z)|. \tag{27}$$

Proof. The bounded closed domain $\bar{G}$ must contain a point ζ for which (27) holds, by Chap. 3, Prob. 12c. If $f(z) \not\equiv$ constant, ζ cannot belong to G, by the maximum modulus principle, and hence ζ must belong to the boundary of G. If $f(z) \equiv$ constant, we can obviously choose ζ in the boundary of G. ∎

c. COROLLARY. *Given a bounded domain G, let $f(z)$ be analytic and nonvanishing in G and continuous in $\bar{G}$. Then the boundary of G contains a point ζ' such that*

$$|f(\zeta')| = \min_{z \in \bar{G}} |f(z)|. \tag{27′}$$

Proof. The bounded closed domain $\bar{G}$ must contain a point ζ' for which (27′) holds, again by Chap. 3, Prob. 12c. If $f(z) \not\equiv$ constant, ζ cannot belong to G, this time by the minimum modulus principle, and hence ζ must belong to the boundary of G. If $f(z) \equiv$ constant, we can obviously choose ζ in the boundary of G. ▮

d. COROLLARY. *Given a bounded domain G, let $f(z)$ be analytic and nonvanishing in G and continuous in $\bar{G}$, and suppose $f(z)$ is constant on the boundary of G. Then $f(z)$ is constant everywhere in $\bar{G}$.*

Proof. Let Γ be the boundary of G. Then by the preceding two corollaries, Γ contains points ζ and ζ' satisfying (27) and (27′). But $|f(\zeta)| = |f(\zeta')|$ by hypothesis, and hence

$$\max_{z \in \bar{G}} |f(z)| = \min_{z \in \bar{G}} |f(z)|.$$

Therefore $|f(z)|$ is constant in $\bar{G}$. It follows from Theorem 10.27 that $f(z)$ is constant in G and hence in $\bar{G}$ (by continuity). ▮

10.33. The maximum and minimum modulus principles in turn imply a number of important properties of harmonic functions:

a. THEOREM. *If $u(x, y)$ is harmonic and nonconstant in a domain G, then $u(x, y)$ has neither a maximum nor a minimum at any point of G.*

Proof. Given any point $(x_0, y_0) \in G$, let K be a neighborhood of (x_0, y_0) contained in G. As in Sec. 5.85, let $f(z)$ be a function which is analytic in K and has $u(x, y)$ as its real part. Then the function

$$g(z) = e^{f(z)}$$

is analytic and nonconstant in K (why?). Moreover, $g(z)$ is nonvanishing in K (cf. Sec. 8.13e) and

$$|g(z)| = e^{u(x,y)}.$$

The function $u(x, y)$ cannot have a maximum at (x_0, y_0), since otherwise $|g(z)|$ would have a maximum at $z_0 = x_0 + iy_0 \in K$, contrary to the maximum modulus principle. Similarly, $u(x, y)$ cannot have a minimum at (x_0, y_0), since otherwise $|g(z)|$ would have a minimum at z_0, contrary to the minimum modulus principle. ▮

b. COROLLARY. *Given a bounded domain G, let $u(x, y)$ be harmonic in G and continuous in $\bar{G}$. Then the boundary of G contains points (ξ, η) and (ξ', η') such that*

$$u(\xi, \eta) = \max_{(x,y) \in \bar{G}} u(x, y), \qquad u(\xi', \eta') = \min_{(x,y) \in \bar{G}} u(x, y).$$

Proof. Virtually the same as that of Corollaries 10.32b and 10.32c. ▮

c. COROLLARY. *Given a bounded domain G, let $u(x, y)$ be harmonic in G and continuous in $\bar{G}$, and suppose $u(x, y)$ is constant on the boundary of G. Then $u(x, y)$ is constant everywhere in G.*

Proof. Virtually the same as that of Corollary 10.32d. ▮

d. COROLLARY. *Given a bounded domain G, let $u_1(x, y)$ and $u_2(x, y)$ be two functions harmonic in G and continuous in $\bar{G}$, and suppose $u_1(x, y) \equiv u_2(x, y)$ on the boundary of G. Then $u_1(x, y) \equiv u_2(x, y)$ everywhere in $\bar{G}$.*

Proof. The function

$$u(x, y) = u_1(x, y) - u_2(x, y)$$

satisfies the conditions of the preceding corollary, with the constant equal to zero. ▮

COMMENTS

10.1. Section 10.1 brims with important results of complex analysis, e.g., the key Cauchy inequalities, Theorem 10.13 on the relation between analyticity and expandibility in power series (whose importance has already been stressed in the comment to Sec. 7.1), Theorem 10.15 on the relation between radius of convergence and location of singular points, and Liouville's remarkable theorem on the intrinsic unboundedness of nontrivial entire functions. The student should settle for nothing less than complete mastery of this material.

10.2. Both Prob. 19 and the "chain of disks" argument used to prove Theorem 10.23 anticipate the theory of "analytic continuation" developed in Sec. 13.4.

10.3. According to (26), if a function is analytic inside and on a circle γ, its value at the center of γ is the average of its values on γ itself. The same is true of a harmonic function (Corollary 13.16). The maximum and minimum modulus principles reveal once again the tight "organic" structure of both analytic and harmonic functions. This is less surprising when we recall that analytic functions must satisfy a system of first-order partial differential equations, namely the Cauchy–Riemann equations (Sec. 4.22), while harmonic functions must satisfy a related second-order partial differential equation, namely Laplace's equation (Sec. 5.81).

PROBLEMS

1. Prove that

$$\ln z = (z-1) - \frac{(z-1)^2}{2} + \frac{(z-1)^3}{3} - \cdots + (-1)^{n-1}\frac{(z-1)^n}{n} - \cdots$$

in the disk $|z-1|<1$, while

$$\ln(1+z) = z - \frac{z^2}{2} + \frac{z^3}{3} - \cdots + (-1)^{n-1}\frac{z^n}{n} - \cdots$$

in the disk $|z|<1$.†

2. What function is represented by the power series

$$\sum_{n=1}^{\infty} \frac{z^n}{n}?$$

3. Given any complex number α, let $z^\alpha = e^{\alpha \ln z}$ as in Chap. 9, Prob. 6. Prove that

$$z^\alpha = 1 + \sum_{n=1}^{\infty} \frac{\alpha(\alpha-1)\cdots(\alpha-n+1)}{n!}(z-1)^n$$

in the disk $|z-1|<1$, while

$$(1+z)^\alpha = 1 + \sum_{n=1}^{\infty} \frac{\alpha(\alpha-1)\cdots(\alpha-n+1)}{n!}z^n \tag{28}$$

in the disk $|z|<1$.

Comment. Formula (28) is the appropriate generalization of the familiar *binomial theorem* to the case where the exponent α is an arbitrary complex number.

4. Find the Taylor expansion of e^z at the point $z = a$.
5. Find the Taylor expansion of $\cos z$ at the point $z = \pi/4$.
6. Prove that

$$\frac{1}{(1-z)^2} = \sum_{n=0}^{\infty} (n+1)z^n \qquad (|z|<1),$$

$$\frac{2}{(1+z)^3} = \sum_{n=0}^{\infty} (-1)^n(n+2)(n+1)z^n \qquad (|z|<1).$$

7. Use the preceding problem to find the Taylor expansion of $1/z^2$ at the point $z=-1$ and of $1/z^3$ at the point $z=1$. Find the same expansions by direct calculation of the Taylor coefficients.
8. Find the Taylor expansion of the function

$$\frac{1}{(z+1)(z-2)}$$

at the point $z=0$.

9. Prove that

$$\frac{1}{1+z+z^2} = \sum_{n=0}^{\infty} (z^{3n} - z^{3n+1})$$

in the disk $|z|<1$.

---

†In Probs. 1-3, $\ln z$ means the single-valued branch of the logarithm such that $\ln 1 = 0$, i.e., the branch $(\ln z)_0$ in the notation of Sec. 9.22.

10. Consider the series

$$f(z) = \sum_{k=1}^{\infty} f_k(z), \tag{29}$$

where every function $f_k(z)$ is analytic in the open disk $|z - z_0| < R$. Suppose the series (29) is uniformly convergent in every closed disk $|z - z_0| \le r < R$. Prove that $f(z)$ has the Taylor expansion

$$f(z) = \sum_{n=0}^{\infty} c_n(z - z_0)^n \tag{30}$$

in the disk $|z - z_0| < R$, with Taylor coefficients

$$c_n = \sum_{k=1}^{\infty} \frac{f_k^{(n)}(z_0)}{k!} \qquad (n = 0, 1, 2, \ldots). \tag{31}$$

Comment. It follows from (31) that the coefficient of a given power of $z - z_0$ in the right-hand side of (30) is the sum of all the coefficients of the same power of $z - z_0$ in the separate Taylor expansions of the functions $f_1(z), \ldots, f_k(z), \ldots$

11. Prove that the series

$$f(z) = \sum_{k=1}^{\infty} \frac{z^k}{1 - z^k} \tag{32}$$

represents an analytic function in the disk $|z| < 1$.

12. Find the Taylor expansion of the function (32) in the unit disk $|z| < 1$.

13. Suppose $f(z)$ is analytic at z_0, with Taylor expansion

$$f(z) = a_0 + a_1(z - z_0) + \cdots + a_k(z - z_0)^k + \cdots, \tag{33}$$

while $\varphi(w)$ is analytic at the point $w_0 = f(z_0) = a_0$, with Taylor expansion

$$\varphi(w) = A_0 + A_1(w - w_0) + \cdots + A_n(w - w_0)^n + \cdots. \tag{34}$$

Suppose we replace w by $f(z)$ in (34), thereby obtaining the formal series expansion

$$\begin{aligned}\varphi(f(z)) = A_0 &+ A_1[a_1(z - z_0) + \cdots + a_k(z - z_0)^k + \cdots] \\ &+ \cdots + A_n[a_1(z - z_0) + \cdots + a_k(z - z_0)^k + \cdots]^n + \cdots\end{aligned} \tag{35}$$

for the composite function $\varphi(f(z))$. Suppose all the algebraic operations called for in (35) are carried out formally, i.e., suppose we raise the series $a_1(z - z_0) + \cdots + a_k(z - z_0)^k + \cdots$ to the indicated powers, afterwards grouping together and adding all the coefficients of each power of $z - z_0$. This gives a new series of the form

$$\varphi(f(z)) = b_0 + b_1(z - z_0) + \cdots + b_n(z - z_0)^n + \cdots, \tag{36}$$

said to be obtained by *substitution* of the series (33) into the series (34). Prove the validity of the expansion (36), i.e., show that $\varphi(f(z))$ is analytic at z_0 and actually has the right-hand side of (36) as its Taylor expansion at z_0.

14. Write the first four terms of the Taylor expansion at $z = 0$ of each of the following functions:

a) $e^{1/(1-z)}$; b) $\sin \dfrac{1}{1 - z}$; c) $e^{z \sin z}$; d) $\sqrt{\cos z}$, where $\sqrt{\cos 0} = 1$.

15. Starting from Cauchy's integral formula, show that

$$f(a) - f(b) = \frac{a-b}{2\pi i} \int_{|z|=R} \frac{dz}{(z-a)(z-b)}$$

if $f(z)$ is an entire function and $|a| < R$, $|b| < R$. Use this to give another proof of Liouville's theorem.

16. Prove the following generalization of Liouville's theorem: If $f(z)$ is an entire function and if the function

$$M(R) = \max_{|z|=R} |f(z)|$$

satisfies the inequality

$$M(R) \leq MR^k,$$

where M is a positive constant and k is a fixed positive integer, then $f(z)$ is a polynomial of degree no higher than k.

17. Is there a function analytic at the point $z = 0$ which takes the following values at the points $z = 1/n$ $(n = 1, 2, \ldots)$:
a) $0, 1, 0, 1, 0, 1, \ldots$; b) $0, \frac{1}{2}, 0, \frac{1}{4}, 0, \frac{1}{6}, \ldots$; c) $\frac{1}{2}, \frac{1}{2}, \frac{1}{4}, \frac{1}{4}, \frac{1}{6}, \frac{1}{6}, \ldots$;
d) $\frac{1}{2}, \frac{2}{3}, \frac{3}{4}, \frac{4}{5}, \ldots$?

18. Is there a function analytic at the point $z = 0$ such that

a) $f\left(\frac{1}{n}\right) = f\left(-\frac{1}{n}\right) = \frac{1}{n^2}$; b) $f\left(\frac{1}{n}\right) = f\left(-\frac{1}{n}\right) = \frac{1}{n^3}$?

19. Give an alternative derivation of the trigonometric identity

$$\cos^2 z + \sin^2 z \equiv 1$$

(Sec. 8.14a), regarding the identity as known for real values of z. Give other examples of the same technique of deducing complex identities from their real counterparts.

20. Prove that if $f(z)$ and $g(z)$ are analytic at a point z_0 and have zeros at z_0 of orders m and n, respectively, then $f(z)g(z)$ is analytic at z_0 and has a zero at z_0 of order $m + n$. How about the function $f(z) + g(z)$?

21. Find the order of the zero at $z = 0$ of the function

$$6 \sin z^3 + z^3(z^6 - 6).$$

22. Find all the zeros of the function

$$\sin \frac{1}{1-z},$$

and show that they have the point $z = 1$ as a limit point. Why is this compatible with the uniqueness theorem for analytic functions?

23. Prove that if $f(z) \not\equiv$ constant is analytic in a simply connected domain G, then any closed Jordan curve C lying in G encloses no more than a finite number of roots of the equation $f(z) = A$ where A is any finite complex number. *Comment.* The roots of the equation $f(z) = A$ are often called "A-points" of $f(z)$. More exactly, z_0 is said to be an *A-point of order m* of $f(z)$ if z_0 is a zero of order m of $f(z) - A$.

24. Let $f(z)$ be analytic in a domain G, and suppose $f'(z)$ vanishes at a point $z_0 \in G$. Prove that $f(z)$ fails to be conformal at z_0.

25. Prove that an analytic function of constant real part is itself constant. More exactly, prove that if $f(z)$ is analytic in a domain G and if Re $f(z)$ is constant in G, then $f(z)$ is constant in G.

26. Give an alternative proof of Theorem 10.27 based on the formula

$$\ln f(z) = \ln |f(z)| + i \arg f(z).$$

27. Given a bounded domain G, let $f(z)$ be analytic and nonconstant in G and continuous in $\bar{G}$, and suppose $|f(z)|$ is constant on the boundary of G. Prove that $f(z)$ has at least one zero at a point of G.

28. Prove the following proposition known as *Schwarz's lemma*: Given a function $f(z)$ which is analytic in the disk $|z| < 1$ and vanishes at $z = 0$, suppose $|f(z)| \leq 1$ for all $|z| < 1$. Then $|f(z)| \leq |z|$ for all $|z| < 1$, where equality is achieved at a nonzero point of the disk only if $f(z) = e^{i\theta}z$ for some real θ.

29. Interpret Schwarz's lemma geometrically.

30. Prove that

$$\int_0^{2\pi} \cos(\cos\theta)\cosh(\sin\theta)\, d\theta = 2\pi.$$

CHAPTER ELEVEN

Laurent Series

11.1. The Laurent Expansion of an Analytic Function

11.11. We begin by studying series that resemble power series except that they involve *negative* powers of the variable z:

THEOREM. *Given a series*

$$c_0 + \frac{c_1}{z-a} + \frac{c_2}{(z-a)^2} + \cdots + \frac{c_n}{(z-a)^n} + \cdots, \tag{1}$$

let

$$l = \overline{\lim_{n\to\infty}} \sqrt[n]{|c_n|}.$$

Then there are just three possibilities:

1) *If $l = 0$ the series is absolutely convergent for all z in the extended plane except $z = a$;*
2) *If $0 < l < \infty$ the series is absolutely convergent for all z outside the circle $|z - a| = l$ and divergent for all z inside the circle $|z - a| = l$;*
3) *If $l = \infty$ the series is divergent for all finite z.*

Proof. We need only apply Theorem 7.24, noting that the substitution

$$\zeta = \frac{1}{z-a} \tag{2}$$

carries the series (1) into the series

$$c_0 + c_1\zeta + c_2\zeta^2 + \cdots + c_n\zeta^n + \cdots \tag{1'}$$

with radius of convergence

$$\frac{1}{l} = \frac{1}{\overline{\lim\limits_{n\to\infty}} \sqrt[n]{|c_n|}}$$

and the points $z = a$, $z = \infty$ into the points $\zeta = \infty$, $\zeta = 0$, while carrying points inside the circle $|z - a| = l$ into points outside the circle $|\zeta| = 1/l$ and vice versa. ∎

11.12. It is clear that the transformation (2) maps every bounded closed domain $\bar{D}$ outside the circle $|z - a| = l$ into a bounded closed domain $\bar{D}'$ inside the circle $|\zeta| = 1/l$. But the series (1′) is uniformly convergent in $\bar{D}'$ by Theorem 7.17, being uniformly convergent in every closed disk $|\zeta| \leq \rho < 1/l$. Therefore the series (1) is uniformly convergent in every bounded closed domain $\bar{D}$ lying outside the circle $|z - a| = l$. Since every term of (1) is analytic outside the circle $|z - a| = l$, it follows from Theorem 6.39 that the sum $f(z)$ of the series (1) is analytic at every (finite) point outside the circle $|z - a| = l$. If $l = 0$ the circle $|z - a| = l$ reduces to the single point $z = a$, and $f(z)$ is analytic for all $z \neq a$.

11.13. We now introduce series of the form

$$\sum_{n=-\infty}^{\infty} c_n(z - a)^n, \tag{3}$$

interpreted as the sum of the two series

$$\sum_{n=0}^{\infty} c_n(z - a)^n, \qquad \sum_{m=1}^{\infty} c_{-m}(z - a)^{-m}. \tag{3′}$$

The series (3), involving both positive and negative powers of $z - a$ and known as a *Laurent series*, is regarded as convergent if and only if both series (3′) converge. The first of the series (3′), called the *regular part* of the Laurent series (3), is an ordinary power series with radius of convergence

$$R = \frac{1}{\overline{\lim\limits_{n\to\infty}} \sqrt[n]{|c_n|}},$$

while the second of the series (3′), called the *principal part* of (3), has the behavior described in Theorem 11.11 involving the number

$$r = \overline{\lim_{n\to\infty}} \sqrt[n]{|c_{-n}|}.$$

Therefore the series (3) is absolutely and uniformly convergent in every bounded closed subdomain of the annulus

$$r < |z - a| < R, \tag{4}$$

provided of course that $r < R$.† It follows from Theorem 6.39 that the func-

†Henceforth, in talking about a Laurent series, we will always assume that the condition $r < R$ is satisfied, so that the series converges in some annulus $r < |z - a| < R$. If $r = 0$ the annulus reduces to the "punctured disk" $0 < |z - a| < R$, while if $R = \infty$ it reduces to the exterior of the circle $|z - a| = r$. If $r = 0$, $R = \infty$ the annulus becomes the whole finite plane with the exception of the point $z = a$. We will continue to use the word "annulus" to describe these "degenerate" cases.

tion defined by

$$f(z) = \sum_{n=-\infty}^{\infty} c_n(z - a)^n \tag{5}$$

is analytic in the annulus (4). Note that if $|z - a| < r$ or $|z - a| > R$, then one of the series (3′) diverges and hence so does the Laurent series (3).

11.14. Next we prove the analogue for Laurent series of a result already proved for power series in Sec. 10.12:

THEOREM. *Let C be any circle $|z - a| = \rho$ with $r < \rho < R$. Then the coefficients of the Laurent series* (5) *are given by*

$$c_n = \frac{1}{2\pi i}\int_C \frac{f(z)}{(z - a)^{n+1}}\,dz \qquad (n = 0, \pm 1, \pm 2, \ldots). \tag{6}$$

Proof. The series (5) is uniformly convergent on C, and hence the same is true of each of the series

$$\frac{1}{2\pi i}\frac{f(z)}{(z - a)^{n+1}} = \sum_{k=-\infty}^{\infty} c_k \frac{(z - a)^k}{(z - a)^{n+1}} \qquad (n = 0, \pm 1, \pm 2, \ldots), \tag{7}$$

since

$$\left|\frac{1}{2\pi i}\frac{1}{(z - a)^{n+1}}\right| = \frac{1}{2\pi\rho^{n+1}}$$

for all $z \in C$ (cf. Chap. 6, Prob. 12). Therefore, by Theorem 6.38, we can integrate (7) term by term along C, obtaining

$$\frac{1}{2\pi i}\int_C \frac{f(z)}{(z - a)^{n+1}}\,dz = \sum_{k=-\infty}^{\infty} c_k \frac{1}{2\pi i}\int_C (z - a)^{k-n-1}\,dz$$

$$= \sum_{k=-\infty}^{\infty} c_k \frac{1}{2\pi}\int_0^{2\pi} \rho^{k-n} e^{i(k-n)\theta}\,d\theta = c_n,$$

where we use the fact that the last integral on the right vanishes if $k \neq n$ and equals 2π if $k = n$. ∎

A series of the form (5) with coefficients related to a given function $f(z)$ by the formulas (6) is called the *Laurent expansion* of $f(z)$, in the annulus $r < |z - a| < R$. Thus we see that every Laurent series is the Laurent expansion of its own sum.

11.15. According to Sec. 11.13, a Laurent series represents an analytic function in its region of convergence (an annulus). We now prove the converse proposition (the analogue of Theorem 10.13 for power series):

THEOREM. *Let K be the annulus $r < |z - a| < R$, and suppose $f(z)$ is analytic in K. Then $f(z)$ has a Laurent expansion in K, i.e., the Laurent series* (5) *with coefficients* (6) *converges to $f(z)$ at every point of K.*

Proof. Given any point $\zeta \in K$, choose numbers r' and R' such that

$$0 < r < r' < |\zeta - a| = \rho < R' < R.$$

Let γ be the circle $|z - a| = r'$ and Γ the circle $|z - a| = R'$, while L is a circle centered at ζ which is small enough to be contained in the annulus $r' < |z - a| < R'$ (see Figure 32). Then the nonintersecting circles γ and L lie inside the circle Γ, and the function $f(z)/(z - \zeta)$ is analytic in the closed domain between the outer circle Γ and the inner circles γ and L. It follows from Sec. 5.43 that

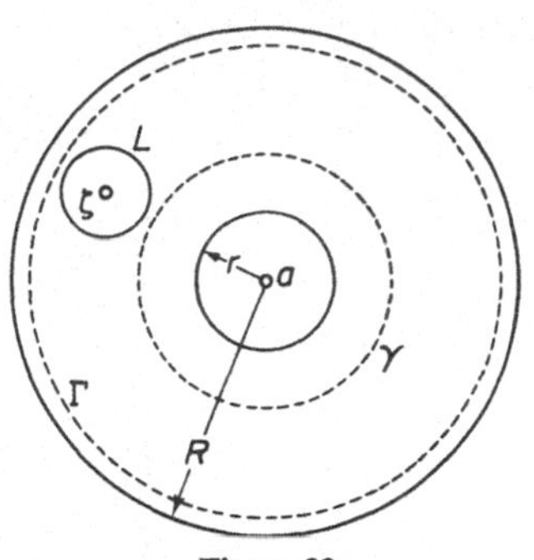

Figure 32

$$\frac{1}{2\pi i}\int_\Gamma \frac{f(z)}{z - \zeta}\,dz = \frac{1}{2\pi i}\int_\gamma \frac{f(z)}{z - \zeta}\,dz + \frac{1}{2\pi i}\int_L \frac{f(z)}{z - \zeta}\,dz.$$

But

$$\frac{1}{2\pi i}\int_L \frac{f(z)}{z - \zeta}\,dz = f(\zeta)$$

by Cauchy's integral formula, since $f(z)$ is analytic inside and on L, and hence

$$f(\zeta) = \frac{1}{2\pi i}\int_\Gamma \frac{f(z)}{z - \zeta}\,dz - \frac{1}{2\pi i}\int_\gamma \frac{f(z)}{z - \zeta}\,dz,$$

or equivalently

$$f(\zeta) = \frac{1}{2\pi i}\int_\Gamma \frac{f(z)}{z - \zeta}\,dz + \frac{1}{2\pi i}\int_\gamma \frac{f(z)}{\zeta - z}\,dz. \tag{8}$$

As we now show, the first integral on the right in (8) leads to the nonnegative powers of $z - a$ in the Laurent series (5), while the second integral leads to the negative powers of $z - a$. To this end, we note that if $z \in \Gamma$, as in the first integral, then

$$\frac{1}{z - \zeta} = \frac{1}{(z - a)\left(1 - \dfrac{\zeta - a}{z - a}\right)} = \sum_{n=0}^{\infty} \frac{(\zeta - a)^n}{(z - a)^{n+1}} \tag{9}$$

just as in the proof of Theorem 10.13, where we use the fact that

$$\left|\frac{\zeta - a}{z - a}\right| = \frac{\rho}{R'} < 1.$$

On the other hand, if $z \in \gamma$ as in the second integral, then

$$\frac{1}{\zeta - z} = \frac{1}{(\zeta - a)\left(1 - \dfrac{z - a}{\zeta - a}\right)} = \sum_{n=0}^{\infty} \frac{(z - a)^n}{(\zeta - a)^{n+1}}$$

$$= \sum_{n=1}^{\infty} \frac{(z - a)^{n-1}}{(\zeta - a)^n} = \sum_{n=1}^{\infty} \frac{(\zeta - a)^{-n}}{(z - a)^{-n+1}}, \tag{10}$$

since this time

$$\left|\frac{z-a}{\zeta-a}\right| = \frac{r'}{\rho} < 1.$$

Substituting (9) and (10) into (8) and integrating term by term,† we get

$$f(\zeta) = \sum_{n=0}^{\infty} c_n(\zeta - a)^n + \sum_{n=1}^{\infty} c_{-n}(\zeta - a)^{-n}, \tag{11}$$

where

$$c_n = \frac{1}{2\pi i}\int_\Gamma \frac{f(z)}{(z-a)^{n+1}}\,dz \qquad (n = 0, 1, 2, \ldots), \tag{12}$$

$$c_{-n} = \frac{1}{2\pi i}\int_\gamma \frac{f(z)}{(z-a)^{-n+1}}\,dz \qquad (n = 1, 2, \ldots). \tag{12'}$$

But we can replace Γ and γ in (12) and (12′) by any circle C of radius ρ centered at ζ provided that $r < \rho < R$ (why?), so that (12) and (12′) can be combined into the single formula (6). To complete the proof, we need only write (11) in the form

$$f(\zeta) = \sum_{n=-\infty}^{\infty} c_n(\zeta - a)^n$$

and then change ζ to z. ∎

11.16. Suppose $|f(z)| \le M$ for all $r < |z - a| < R$. Then, applying Theorem 5.23 to (6), we get *Cauchy's inequalities*

$$|c_n| \le \frac{M}{2\pi}\frac{2\pi\rho}{\rho^{n+1}} = \frac{M}{\rho^n} \qquad (n = 0, \pm 1, \pm 2, \ldots), \tag{13}$$

valid for all $r < \rho < R$. We have already proved (13) for the case of nonnegative $n = 0, 1, 2, \ldots$ (and $r = 0$) in Sec. 10.12.

11.17. Example. The function

$$f(z) = \frac{1}{z(1-z)}$$

has the Laurent expansion

$$f(z) = \frac{1}{z} + \frac{1}{1-z} = \frac{1}{z} + \sum_{n=0}^{\infty} z^n$$

in the annulus $0 < |z| < 1$ and the Laurent expansion

$$f(z) = \frac{1}{1 + (z-1)} - \frac{1}{z-1} = -\frac{1}{z-1} + \sum_{n=0}^{\infty}(-1)^n(z-1)^n$$

in the annulus $0 < |z - 1| < 1$.

---

†Justify this in the usual way, invoking the uniform convergence of the resulting series on the circles Γ and γ respectively.

11.2. Isolated Singular Points

11.21. Given a (finite) point z_0, let $f(z)$ be a single-valued analytic function defined at every point of some neighborhood of z_0 except at the point z_0 itself. Then z_0 is called an *isolated singular point* of $f(z)$.† Suppose $f(z)$ has an isolated singular point at z_0. Then, according to Theorem 11.15, $f(z)$ has a Laurent expansion

$$f(z) = \sum_{n=-\infty}^{\infty} c_n(z - z_0)^n \tag{14}$$

in some "deleted neighborhood" of z_0, i.e., in some annulus of the form $0 < |z - z_0| < R$‡. There are now just three possibilities:

1) The series (14) contains no negative powers of $z - z_0$, in which case z_0 is called a *removable singular point*;
2) The series (14) contains only a finite number of negative powers of $z - z_0$, in which case z_0 is called a *pole*;
3) The series (14) contains infinitely many negative powers of $z - z_0$, in which case z_0 is called an *essential singular point*.

We now analyze the behavior of $f(z)$ at each of these three kinds of isolated singular points.

11.22. Removable singular points. If z_0 is a removable singular point of $f(z)$, the right-hand side of (14) reduces to an ordinary power series, whose sum $\varphi(z)$ is analytic in some "full neighborhood" $|z - z_0| < R$ (recall Theorem 7.19). Obviously $\varphi(z) = f(z)$ if $z \neq z_0$ and $\varphi(z_0) = c_0$. Hence we can make $f(z)$ analytic at z_0, thereby changing z_0 from a singular point of $f(z)$ to a regular point of $f(z)$, by the simple expedient of setting

$$f(z_0) = c_0$$

(recall that $f(z)$ is originally undefined at z_0). It is in this sense that the singular point z_0 is "removable." Clearly if z_0 is a removable singular point of $f(z)$, then

$$\lim_{z \to z_0} f(z) = c_0.$$

In particular, there exist positive numbers M and δ such that $|f(z)| \leq M$ whenever $0 < |z - z_0| < \delta$. In other words, if z_0 is a removable singular point of $f(z)$, then $f(z)$ is bounded in some deleted neighborhood of z_0 (see Chap. 3, Prob. 12a).

11.23. Poles. Next suppose z_0 is a pole of $f(z)$, i.e., suppose the series (14) contains only a finite number of negative powers of $f(z)$. Let m be the

---

†Cf. Sec. 10.14 and the use of the word "isolated" in the statement of Theorem 10.26.

‡In this case, we often call (14) the Laurent expansion of $f(z)$ *at the point* z_0.

highest power of $1/(z - z_0)$ appearing in (14), so that

$$f(z) = \sum_{n=0}^{\infty} c_n(z - z_0)^n + \frac{c_{-1}}{z - z_0} + \frac{c_{-2}}{(z - z_0)^2} + \cdots + \frac{c_{-m}}{(z - z_0)^m}, \quad (14')$$

where $c_{-m} \neq 0$. The point z_0 is then called a *pole of order m* of $f(z)$. In particular, z_0 is said to be a *simple pole* if $m = 1$ and a *multiple pole* if $m > 1$. Multiplying both sides of (14') by $(z - z_0)^m$, we get

$$(z - z_0)^m f(z) = \sum_{n=0}^{\infty} c_n(z - z_0)^{n+m} + c_{-1}(z - z_0)^{m-1} + c_{-2}(z - z_0)^{m-2} + \cdots + c_{-m}, \quad (15)$$

where the right-hand side of (15) is just an ordinary power series, whose constant term c_{-m} is nonzero. Therefore the point z_0 is a removable singular point of the function $(z - z_0)^m f(z)$. Moreover

$$\lim_{z \to z_0} (z - z_0)^m f(z) = c_{-m} \neq 0,$$

and hence

$$\lim_{z \to z_0} f(z) = \lim_{z \to z_0} \frac{c_{-m}}{(z - z_0)^m} = \infty,$$

i.e., $f(z)$ approaches infinity as $z \longrightarrow z_0$ (cf. Chap. 3, Prob. 3).

11.24. Before considering the case of essential singular points, we first examine the relation between zeros and poles:

a. THEOREM. *Let z_0 be a zero of order m of a function $f(z)$ analytic at z_0. Then $1/f(z)$ is analytic in a deleted neighborhood of z_0, with a pole of order m at z_0.*

Proof. By Sec. 10.25, $f(z)$ has a power series expansion of the form

$$f(z) = c_m(z - z_0)^m + c_{m+1}(z - z_0)^{m+1} + \cdots \qquad (c_m \neq 0)$$

at z_0, or equivalently

$$f(z) = (z - z_0)^m \varphi(z) \quad (16)$$

where $\varphi(z)$ is analytic and nonzero at z_0. It follows from (16) that

$$\frac{1}{f(z)} = \frac{1}{(z - z_0)^m} \frac{1}{\varphi(z)} = \frac{\psi(z)}{(z - z_0)^m}, \quad (17)$$

where the function $\psi(z) = 1/\varphi(z)$ is itself analytic and nonzero at z_0 (why?). Therefore $1/f(z)$ is analytic in a deleted neighborhood of z_0. Writing

$$\psi(z) = \psi(z_0) + \psi'(z_0)(z - z_0) + \cdots, \quad (18)$$

we have

$$\frac{1}{f(z)} = \frac{\psi(z_0)}{(z - z_0)^m} + \frac{\psi'(z_0)}{(z - z_0)^{m-1}} + \cdots$$

in a deleted neighborhood of z_0, i.e., $1/f(z)$ has a pole of order m at z_0. ∎

b. THEOREM. *Let z_0 be a pole of order m of a function $f(z)$ analytic in a deleted neighborhood of z_0. Then $1/f(z)$ is analytic at z_0 provided that $1/f(z_0) = 0$, with a zero of order m at z_0.*

Proof. By Sec. 11.23, $f(z)$ can be written in the form

$$f(z) = \frac{\varphi(z)}{(z - z_0)^m}, \tag{19}$$

where $\varphi(z)$ approaches a nonzero limit α as $z \to z_0$ and hence can be regarded as a function which is analytic and nonzero at z_0 if we set $\varphi(z_0) = \alpha$. It follows from (19) that

$$\frac{1}{f(z)} = (z - z_0)^m \frac{1}{\varphi(z)} = (z - z_0)^m \psi(z),$$

where $\psi(z)$ is itself analytic and nonzero at z_0. Therefore $1/f(z)$ is analytic at z_0 provided we set $1/f(z_0) = 0$. Substituting from (18), we get

$$\frac{1}{f(z)} = \psi(z_0)(z - z_0)^m + \psi'(z_0)(z - z_0)^{m+1} + \cdots \qquad (\psi(z_0) \neq 0)$$

in a neighborhood of z_0, i.e., $1/f(z)$ has a zero of order m at z_0. ▮

11.25. COROLLARY. *If $f(z)$ is nonvanishing in a deleted neighborhood of z_0 and has an essential singular point at z_0, then $1/f(z)$ also has an essential singular point at z_0.*

Proof. If z_0 is not an essential singular point of

$$\varphi(z) = \frac{1}{f(z)},$$

then z_0 is either a pole or a removable singular point of $\varphi(z)$. In the first case, it follows from Theorem 11.24b that z_0 is a zero of $f(z)$, contrary to hypothesis. In the second case, either

$$\lim_{z \to z_0} \varphi(z) = 0 \tag{20}$$

or

$$\lim_{z \to z_0} \varphi(z) \neq 0. \tag{20'}$$

If (20) holds, then by Theorem 11.24a, $f(z)$ has a pole at z_0 contrary to hypothesis, while if (20′) holds, $f(z)$ has a removable singular point at z_0, again contrary to hypothesis. ▮

11.26. The behavior of a function at an essential singular point is described by the following remarkable

THEOREM *(Casorati–Weierstrass). If z_0 is an essential singular point of $f(z)$, then given any complex number A (finite or infinite), there exists a sequence* *of points z_n converging to z_0 such that*

$$\lim_{n\to\infty} f(z_n) = A. \tag{21}$$

Proof. First let $A = \infty$. Suppose $f(z)$ is bounded in a deleted neighborhood of z_0, so that $|f(z)| \leq M$ for all z in some annulus $0 < |z - z_0| < R$. Then, by Cauchy's inequalities (Sec. 11.16),

$$|c_n| \leq \frac{M}{\rho^n} \qquad (n = 0, \pm 1, \pm 2, \ldots)$$

for all $0 < \rho < R$. Letting $\rho \longrightarrow 0$ in the case $n < 0$, we find that

$$c_n = 0 \qquad (n = -1, -2, \ldots).$$

Therefore the Laurent series (14) contains no negative powers of $z - z_0$, i.e., z_0 is a removable singular point rather than an essential singular point of $f(z)$. This contradiction shows that $f(z)$ cannot be bounded in any deleted neighborhood of z_0. Hence, given any positive integer n, there exists a point z_n such that

$$0 < |z_n - z_0| < \frac{1}{n}, \qquad |f(z_n)| > n.$$

But then the sequence z_n converges to z_0 and

$$\lim_{n\to\infty} f(z_n) = \infty,$$

as required.

Now let A be any finite complex number. If every deleted neighborhood of z_0 contains a point z such that $f(z) = A$, there is obviously nothing more to prove. Thus suppose z_0 has a deleted neighborhood K in which $f(z) \neq A$. Then the function

$$\varphi(z) = \frac{1}{f(z) - A}$$

is analytic in K and has an essential singular point at z_0, by Corollary 11.25. Hence, by the first part of the proof, there exists a sequence z_n converging to z_0 such that

$$\lim_{n\to\infty} \varphi(z_n) = \infty. \tag{21'}$$

But (21′) is equivalent to (21). ∎

If z_0 is an essential singular point of $f(z)$, then

$$\lim_{z\to z_0} f(z) \tag{22}$$

cannot exist. In fact, suppose (22) exists and equals some number A_0 (finite or infinite). Then $f(z)$ must be near A_0 for every z sufficiently near z_0, contrary to Theorem 11.26.

11.27. Example. The function

$$f(z) = e^{1/z} = 1 + \frac{1}{z} + \frac{1}{2!}\frac{1}{z^2} + \frac{1}{3!}\frac{1}{z^3} + \cdots \tag{23}$$

obviously has an essential singular point at $z = 0$. If $A = \infty$, the sequence $z_n = 1/n$ $(z_n \to 0)$ satisfies the condition (21), since

$$\lim_{n\to\infty} f(z_n) = \lim_{n\to\infty} e^n = \infty.$$

If $A = 0$, the sequence $z_n = -1/n$ $(z_n \to 0)$ satisfies (21), since

$$\lim_{n\to\infty} f(z_n) = \lim_{n\to\infty} e^{-n} = 0.$$

On the other hand, if $A \neq \infty$, $A \neq 0$, then, solving the equation

$$e^{1/z} = A,$$

we get

$$z = \frac{1}{\ln A}. \tag{24}$$

Let $(\ln z)_0$ be the branch of the logarithm such that $0 \leq \arg z < 2\pi$ (cf. Sec. 9.22). Then (24) can be written as

$$z = \frac{1}{(\ln A)_0 + 2k\pi i}, \tag{24'}$$

where k is any integer. Choosing

$$z_n = \frac{1}{(\ln A)_0 + 2n\pi i} \qquad (n = 1, 2, \ldots),$$

we get a sequence z_n which converges to zero and satisfies the condition (21), in fact the much stronger condition

$$f(z_n) = A \qquad (n = 1, 2, \ldots). \tag{25}$$

11.3. Residues

11.31. Definition. Let z_0 be an isolated singular point of a function $f(z)$. Then by the *residue of* $f(z)$ *at* z_0, denoted by

$$\operatorname*{Res}_{z=z_0} f(z),$$

is meant the coefficient c_{-1} in the Laurent expansion

$$f(z) = \sum_{n=-\infty}^{\infty} c_n(z - z_0)^n.$$

Note that the residue of $f(z)$ at z_0 may be zero or nonzero if z_0 is a pole or an essential singular point, but is automatically zero if z_0 is a removable singular point.

11.32. THEOREM ***(Residue theorem).*** *If $f(z)$ is analytic inside and on a piecewise smooth closed Jordan curve C, except for isolated singular points $z_1, \ldots, z_N$ lying inside C, then*

$$\int_C f(z)\,dz = 2\pi i \sum_{k=1}^{N} \operatorname*{Res}_{z=z_k} f(z). \tag{26}$$

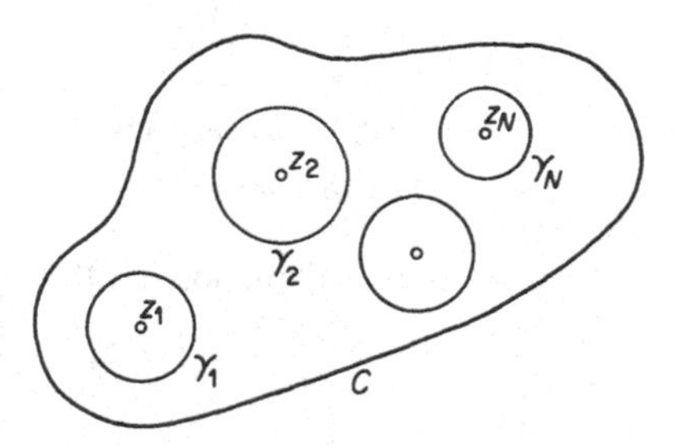

Figure 33

Proof. As in Figure 33, let $\gamma_1, \ldots, \gamma_N$ be circles centered at the points $z_1, \ldots, z_N$, respectively, which are so small that they all lie inside C and do not intersect each other. Then, by Sec. 5.43,

$$\int_C f(z)\,dz = \sum_{k=1}^{N} \int_{\gamma_k} f(z)\,dz, \tag{27}$$

where the curves $C, \gamma_1, \ldots, \gamma_N$ are all traversed in the positive (counterclockwise) direction. Suppose the Laurent expansion of $f(z)$ at z_k is

$$f(z) = \sum_{n=-\infty}^{\infty} c_n^{(k)}(z - z_k)^n \qquad (k = 1, \ldots, N). \tag{28}$$

Then integrating (28) term by term along γ_k, which is permissible because of the uniform convergence of the series on γ_k, we obtain

$$\begin{aligned}\int_{\gamma_k} f(z)\,dz &= \int_{\gamma_k} \sum_{n=-\infty}^{\infty} c_n^{(k)}(z - z_k)^n\,dz = \sum_{n=-\infty}^{\infty} c_n^{(k)} \int_{\gamma_k} (z - z_k)^n\,dz \\ &= \sum_{n=-\infty}^{\infty} c_n^{(k)} \int_0^{2\pi} (r_k e^{i\theta})^n d(r_k e^{i\theta}) = \sum_{n=-\infty}^{\infty} i r_k^{n+1} \int_0^{2\pi} e^{i(n+1)\theta}\,d\theta, \end{aligned} \tag{29}$$

where r_k is the radius of γ_k. But

$$\int_0^{2\pi} e^{i(n+1)\theta}\,d\theta = \begin{cases} 2\pi & \text{if } n = -1, \\ 0 & \text{otherwise.} \end{cases}$$

Therefore (29) reduces to

$$\int_{\gamma_k} f(z)\,dz = 2\pi i c_{-1}^{(k)} \qquad (k = 1, \ldots, N). \tag{30}$$

Substituting (30) into (27), we finally get

$$\int_C f(z)\,dz = \sum_{k=1}^{N} \int_{\gamma_k} f(z)\,dz = 2\pi i \sum_{k=1}^{N} c_{-1}^{(k)} = 2\pi i \sum_{k=1}^{N} \operatorname*{Res}_{z=z_k} f(z). \qquad \blacksquare$$

11.33. Example. Evaluate the integral

$$\int_{|z|=2} \frac{e^z}{(z-1)^n}\,dz.$$

Solution. The integrand has only one singular point inside the circle $|z| = 2$, namely a pole of order n at $z = 1$. Therefore, by the residue theorem,

$$\int_{|z|=2} \frac{e^z}{(z-1)^n}\,dz = 2\pi i \operatorname*{Res}_{z=1} \frac{e^z}{(z-1)^n}. \tag{31}$$

Expanding $e^z/(z-1)^n$ in Laurent series in a deleted neighborhood of $z = 1$, we get

$$\begin{aligned}\frac{e^z}{(z-1)^n} &= e\frac{e^{z-1}}{(z-1)^n}\\ &= \frac{e}{(z-1)^n}\left[1 + (z-1) + \cdots + \frac{(z-1)^{n-1}}{(n-1)!} + \frac{(z-1)^n}{n!} + \cdots\right]\\ &= e\left[\frac{1}{(z-1)^n} + \frac{1}{(z-1)^{n-1}} + \cdots + \frac{1}{(n-1)!}\frac{1}{z-1} + \frac{1}{n!} + \cdots\right],\end{aligned}$$

and hence

$$\operatorname*{Res}_{z=1} \frac{e^z}{(z-1)^n} = \frac{e}{(n-1)!}. \tag{32}$$

It follows from (31) that

$$\int_{|z|=2} \frac{e^z}{(z-1)^n}\,dz = \frac{2\pi i e}{(n-1)!}.$$

11.34. Next we show how to calculate residues at a pole without making explicit use of a Laurent series as in the above example. First let z_0 be a simple pole of $f(z)$, so that $f(z)$ has a Laurent expansion of the form

$$f(z) = \frac{c_{-1}}{z-z_0} + c_0 + c_1(z-z_0) + \cdots$$

in a deleted neighborhood of z_0. Then

$$(z-z_0)f(z) = c_{-1} + c_0(z-z_0) + c_1(z-z_0)^2 + \cdots,$$

where the right-hand side is an ordinary power series and hence is continuous at z_0. It follows that

$$c_{-1} = \operatorname*{Res}_{z=z_0} f(z) = \lim_{z\to z_0} (z-z_0)f(z). \tag{33}$$

The calculation is particularly simple if $f(z)$ has the form

$$f(z) = \frac{\varphi(z)}{\psi(z)},$$

where $\varphi(z_0) \neq 0$ and $\psi(z)$ has a simple zero at z_0, i.e., $\psi(z_0) = 0$, $\psi'(z_0) \neq 0$.

Then z_0 is a simple pole of $f(z)$, and hence, by (33),

$$\operatorname*{Res}_{z=z_0} f(z) = \operatorname*{Res}_{z=z_0} \frac{\varphi(z)}{\psi(z)} = \lim_{z\to z_0} \frac{(z-z_0)\varphi(z)}{\psi(z)}$$

$$= \lim_{z\to z_0} \frac{\varphi(z)}{\dfrac{\psi(z)-\psi(z_0)}{z-z_0}} = \frac{\varphi(z_0)}{\psi'(z_0)}. \tag{34}$$

Now let z_0 be a pole of order $m > 1$ of $f(z)$. Then the Laurent expansion of $f(z)$ at z_0 is of the form

$$f(z) = \frac{c_{-m}}{(z-z_0)^m} + \cdots + \frac{c_{-1}}{z-z_0} + c_0 + c_1(z-z_0) + \cdots,$$

and hence

$$(z-z_0)^m f(z) = c_{-m} + \cdots + c_{-1}(z-z_0)^{m-1}$$
$$+ c_0(z-z_0)^m + c_1(z-z_0)^{m+1} + \cdots. \tag{35}$$

Differentiating (35) $m-1$ times, we get

$$\frac{d^{m-1}}{dz^{m-1}}[(z-z_0)^m f(z)] = (m-1)!c_{-1} + \frac{m!}{1!}c_0(z-z_0)$$
$$+ \frac{(m+1)!}{2!}c_1(z-z_0)^2 + \cdots,$$

which implies

$$(m-1)!c_{-1} = \lim_{z\to z_0} \frac{d^{m-1}}{dz^{m-1}}[(z-z_0)^m f(z)]$$

or

$$c_{-1} = \operatorname*{Res}_{z=z_0} f(z) = \lim_{z\to z_0} \frac{1}{(m-1)!} \frac{d^{m-1}}{dz^{m-1}}[(z-z_0)^m f(z)]. \tag{36}$$

Note that (36) reduces to (33) if $m = 1$. Applying (36) to Example 11.33, we find that

$$\operatorname*{Res}_{z=1} \frac{e^z}{(z-1)^n} = \lim_{z\to 1} \frac{1}{(n-1)!} \frac{d^{n-1} e^z}{dz^{n-1}} = \lim_{z\to 1} \frac{e^z}{(n-1)!} = \frac{e}{(n-1)!},$$

in keeping with (32).

COMMENTS

11.1. The motivation for the considerations of Sec. 11.1 is the desire to generalize the theory of Sec. 10.1 to allow for functions that fail to be analytic at certain isolated points. This leads us to consider the case where $f(z)$ is analytic merely in an *annulus* K rather than in a full disk. Then Theorem 11.15 (the analogue of Theorem 10.13) asserts that $f(z)$ is the sum of a Laurent series (rather than a power series) convergent in K, the price for making the region of analyticity of $f(z)$ an annulus rather than a disk being to replace the Taylor series of $f(z)$ at a by an expansion involving *negative* powers of $z - a$ (where a is the center of K). This complication notwithstanding, Theorem 11.15 remains one of the central results of complex analysis, since it is the point of departure for treating functions with isolated singular points (Sec. 11.21).

11.2. We call attention to the alternative characterization of isolated singular points given in Prob. 15, in terms of the limiting behavior of the function $f(z)$ at a given singular point. Equation (25) is no accident, in view of the following remarkable proposition known as *Picard's theorem* (proved, for example, in A.I. Markushevich, *op. cit.*, Volume III, Sec. 51): If z_0 is an essential singular point of $f(z)$, then given any complex number $A \neq \infty$ with the possible exception of a single value $A = A_0$, there exists a sequence of points z_n which converges to z_0 and satisfies (25). Note that in the case of the function (23), the number $A_0 = 0$ is the exceptional value allowed for in Picard's theorem, since $e^{1/z}$ never vanishes (Sec. 8.13e).

11.3. The importance of Definition 11.31 and Theorem 11.32 (the residue theorem) can hardly be exaggerated. Chapter 12 is devoted in its entirety to some of the manifold applications of residue theory.

PROBLEMS

1. Prove that there is a unique Laurent series in the variable $z - z_0$ with a given sum on a circle with center z_0.
2. Find the Laurent expansion of the function

$$\frac{1}{(z-a)(z-b)} \qquad (0 < |a| < |b|)$$

in
a) A deleted neighborhood of the point $z = 0$;
b) A deleted neighborhood of the point $z = a$;
c) The annulus $|a| < |z| < |b|$;
d) The domain $|z| > |b|$.

3. Find the Laurent expansion of the function $e^{z+(1/z)}$ in the domain $0 < |z| < \infty$.
4. Find the principal part of the Laurent expansion of each of the following functions at the indicated point z_0:

$$\text{(a)}\ \frac{z}{(z+2)^2}\ \ (z_0 = -2);\quad \text{(b)}\ \frac{e^z+1}{e^z-1}\ \ (z_0 = 2\pi i);\quad \text{(c)}\ \frac{z-1}{\sin^2 z}\ \ (z_0 = 0);$$

$$\text{(d)}\ \frac{e^{iz}}{z^2+b^2}\ \ (z_0 = ib, b > 0);\quad \text{(e)}\ \frac{1}{\sin \pi z}\ \ (z_0 = n);\quad \text{(f)}\ \cot \pi z\ \ (z_0 = n)$$

(n an arbitrary integer).

5. Can the function $\ln (1/(1-z))$ be expanded in Laurent series in the domain $|z| > 1$?

6. Let $f(z)$ be the branch of the function $\sqrt{1+z^2}$ defined in the z-plane cut along the segment joining the points $-i$ and i such that $f(-\frac{3}{4}) = \frac{5}{4}$. Find the Laurent expansion of $f(z)$ in the domain $|z| > 1$.

7. Let $f(z)$ be a rational function, i.e., a ratio

$$f(z) = \frac{a_0 + a_1 z + \cdots + a_m z^m}{b_0 + b_1 z + \cdots + b_n z^n} \quad (a_m \neq 0,\ b_n \neq 0) \tag{37}$$

of two polynomials. Describe the finite singular points of $f(z)$, assuming that the numerator and denominator have no common zeros.

8. Suppose $f(z)$ has a pole of order m at the point z_0. Prove that the nth derivative $f^{(n)}(z)$ has a pole of order $m + n$ at z_0.

9. Suppose $f(z)$ and $g(z)$ have poles at the point z_0, of orders m and n respectively. Describe the behavior of each of the following functions at z_0:

a) $f(z) + g(z)$; b) $f(z)g(z)$; c) $\dfrac{f(z)}{g(z)}$,

10. Describe all (finite) singular points of each of the following functions:

a) $\dfrac{1}{z - z^3}$; b) $\dfrac{z^4}{1 + z^4}$; c) $\dfrac{z^5}{(1 - z)^2}$; d) $\dfrac{1}{z(z^2 + 4)^2}$; e) $\dfrac{e^z}{1 + z^2}$;

f) $\dfrac{z^2 + 1}{e^z}$; g) ze^{-z}; h) $\dfrac{1}{e^z - 1} - \dfrac{1}{z}$.

11. Do the same for each of the following functions:

a) $\dfrac{z}{\sin z}$; b) $\dfrac{\cos z}{z^2}$; c) $\dfrac{1}{\sin \dfrac{1}{z}}$; d) $\tan^2 z$; e) $\cot z - \dfrac{1}{z}$;

f) $\cot \dfrac{1}{z}$; g) $e^{\cot (1/z)}$; h) $\sin \left(\dfrac{1}{\cos \dfrac{1}{z}} \right)$.

12. Prove that if z_0 is an essential singular point of $f(z)$ and if K is any deleted neighborhood of z_0, then every point of the extended w-plane is a limit point of the image of K under the mapping $w = f(z)$.

13. Show that Theorem 11.26 remains valid if z_0 is a limit point of poles.

14. Verify Picard's theorem for the function $\sin (1/z)$ at the point $z = 0$. Is there an exceptional point?

15. Consider the expression

$$\lim_{z \to z_0} f(z), \tag{38}$$

where z_0 is an isolated singular point of $f(z)$. It has been shown that
a) (38) exists and is finite if z_0 is a removable singular point;
b) (38) exists and is infinite if z_0 is a pole;
c) (38) fails to exist if z_0 is an essential singular point.
Prove that conversely
a′) z_0 is a removable singular point if (38) exists and is finite;
b′) z_0 is a pole if (38) exists and is infinite;
c′) z_0 is an essential singular point if (38) fails to exist.

16. Suppose $f(z)$ is analytic in a deleted neighborhood of infinity (cf. Sec. 2.44), with a Laurent expansion of the form

$$f(z) = \cdots + \frac{c_{-n}}{z^n} + \cdots + \frac{c_{-1}}{z} + c_0 + c_1 z + \cdots + c_n z^n + \cdots \qquad (R < z < \infty). \tag{39}$$

Then the point at infinity (∞) is called an *isolated singular point* of $f(z)$, more exactly

a) A *removable singular point* if the series (39) contains no *positive* powers of z;

b) A *pole of order m* if (39) contains only a finite number of positive powers of z, the highest positive power being z^m;

c) An *essential singular point* if (39) contains infinitely many positive powers of z.

Prove that the nature of the singular point of $f(z)$ at $z = \infty$ is precisely the same as that of the function $\varphi(\zeta) = f(1/\zeta)$ at $\zeta = 0$. Show that the six assertions in Prob. 15 remain valid for $z_0 = \infty$ if we replace (38) by

$$\lim_{z \to \infty} f(z). \tag{38'}$$

17. Show that Theorem 11.26 remains valid if $z_0 = \infty$.

18. Prove that the rational function (37) has a removable singular point at ∞ if $m \leq n$ and a pole of order $m - n$ at ∞ if $m > n$.

19. Investigate the behavior at infinity of each of the functions in Prob. 10.

20. Do the same for each of the functions in Prob. 11.

21. Find the residues of each of the following functions $f(z)$ at all of its (finite) isolated singular points:

a) $\dfrac{1}{z^3 - z^5}$; b) $\dfrac{z^2}{(z^2+1)^2}$; c) $\dfrac{\sin 2z}{(z+1)^3}$; d) $\dfrac{e^z}{z^2+9}$; e) $\dfrac{1}{\sin z}$;

f) $\sin \dfrac{z}{z+1}$; g) $\sin z \sin \dfrac{1}{z}$; h) $z^n \sin \dfrac{1}{z}$ (n an integer).

22. Find

$$\operatorname*{Res}_{z=z_0} f(z)g(z)$$

if $f(z)$ is analytic at z_0 and

a) $g(z)$ has a simple pole at z_0 with residue c_{-1};

b) $g(z)$ has a pole of order m at z_0 with principal part

$$\frac{c_{-1}}{z - z_0} + \cdots + \frac{c_{-m}}{(z - z_0)^m}.$$

23. Evaluate the integral

$$\int_{|z|=2} \frac{z^3}{z^4 - 1}\, dz.$$

24. Evaluate the integral

$$\int_C \frac{dz}{z^4 + 1},$$

where C is the circle $x^2 + y^2 = 2x$.

25. Evaluate the integral

$$\int_{|z|=1} z^n e^{2/z}\, dz,$$

where n is an integer.

26. Evaluate the integral

$$\int_C \frac{dz}{(z-1)(z-2)^2},$$

where C is the circle $|z-2| = \frac{1}{2}$.

27. Evaluate the integral

$$\int_{|z|=n} \tan \pi z.$$

28. Suppose $f(z)$ is analytic in a deleted neighborhood of infinity, with a Laurent expansion of the form

$$f(z) = \cdots + \frac{c_{-n}}{z^n} + \cdots + \frac{c_{-1}}{z} + c_0 + c_1 z + \cdots + c_n z^n + \cdots \qquad (R < |z| < \infty).$$

Then by the *residue of* $f(z)$ *at infinity*, denoted by

$$\operatorname*{Res}_{z=\infty} f(z),$$

is meant the number $-c_{-1}$ (note the minus sign!). Find

$$\operatorname*{Res}_{z=\infty} f^2(z).$$

29. Suppose $f(z)$ is analytic at every point of the finite plane except at isolated singular points $z_1, \ldots, z_n$. Prove that

$$\operatorname*{Res}_{z=z_1} f(z) + \cdots + \operatorname*{Res}_{z=z_n} f(z) + \operatorname*{Res}_{z=\infty} f(z) = 0.$$

30. Avoiding excessive calculation, show that

$$\int_{|z|=2} \frac{dz}{(z-3)(z^5-1)} = -\frac{\pi i}{121}.$$

Give a quick calculation of the integral in Prob. 23.

31. Use residues to show that

$$\int_0^{2\pi} \frac{dx}{1 - 2p\cos x + p^2} = \frac{2\pi}{1-p^2} \qquad (0 < p < 1).$$

32. Prove that

$$\int_0^{2\pi} \frac{dx}{(p + q\cos x)^2} = \frac{2\pi p}{(p^2 - q^2)^{3/2}} \qquad (p > q > 0).$$

33. Prove that

$$\int_0^{\pi} \cot(x-a)\, dx = \begin{cases} \pi i & \text{if } \operatorname{Im} a > 0, \\ -\pi i & \text{if } \operatorname{Im} a < 0 \end{cases}$$

(the integral diverges if $\operatorname{Im} a = 0$).

CHAPTER TWELVE

Applications of Residues

12.1. Logarithmic Residues and the Argument Principle

12.11. By the *logarithmic residue* of a function $f(z)$ at a point a is meant the quantity

$$\operatorname*{Res}_{z=a} \frac{f'(z)}{f(z)};$$

i.e., the residue of the logarithmic derivative

$$\frac{d \ln f(z)}{dz} = \frac{f'(z)}{f(z)}$$

at a.† Suppose a is a zero of order α of $f(z)$. Then the Taylor expansion of $f(z)$ at a is of the form

$$f(z) = c_\alpha(z-a)^\alpha + c_{\alpha+1}(z-a)^{\alpha+1} + \cdots \qquad (c_\alpha \neq 0),$$

and hence

$$f'(z) = \alpha c_\alpha (z-a)^{\alpha-1} + (\alpha+1)c_{\alpha+1}(z-a)^\alpha + \cdots.$$

Therefore

$$\frac{f'(z)}{f(z)} = \frac{\alpha c_\alpha (z-a)^{\alpha-1} + (\alpha+1)c_{\alpha+1}(z-a)^\alpha + \cdots}{c_\alpha(z-a)^\alpha + c_{\alpha+1}(z-a)^{\alpha+1} + \cdots},$$

which implies

$$\frac{f'(z)}{f(z)} = \frac{1}{z-a} \, \frac{\alpha + (\alpha+1)\frac{c_{\alpha+1}}{c_\alpha}(z-a) + \cdots}{1 + \frac{c_{\alpha+1}}{c_\alpha}(z-a) + \cdots}$$

$$= \frac{\alpha}{z-a} + c_0' + c_1'(z-a) + \cdots$$

---

†In writing $d \ln f(z)/dz$, we rely on the fact that the branches of $\ln f(z)$ all have the same derivative (cf. Chap. 9, Prob. 8).

for suitable $c_0', c_1', \ldots$ But then the logarithmic residue of $f(z)$ at a is just

$$\operatorname*{Res}_{z=a} \frac{f'(z)}{f(z)} = \alpha, \tag{1}$$

i.e., the order of the zero at a.

Similarly, if b is a pole of order $f(z)$ of order β, then the Laurent expansion of $f(z)$ at b is of the form

$$f(z) = \frac{c_{-\beta}}{(z-b)^\beta} + \frac{c_{-\beta+1}}{(z-b)^{\beta-1}} + \cdots \qquad (c_{-\beta} \neq 0),$$

and hence

$$f'(z) = -\frac{\beta c_{-\beta}}{(z-b)^{\beta+1}} - \frac{(\beta-1)c_{-\beta+1}}{(z-b)^\beta} + \cdots.$$

Therefore

$$\begin{aligned}\frac{f'(z)}{f(z)} &= \frac{1}{z-b}\,\frac{-\beta c_{-\beta} - (\beta-1)c_{-\beta+1}(z-b) + \cdots}{c_{-\beta} + c_{-\beta+1}(z-b) + \cdots} \\ &= \frac{-\beta}{z-b} + c_0' + c_1'(z-b) + \cdots\end{aligned}$$

for suitable $c_0', c_1', \ldots$, so that the logarithmic residue of $f(z)$ at b is just

$$\operatorname*{Res}_{z=b} \frac{f'(z)}{f(z)} = -\beta, \tag{2}$$

i.e., the *negative* of the order of the pole at b.

12.12. THEOREM. *Given a piecewise smooth closed Jordan curve C, suppose $f(z)$ is analytic inside and on C except for poles inside C at the points $b_1, \ldots, b_n$. Moreover, suppose $f(z)$ has zeros $a_1, \ldots, a_m$ inside C, but none on C itself. Then*

$$\frac{1}{2\pi i}\int_C \frac{f'(z)}{f(z)}\,dz = \sum_{k=1}^{m} \alpha_k - \sum_{k=1}^{n} \beta_k, \tag{3}$$

where α_k is the order of a_k and β_k the order of b_k.

Proof. Since the only singular points of $f'(z)/f(z)$ inside C are the poles and zeros of $f(z)$, (3) is an immediate consequence of the residue theorem and formulas (1) and (2). ∎

12.13. Let N be the total number of zeros and P the total number of poles of $f(z)$ inside C, each counted a number of times equal to its order. Then (3) implies

$$\begin{aligned}N - P &= \frac{1}{2\pi i}\int_C \frac{f'(z)}{f(z)}\,dz = \frac{1}{2\pi i}\int_C \frac{d\ln f(z)}{dz}\,dz \\ &= \frac{1}{2\pi i}\int_C d\ln f(z) = \frac{1}{2\pi i}\Delta_C \ln f(z),\end{aligned} \tag{4}$$

where $\Delta_C \ln f(z)$ denotes the change in $\ln f(z)$ as the point z makes one circuit around C in the positive (counterclockwise) direction. But

$$\ln f(z) = \ln |f(z)| + i \arg f(z)$$

(see Sec. 9.14b), and there is obviously no change in $|f(z)|$ as z makes a circuit around C. It follows that

$$\Delta_C \ln f(z) = i\,\Delta_C \arg f(z), \tag{5}$$

where this time $\Delta_C \arg f(z)$ is the change in $\arg f(z)$ as z makes one circuit around C. Substituting (5) into (4), we get

$$N - P = \frac{1}{2\pi} \Delta_C \arg f(z), \tag{6}$$

an important result known as the *argument principle.*

12.14. Formula (6) has a simple geometric interpretation: As z traverses C once in the positive direction, the image point $w = f(z)$ traverses some closed curve Γ in the w-plane, where Γ need not be a Jordan curve (see Figure 34). Suppose $w = f(z)$ "winds around" the origin of the w-plane n_+

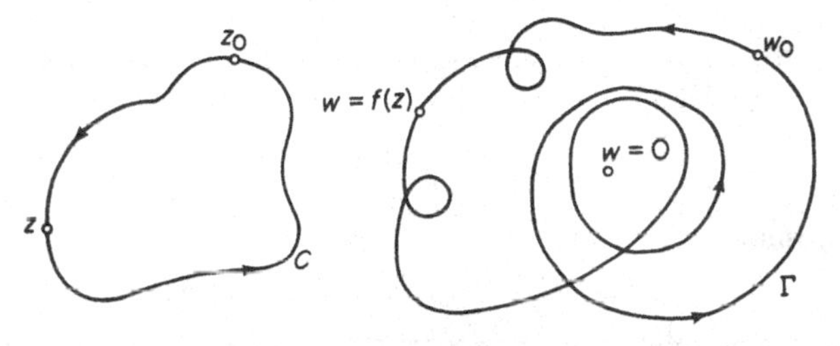

Figure 34

times in the positive direction and n_- times in the negative direction as z makes one circuit around C in the positive direction, and let

$$\nu = n_+ - n_-$$

($n_+ = 3$, $n_- = 0$, $\nu = 3$ in the figure). Then

$$\Delta_C \arg f(z) = 2\nu\pi$$

for the same reason as in Sec. 9.24b, so that (6) becomes simply

$$N - P = \nu. \tag{6'}$$

In other words, the difference between the total number of zeros and the total number of poles of $f(z)$ inside C equals the net number of times the point $w = f(z)$ winds around the origin $w = 0$ in the positive direction as z makes one circuit around C in the positive direction.

12.2. Rouché's Theorem and Its Implications

12.21. THEOREM *(Rouché). Suppose $f(z)$ and $g(z)$ are analytic inside and on a piecewise smooth closed Jordan curve C, and suppose*

$$|f(z)| > |g(z)| \tag{7}$$

at every point of C. Then $f(z)$ and $f(z) + g(z)$ have the same number of zeros inside C.

Proof. Since $f(z)$ cannot vanish on C, because of (7), we have

$$\begin{aligned}\Delta_C \arg\,[f(z) + g(z)] &= \Delta_C \arg\left\{f(z)\left[1 + \frac{g(z)}{f(z)}\right]\right\}\\ &= \Delta_C \arg f(z) + \Delta_C \arg\left[1 + \frac{g(z)}{f(z)}\right].\end{aligned} \tag{8}$$

But

$$\left|\frac{g(z)}{f(z)}\right| < 1$$

for all $z \in C$, and hence the variable point

$$w = 1 + \frac{g(z)}{f(z)}$$

stays in the disk $|w - 1| < 1$ as z describes the curve C. Therefore w cannot wind around the origin, i.e.,

$$\Delta_C \arg\left[1 + \frac{g(z)}{f(z)}\right] = 0,$$

so that (8) implies

$$\Delta_C \arg\,[f(z) + g(z)] = \Delta_C \arg f(z).$$

The theorem now follows at once from the argument principle. ∎

12.22. Example. How many zeros does the function

$$z^8 - 4z^5 + z^2 - 1 \tag{9}$$

have inside the unit circle $|z| = 1$?

Solution. Writing (9) in the form $f(z) + g(z)$, where

$$f(z) = -4z^5, \qquad g(z) = z^8 + z^2 - 1,$$

we see that $|f(z)| > |g(z)|$ on the circle $|z| = 1$, since

$$|f(z)| = |4z^5| = 4, \qquad |g(z)| = |z^8 + z^2 - 1| \leq |z^8| + |z^2| + 1 = 3$$

if $|z| = 1$. Hence, by Rouché's theorem, the function (9) has the same number of zeros inside the circle $|z| = 1$ as the function $f(z) = -4z^5$, namely 5, since $f(z)$ obviously has 5 zeros inside $|z| = 1$, namely a zero of order 5 at the origin.

12.23. Rouché's theorem leads to a particularly elegant proof of the following key

THEOREM *(Fundamental theorem of algebra). Every polynomial*

$$P(z) = a_0 + a_1 z + \cdots + a_{n-1}z^{n-1} + a_n z^n \qquad (a_n \neq 0)$$

of degree $n \geq 1$ has precisely n zeros.†

Proof. Choosing

$$f(z) = a_n z^n, \qquad g(z) = a_0 + a_1 z + \cdots + a_{n-1}z^{n-1},$$

we have

$$|f(z)| = |a_n|R^n, \qquad |g(z)| \leq |a_0| + |a_1|R + \cdots + |a_{n-1}|R^{n-1}$$

on the circle $|z| = R$. But

$$\lim_{z\to\infty}\left|\frac{f(z)}{g(z)}\right| \geq \lim_{z\to\infty}\frac{|a_n|R^n}{|a_0| + |a_1|R + \cdots + |a_{n-1}|R^{n-1}} = \infty,$$

and hence $|f(z)| > |g(z)|$ on the circle $|z| = R$ for all sufficiently large R. Hence by Rouché's theorem, $P(z)$ has the same number of zeros inside the circle $|z| = R$ as the function $f(z) = a_n z^n$, i.e., precisely n zeros (a zero of order n at $z = 0$). Moreover the circle $|z| = R$ encloses all the zeros of $P(z)$ if R is sufficiently large, since

$$\lim_{z\to\infty} P(z) = \infty.$$

It follows that $P(z)$ has precisely n zeros (in the finite plane).

12.24. Next we use Rouché's theorem to prove an important proposition anticipated in Sec. 9.11:

THEOREM. *If $f(z)$ is univalent in a domain G, then the derivative $f'(z)$ is nonvanishing in G.*

Proof. Given any point $z_0 \in G$, suppose $f'(z_0) = 0$. Then $f(z)$ has a Taylor expansion of the form

$$f(z) = c_0 + c_k(z - z_0)^k + c_{k+1}(z - z_0)^{k+1} + \cdots \qquad (c_k \neq 0, k \geq 2)$$

in some closed disk $|z - z_0| \leq r$ contained in G. The radius of this disk can clearly be chosen so small that $f'(z)$ does not vanish if $0 < |z - z_0| \leq r$, while the sum $\varphi(z)$ of the series

$$c_k + c_{k+1}(z - z_0) + \cdots$$

does not vanish if $0 \leq |z - z_0| \leq r$. In fact, the zero of $f'(z)$ at z_0 is isolated,

†Recall from Sec. 12.13 that each multiple zero is counted a number of times equal to its order.

by Theorem 10.26, while $\varphi(z)$ is nonvanishing in a neighborhood of z_0 by the continuity of $\varphi(z)$, since $\varphi(z_0) = c_k \neq 0$.

Now let†

$$\mu = \min_{|z-z_0|=r} |c_k(z-z_0)^k + c_{k+1}(z-z_0)^{k+1} + \cdots|,$$

and let $-a \neq 0$ be any number of modulus less than μ. Then by Rouché's theorem, the function

$$f(z) - (c_0 + a) = -a + c_k(z-z_0)^k + c_{k+1}(z-z_0)^{k+1} + \cdots$$

has precisely the same number of zeros inside the circle $|z - z_0| = r$ as the function

$$c_k(z-z_0)^k + c_{k+1}(z-z_0)^{k+1} + \cdots = (z-z_0)^k[c_k + c_{k+1}(z-z_0) + \cdots],$$

i.e., precisely k zeros, where each zero is simple since $[f(z) - (c_0 + a)]' = f'(z)$ does not vanish if $0 < |z - z_0| \leq r$. But then $f(z)$ takes the value $c_0 + a$ at $k \geq 2$ distinct points (inside $|z - z_0| = r$), which is impossible since $f(z)$ is univalent in G by hypothesis. This contradiction shows that $f'(z_0) \neq 0$ for all $z_0 \in G$. ∎

12.25. Corollary. *If $f(z)$ is univalent in a domain G, then $f(z)$ is conformal at every point of G.*

Proof. An immediate consequence of the above theorem and the last assertion of Sec. 4.33. ∎

12.3. Evaluation of Improper Real Integrals

We have already used residues to evaluate various complex integrals.‡ The residue method is also a powerful tool for evaluating improper real integrals (with infinite limits), as we now illustrate by a variety of examples:

12.31. Example. Evaluate the integral

$$\int_{-\infty}^{\infty} \frac{dx}{(x^2 + a^2)^3} \qquad (a > 0).$$

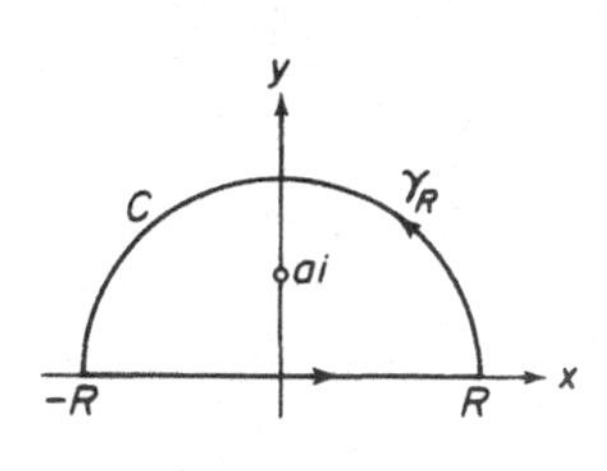

Figure 35

Solution. Consider the function

$$f(z) = \frac{1}{(z^2 + a^2)^3}$$

and the contour C shown in Figure 35, consisting of the segment $-R \leq x \leq R$ of the real axis and a semicircular arc γ_R

†Note that $\mu > 0$ (why?).

‡See Chap. 11, Probs. 23–27, 30–33, and Example 11.33.

of radius $R > a$ in the upper half-plane. The function $f(z)$ has only one singular point inside or on C, namely a pole of order 3 at $z = ai$, with residue

$$\operatorname*{Res}_{z=ai} f(z) = \frac{1}{2!} \lim_{z \to ai} \frac{d^2}{dz^2} \frac{(z - ai)^3}{(z^2 + a^2)^3}$$

$$= \frac{1}{2}\left[\frac{d^2}{dz^2}\frac{1}{(z + ai)^3}\right]_{z=ai} = \frac{1}{2}\frac{3 \cdot 4}{(2ai)^5} = \frac{3}{16a^5 i},$$

by formula (36), p. 166. Hence, by the residue theorem,

$$\int_C f(z)\,dz = \int_{-R}^{R} f(x)\,dx + \int_{\gamma_R} f(z)\,dz = 2\pi i \operatorname*{Res}_{z=ai} f(z) = \frac{3\pi}{8a^5}. \tag{10}$$

But

$$\frac{1}{|z^2 + a^2|} = \frac{1}{|z^2 - (-a^2)|} \leq \frac{1}{||z^2| - |a^2||} = \frac{1}{R^2 - a^2}$$

if $z \in \gamma_R$ (cf. Sec. 1.38), and hence

$$\left|\int_{\gamma_R} f(z)\,dz\right| \leq \frac{\pi R}{(R^2 - a^2)^3},$$

by Theorem 5.23. It follows that

$$\lim_{R \to \infty} \int_{\gamma_R} f(z)\,dz = 0.$$

Therefore, taking the limit as $R \to \infty$ in (10), we get

$$\lim_{R \to \infty} \int_{-R}^{R} f(x)\,dx = \int_{-\infty}^{\infty} f(x)\,dx = \frac{3\pi}{8a^5},$$

i.e.,

$$\int_{-\infty}^{\infty} \frac{dx}{(x^2 + a^2)^3} = \frac{3\pi}{8a^5} \qquad (a > 0).$$

12.32. Example. Evaluate the integral

$$\int_0^{\infty} \frac{\cos x}{x^2 + a^2}\,dx \qquad (a > 0).$$

Solution. Let the contour C be the same as in the preceding example, but this time choose the function

$$f(z) = \frac{e^{iz}}{z^2 + a^2},$$

whose real part coincides with the integrand on the real axis. On the semicircle γ_R we have $|e^{iz}| = e^{-y} \leq 1$ since $y = \operatorname{Im} z \geq 0$. Therefore

$$\left|\int_{\gamma_R} f(z)\,dz\right| \leq \frac{\pi R}{R^2 - a^2}$$

and

$$\lim_{R \to \infty} \int_{\gamma_R} f(z)\, dz = 0,$$

just as before. The function $f(z)$ again has only one singular point inside or on C, namely a simple pole at $z = ai$, with residue

$$\operatorname*{Res}_{z=ai} f(z) = \left[\frac{e^{iz}}{(z^2 + a^2)'}\right]_{z=ai} = \left[\frac{e^{iz}}{2z}\right]_{z=ai} = \frac{e^{-a}}{2ai},$$

by formula (34), p. 166. Therefore, by the residue theorem,

$$\int_C f(z)\, dz = \int_{-R}^{R} \frac{e^{ix}}{x^2 + a^2}\, dx + \int_{\gamma_R} f(x)\, dz = 2\pi i \frac{e^{-a}}{2ai} = \frac{\pi e^{-a}}{a},$$

which gives

$$\int_{-R}^{R} \frac{\cos x}{x^2 + a^2}\, dx + \operatorname{Re} \int_{\gamma_R} f(z)\, dz = \frac{\pi e^{-a}}{a} \qquad (11)$$

after taking real parts. Taking the limit as $R \to \infty$ in (11), we get

$$\lim_{R \to \infty} \int_{-R}^{R} \frac{\cos x}{x^2 + a^2}\, dx = \int_{-\infty}^{\infty} \frac{\cos x}{x^2 + a^2}\, dx = \frac{\pi e^{-a}}{a},$$

and hence finally†

$$\int_0^{\infty} \frac{\cos x}{x^2 + a^2}\, dx = \frac{\pi e^{-a}}{2a} \qquad (a > 0).$$

12.33. Example. Evaluate the integral

$$\int_0^{\infty} \frac{\sin x}{x}\, dx.$$

Solution. Consider the function

$$f(z) = \frac{e^{iz}}{z},$$

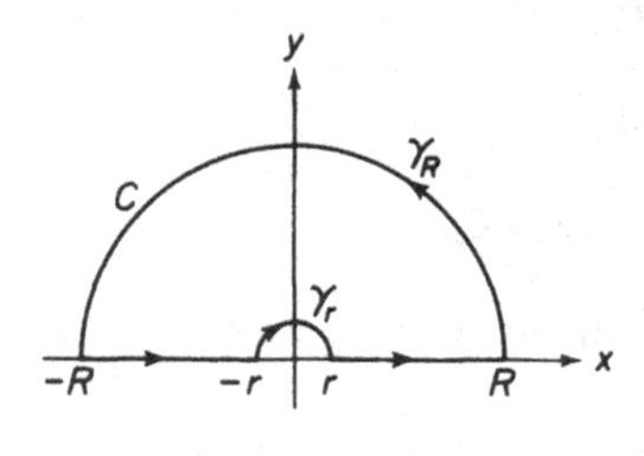

Figure 36

whose imaginary part coincides with the integrand on the real axis. We can no longer integrate $f(z)$ along the contour shown in Figure 35, since $f(z)$ becomes infinite at the origin. Instead we choose C to be the "indented" contour shown in Figure 36, where γ_R is a semicircular arc of radius R, as before, and γ_r is a semicircular arc of radius $r < R$

---

†If $f(x)$ is *even*, i.e., if $f(-x) \equiv f(x)$, then obviously

$$\int_{-\infty}^{\infty} f(x)\, dx = 2 \int_0^{\infty} f(x)\, dx.$$

(also in the upper half-plane). Since $f(z)$ is analytic inside and on C, we have

$$\int_C f(z)\,dz = \int_{-R}^{-r} \frac{e^{ix}}{x}\,dx + \int_{\gamma_r} \frac{e^{iz}}{z}\,dz + \int_r^R \frac{e^{ix}}{x}\,dx + \int_{\gamma_R} \frac{e^{iz}}{z}\,dz = 0, \tag{12}$$

by Cauchy's integral theorem. (Note that the residue theorem reduces to Cauchy's integral theorem if there are no singular points inside C.) Taking the limit of (12) as both $r \longrightarrow 0$ and $R \longrightarrow \infty$, we get

$$\int_{-\infty}^{0} \frac{e^{ix}}{x}\,dx + \lim_{r\to 0} \int_{\gamma_r} \frac{e^{iz}}{z}\,dz + \int_0^{\infty} \frac{e^{ix}}{x}\,dx + \lim_{R\to\infty} \int_{\gamma_R} \frac{e^{iz}}{z}\,dz = 0. \tag{13}$$

To calculate the fourth integral on the left, we integrate by parts,† obtaining

$$\int_{\gamma_R} \frac{e^{iz}}{z}\,dz = \int_{\gamma_R} \frac{d(e^{iz})}{iz} = \frac{e^{iz}}{iz}\Big|_{-R}^{R} + \int_{\gamma_R} \frac{e^{iz}}{iz^2}\,dz = \frac{e^{iR} + e^{-iR}}{iR} + \frac{1}{i}\int_{\gamma_R} \frac{e^{iz}}{z^2}\,dz. \tag{14}$$

But then

$$\left|\int_{\gamma_R} \frac{e^{iz}}{z}\,dz\right| \le \left|\frac{e^{iR} + e^{-iR}}{iR}\right| + \left|\frac{1}{i}\int_{\gamma_R} \frac{e^{iz}}{z^2}\,dz\right| \le \frac{2}{R} + \frac{\pi R}{R^2} \longrightarrow 0$$

as $R \longrightarrow \infty$, since $|e^{iz}| \le 1$ on γ_R, and hence

$$\lim_{R\to\infty} \int_{\gamma_R} \frac{e^{iz}}{z}\,dz = 0. \tag{15}$$

On the other hand, to evaluate

$$\lim_{r\to 0} \int_{\gamma_r} \frac{e^{iz}}{z}\,dz,$$

we note that the Laurent expansion of e^{iz}/z at $z = 0$ is

$$\frac{e^{iz}}{z} = \frac{1 + iz + \dfrac{(iz)^2}{2!} + \dfrac{(iz)^3}{3!} + \cdots}{z} = \frac{1}{z} + P(z),$$

where $P(z)$, the regular part of the expansion, is analytic at $z = 0$. Therefore

$$\lim_{r\to 0} \int_{\gamma_r} \frac{e^{iz}}{z}\,dz = \lim_{r\to 0} \int_{\gamma_r} \frac{dz}{z} + \lim_{r\to 0} \int_{\gamma_r} P(z)\,dz = \lim_{r\to 0} \int_{\gamma_r} \frac{dz}{z},$$

since $|P(z)| \le M$, say, in a neighborhood of $z = 0$, so that

$$\left|\int_{\gamma_r} P(z)\,dz\right| \le M\pi r \longrightarrow 0$$

†Without the preliminary integration by parts, we can only use Theorem 5.23 to deduce that the left-hand side of (14) is of modulus less than $\pi R/R = \pi$, which is too crude an estimate for our purposes.

as $r \to 0$. But

$$\int_{\gamma_r} \frac{dz}{z} = \int_\pi^0 \frac{ire^{i\theta}\, d\theta}{re^{i\theta}} = -\pi i,$$

and hence

$$\lim_{r\to 0} \int_{\gamma_r} \frac{e^{iz}}{z}\, dz = -\pi i. \tag{16}$$

Substituting (15) and (16) into (13), we finally get

$$\int_{-\infty}^0 \frac{\sin x}{x}\, dx + \int_0^\infty \frac{\sin x}{x}\, dx = \pi,$$

after taking imaginary parts, or equivalently

$$\int_0^\infty \frac{\sin x}{x}\, dx = \frac{\pi}{2},$$

since the integrand is even.

12.34. Example. Evaluate the *Fresnel integrals*

$$\int_0^\infty \cos^2 x\, dx, \qquad \int_0^\infty \sin^2 x\, dx.$$

Solution. This time we choose

$$f(z) = e^{iz^2}, \tag{17}$$

since

$$\operatorname{Re} f(x) = \cos x^2, \qquad \operatorname{Im} f(x) = \sin x^2.$$

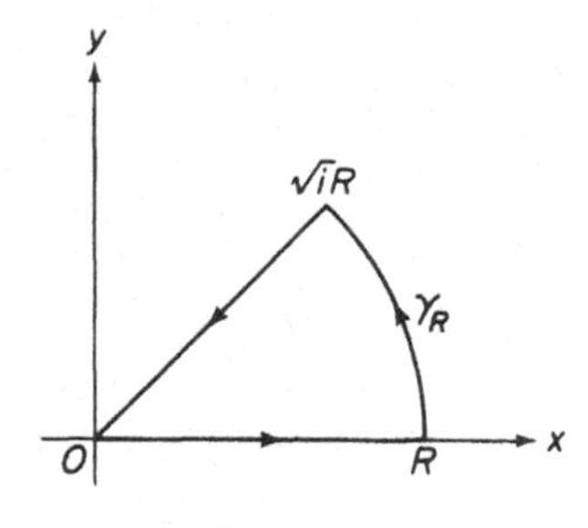

Figure 37

On the ray bisecting the first quadrant of the z-plane, we have $z = \sqrt{i}\, r$ ($r \geq 0$), so that (17) becomes

$$f(\sqrt{i}\, r) = e^{-r^2},$$

the integrand of the familiar integral

$$\int_0^\infty e^{-r^2}\, dr = \frac{\sqrt{\pi}}{2} \tag{18}$$

(see Prob. 12). To exploit this fact, we use the contour of integration C shown in Figure 37. Since $f(z)$ is analytic inside and on C, we have

$$\int_0^R e^{ix^2}\, dx + \int_{\gamma_R} e^{iz^2}\, dz + \int_R^0 e^{-r^2} \sqrt{i}\, dr = 0, \tag{19}$$

by Cauchy's integral theorem (since $z = \sqrt{i}\, r$, $R \geq r \geq 0$ on the segment joining the point $\sqrt{i}\, R$ to the origin). Taking the limit of (19) as $R \to \infty$, we get

$$\int_0^\infty e^{ix^2}\,dx + \lim_{R\to\infty}\int_{\gamma_R} e^{iz^2}\,dz - \sqrt{i}\int_0^\infty e^{-r^2}\,dr = 0. \tag{20}$$

But

$$\int_{\gamma_R} e^{iz^2}\,dz = \int_{\gamma_R}\frac{d(e^{iz^2})}{2iz} = \frac{e^{iz^2}}{2iz}\bigg|_R^{\sqrt{i}R} + \frac{1}{2i}\int_{\gamma_R}\frac{e^{iz^2}}{z^2}\,dz$$

after integration by parts. The modulus of the first term on the right satisfies the inequality

$$\left|\frac{e^{-R^2}}{2i\sqrt{i}\,R} - \frac{e^{iR^2}}{2iR}\right| \le \frac{e^{-R^2}}{2R} + \frac{1}{2R},$$

and hence approaches zero as $R \to \infty$. As for the second term, its integrand is of modulus

$$\left|\frac{e^{iz^2}}{z^2}\right| = \left|\frac{e^{iR^2(\cos 2\theta + i\sin 2\theta)}}{z^2}\right| = \frac{e^{-R^2\sin 2\theta}}{R^2},$$

where we write $z = R(\cos\theta + i\sin\theta)$ on the arc γ_R. But on γ_R we have

$$\sin 2\theta \ge 0, \qquad e^{-R^2\sin 2\theta} \le 1$$

and hence

$$\left|\frac{e^{iz^2}}{z^2}\right| \le \frac{1}{R^2},$$

so that

$$\left|\int_{\gamma_R}\frac{e^{iz^2}}{z^2}\,dz\right| \le \frac{1}{R^2}\frac{\pi R}{4} = \frac{\pi}{4R} \to 0$$

as $R \to \infty$. Thus

$$\lim_{R\to\infty}\int_{\gamma_R} e^{iz^2}\,dz = 0,$$

and (20) becomes

$$\int_0^\infty e^{ix^2}\,dx = \sqrt{i}\int_0^\infty e^{-r^2}\,dr = \sqrt{i}\,\frac{\sqrt{\pi}}{2} = \frac{1+i}{\sqrt{2}}\frac{\sqrt{\pi}}{2} \tag{21}$$

after using (18). Taking real and imaginary parts of (21), we finally get

$$\int_0^\infty \cos^2 x\,dx = \int_0^\infty \sin^2 x\,dx = \frac{1}{2}\sqrt{\frac{\pi}{2}}.$$

12.35. Example. Evaluate the integral

$$\int_{-\infty}^\infty \frac{e^{ax}}{1+e^x}\,dx \qquad (0 < a < 1).$$

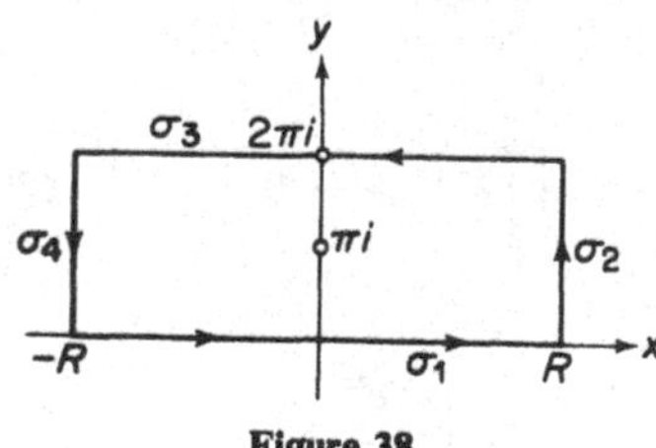

Figure 38

Solution. Consider the function

$$f(x) = \frac{e^{az}}{1 + e^z}$$

and the rectangular contour C shown in Figure 38, consisting of the line segments $\sigma_1, \sigma_2, \sigma_3, \sigma_4$. By the residue theorem,

$$\int_{\sigma_1} f(z)\,dz + \int_{\sigma_2} f(z)\,dz + \int_{\sigma_3} f(z)\,dz + \int_{\sigma_4} f(z)\,dz$$
$$= 2\pi i \operatorname*{Res}_{z=\pi i} \frac{e^{az}}{1+e^z} = 2\pi i \left[\frac{e^{az}}{(1+e^z)'}\right]_{z=\pi i} = -2\pi i e^{a\pi i}. \tag{22}$$

Clearly

$$\int_{\sigma_1} f(z)\,dz = \int_{-R}^{R} \frac{e^{ax}}{1+e^x}\,dx,$$

$$\int_{\sigma_3} f(z)\,dz = \int_{R}^{-R} \frac{e^{a(x+2\pi i)}}{1+e^{x+2\pi i}}\,dx = -e^{2a\pi i}\int_{-R}^{R} \frac{e^{ax}}{1+e^x}\,dx,$$

while

$$|f(z)| = \left|\frac{e^{a(R+iy)}}{1+e^{R+iy}}\right| \le \frac{e^{aR}}{e^R - 1} = \frac{e^{(a-1)R}}{1-e^{-R}} \qquad (z \in \sigma_2),$$

$$|f(z)| = \left|\frac{e^{a(-R+iy)}}{1+e^{-R+iy}}\right| \le \frac{e^{-aR}}{1-e^{-R}} \qquad (z \in \sigma_4).$$

Therefore

$$\left|\int_{\sigma_2} f(z)\,dz\right| \le 2\pi\frac{e^{(a-1)R}}{1-e^{-R}} \longrightarrow 0, \qquad \left|\int_{\sigma_4} f(z)\,dz\right| \le 2\pi\frac{e^{-aR}}{1-e^{-R}} \longrightarrow 0$$

as $R \longrightarrow \infty$ (recall that $0 < a < 1$). Hence, taking the limit as $R \longrightarrow \infty$ in (22), we get

$$\lim_{R\to\infty}\int_{\sigma_1} f(z)\,dz + \lim_{R\to\infty}\int_{\sigma_3} f(z)\,dz$$
$$= (1 - e^{2a\pi i})\int_{-\infty}^{\infty} \frac{e^{ax}}{1+e^x}\,dx = -2\pi i e^{a\pi i},$$

or equivalently

$$\int_{-\infty}^{\infty} \frac{e^{ax}}{1+e^x}\,dx = -2\pi i\frac{e^{a\pi i}}{1-e^{2a\pi i}} = \frac{\pi}{\sin a\pi} \qquad (0 < a < 1).$$

12.4. Integrals Involving Multiple-Valued Functions

The technique of the preceding section often leads to contour integrals involving multiple-valued functions. These are easily handled with a little extra care, as shown by the following two examples:

12.41. Example. Evaluate the integral

$$\int_0^\infty \frac{\ln x}{(x^2+1)^2}\,dx.$$

Solution. Let C be the same contour as in Figure 36 (with $R > 1$), and let

$$f(z) = \frac{\ln z}{(z^2+1)^2},$$

where we choose $\ln z$ to be the branch of the logarithm satisfying the condition

$$-\pi < \operatorname{Im} \ln z = \arg z \leq \pi.$$

The function $f(z)$ is analytic at every point of C and its interior except at the point $z = i$, where it has a pole of order 2 with residue

$$\operatorname*{Res}_{z=i} f(z) = \left\{\frac{d}{dz}\left[\frac{\ln z}{(z+i)^2}\right]\right\}_{z=i} = \frac{\pi + 2i}{8}.$$

Hence, by the residue theorem,

$$\int_{-R}^{-r} f(z)\,dz + \int_{\gamma_r} f(z)\,dz + \int_r^R f(z)\,dz + \int_{\gamma_R} f(z)\,dz$$
$$= 2\pi i \frac{\pi + 2i}{8} = \frac{\pi^2 i}{4} - \frac{\pi}{2}. \tag{23}$$

If $z = Re^{i\theta}$ $(0 \leq \theta \leq \pi)$, then

$$|\ln z| = |\ln|z| + i \arg z| = \sqrt{\ln^2 R + \theta^2} \leq \sqrt{\ln^2 R + \pi^2} \leq 2 \ln R$$

for sufficiently large R, and hence

$$\left|\int_{\gamma_R} f(z)\,dz\right| \leq \frac{2 \ln R}{(R^2-1)^2}\pi R \longrightarrow 0$$

as $R \longrightarrow \infty$, while if $z = re^{i\theta}$ $(\pi \geq \theta \geq 0)$, then

$$|\ln z| \leq 2 \ln \frac{1}{r}$$

for sufficiently small r, and hence

$$\left|\int_{\gamma_r} f(z)\,dz\right| \leq \frac{2 \ln \frac{1}{r}}{(1-r^2)^2}\pi r \longrightarrow 0$$

as $r \longrightarrow 0$ (cf. Prob. 13). Hence, taking the limit as $r \longrightarrow 0$, $R \longrightarrow \infty$ in (23), we get

$$\int_{-\infty}^0 \frac{\ln x}{(x^2+1)^2}\,dx + \int_0^\infty \frac{\ln x}{(x^2+1)^2}\,dx = \frac{\pi^2 i}{4} - \frac{\pi}{2}. \tag{24}$$

But $\ln(-x) = \ln x + \pi i$, and hence

$$\int_{-\infty}^{0} \frac{\ln x}{(x^2+1)^2}\,dx = \int_0^\infty \frac{\ln x}{(x^2+1)^2}\,dx + \pi i \int_0^\infty \frac{dx}{(x^2+1)^2}.$$

Thus (24) becomes

$$2\int_0^\infty \frac{\ln x}{(x^2+1)^2}\,dx + \pi i \int_0^\infty \frac{dx}{(x^2+1)^2} = \frac{\pi^2 i}{4} - \frac{\pi}{2}, \tag{25}$$

which implies

$$\int_0^\infty \frac{\ln x}{(x^2+1)^2}\,dx = -\frac{\pi}{4}$$

after taking real parts.†

12.42. Example. Evaluate the integral

$$\int_0^\infty \frac{x^{a-1}}{1+x}\,dx \qquad (0 < a < 1).$$

Solution. Let C be the contour shown in Figure 39, made up of arcs γ_r, γ_R of the circles $|z| = r$, $|z| = R$ and segments σ_1, σ_2 of the rays $\arg z = \epsilon$, $\arg z = 2\pi - \epsilon$, and let

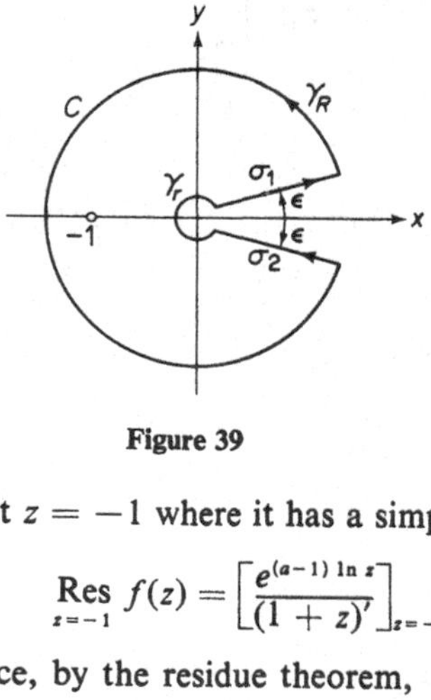

$$f(z) = \frac{z^{a-1}}{1+z} = \frac{e^{(a-1)\ln z}}{1+z}, \tag{26}$$

where this time we choose $\ln z$ to be the branch of the logarithm satisfying the condition

$$0 \le \operatorname{Im} \ln z = \arg z < 2\pi. \tag{27}$$

Figure 39

The function $f(z)$ is single-valued and analytic inside and on C, except at the point $z = -1$ where it has a simple pole with residue

$$\operatorname*{Res}_{z=-1} f(z) = \left[\frac{e^{(a-1)\ln z}}{(1+z)'}\right]_{z=-1} = e^{(a-1)\ln(-1)} = e^{(a-1)\pi i} = -e^{a\pi i}.$$

Hence, by the residue theorem,

$$\int_{\sigma_1} f(z)\,dz + \int_{\gamma_R} f(z)\,dz + \int_{\sigma_2} f(z)\,dz + \int_{\gamma_r} f(z)\,dz = -2\pi i e^{a\pi i}$$

†Taking imaginary parts of (25), we get the elementary integral

$$\int_0^\infty \frac{dx}{(x^2+1)^2} = \frac{\pi}{4}.$$

or equivalently

$$\int_r^R f(\rho e^{i\epsilon})\,d(\rho e^{i\epsilon}) + \int_{\gamma_R} f(z)\,dz + \int_R^r f(\rho e^{i(2\pi-\epsilon)})\,d(\rho e^{i(2\pi-\epsilon)}) + \int_{\gamma_r} f(z)\,dz = -2\pi i e^{a\pi i}. \tag{28}$$

Using (26), we can write (28) in the form

$$\begin{aligned} e^{i\epsilon}\int_r^R \frac{\rho^{a-1}e^{i(a-1)\epsilon}}{1+\rho e^{i\epsilon}}\,d\rho &+ \int_{\gamma_R} f(z)\,dz \\ &- e^{i(2\pi-\epsilon)}\int_r^R \frac{\rho^{a-1}e^{i(a-1)(2\pi-\epsilon)}}{1+\rho e^{i(2\pi-\epsilon)}}\,d\rho + \int_{\gamma_r} f(z)\,dz \\ = e^{i\epsilon+i(a-1)\epsilon}&\int_r^R \frac{\rho^{a-1}}{1+\rho e^{i\epsilon}}\,d\rho + \int_{\gamma_R} f(z)\,dz \\ &- e^{-i\epsilon-i(a-1)\epsilon}e^{2(a-1)\pi i}\int_r^R \frac{\rho^{a-1}}{1+\rho e^{-i\epsilon}}\,d\rho + \int_{\gamma_r} f(z)\,dz = -2\pi i e^{a\pi i}. \end{aligned} \tag{29}$$

Moreover

$$|f(z)| \le \frac{R^{a-1}}{R-1} \qquad (z \in \gamma_R,\ R > 1),$$

$$|f(z)| \le \frac{r^{a-1}}{1-r} \qquad (z \in \gamma_r,\ r < 1),$$

so that

$$\left|\int_{\gamma_R} f(z)\,dz\right| \le \frac{R^{a-1}}{R-1}(2\pi - 2\epsilon)R,$$

$$\left|\int_{\gamma_r} f(z)\,dz\right| \le \frac{r^{a-1}}{1-r}(2\pi - 2\epsilon)r,$$

and hence

$$\lim_{R\to\infty}\lim_{\epsilon\to 0}\int_{\gamma_R} f(z)\,dz = 0,$$

$$\lim_{r\to 0}\lim_{\epsilon\to 0}\int_{\gamma_r} f(z)\,dz = 0$$

(recall that $0 < a < 1$). Therefore, taking the limit first as $\epsilon \to 0$ and then as $r \to 0$, $R \to \infty$ in (29), we get

$$\int_0^\infty \frac{\rho^{a-1}}{1+\rho}\,d\rho - e^{2(a-1)\pi i}\int_0^\infty \frac{\rho^{a-1}}{1+\rho}\,d\rho = -2\pi i e^{a\pi i}, \tag{30}$$

i.e.,

$$\int_0^\infty \frac{\rho^{a-1}}{1+\rho}\,d\rho = -2\pi i\frac{e^{a\pi i}}{1-e^{2(a-1)\pi i}}.$$

But

$$-2\pi i \frac{e^{a\pi i}}{1 - e^{2(a-1)\pi i}} = 2\pi i \frac{1}{e^{a\pi i} - e^{-a\pi i}} = \frac{\pi}{\sin a\pi},$$

and hence finally

$$\int_0^\infty \frac{x^{a-1}}{1+x}\,dx = \frac{\pi}{\sin a\pi} \qquad (0 < a < 1). \tag{31}$$

Note that the substitution $x = e^t$ immediately reduces (31) to the integral evaluated in Example 12.35.

COMMENTS

12.1. The kind of reasoning involved in the argument principle is typical of "geometric function theory," a subject standing at the junction of complex analysis and the geometry (and topology) of curves and other sets in the complex plane. The considerations of Secs. 13.3 and 13.5 belong to this category.

12.2. There is another version of the fundamental theorem of algebra in which it is asserted only that $P(z)$ has at least one zero. An elementary argument then shows (give the details) that $P(z)$ has precisely n zeros. For another proof of the fundamental theorem of algebra, see Prob. 8.

12.3. It will be recalled from advanced calculus that the improper real integrals

$$\int_a^\infty f(x)\,dx, \qquad \int_{-\infty}^a f(x)\,dx, \qquad \int_{-\infty}^\infty f(x)\,dx$$

are defined as the limits

$$\lim_{X\to\infty}\int_a^X f(x)\,dx, \qquad \lim_{X\to-\infty}\int_X^a f(x)\,dx, \qquad \lim_{\substack{X\to-\infty\\ X'\to\infty}}\int_X^{X'} f(x)\,dx,$$

respectively, where in the last case X and X' approach $-\infty$ and ∞ *independently*. In this regard, see also Prob. 21.

12.4. In setting

$$\lim_{\epsilon\to 0}\int_r^R \frac{\rho^{a-1}}{1+\rho e^{\pm i\epsilon}}\,d\rho = \int_r^R \frac{\rho^{a-1}}{1+\rho}\,d\rho$$

in one of the steps leading to (30), we rely on the following theorem of advanced calculus: If

$$I(y) = \int_a^b f(x, y)\,dx,$$

where $f(x, y)$ is a continuous function of two variables in the rectangle $a \le x \le b, \alpha \le y \le \beta$, then $I(y)$ is a continuous function of y in the interval $\alpha \le y \le \beta$.

PROBLEMS

1. With the same notation as in Theorem 12.12, prove that

$$\frac{1}{2\pi i}\int_C \varphi(z)\frac{f'(z)}{f(z)}\,dz = \sum_{k=1}^{m} \alpha_k\varphi(a_k) - \sum_{k=1}^{n} \beta_k\varphi(b_k)$$

if $\varphi(z)$ is analytic inside and on C.

2. Recalling the definition of A-points from Chap. 10, Prob. 23, prove the following generalization of Theorem 12.12: Given a piecewise smooth closed Jordan curve C, suppose $f(z)$ is analytic inside and on C except for poles inside C at the points $b_1, \ldots, b_m$. Moreover, suppose $f(z)$ has A-points $a_1, \ldots, a_m$ inside C, but none on C itself. Then

$$\frac{1}{2\pi i}\int_C \frac{f'(z)}{f(z) - A}\,dz = \sum_{k=1}^{m} \alpha_k - \sum_{k=1}^{n} \beta_k,$$

where α_k is the order of a_k and β_k the order of b_k. State and prove the corresponding generalization of the argument principle.

3. Find the "winding numbers" n_+, n_- and ν (see Sec. 12.14) for the curve Γ shown in Figure 40.

4. How many zeros does each of the following functions have inside the unit circle $|z| = 1$:
a) $z^7 - 5z^4 + z^2 - 2$; b) $2z^5 - z^3 + 3z^2 - z + 8$;
c) $z^9 - 2z^6 + z^2 - 8z - 2$?

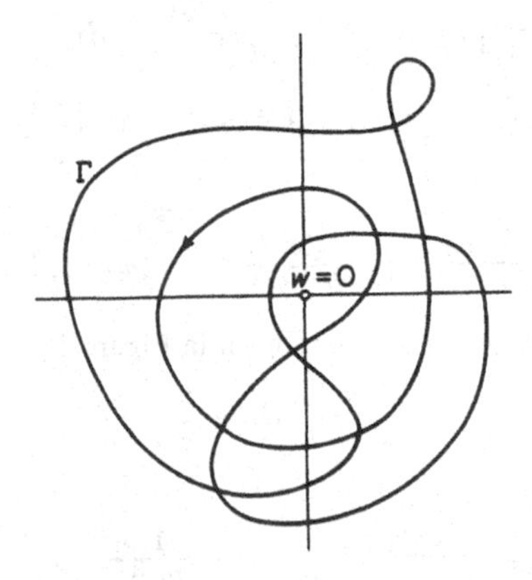

Figure 40

5. How many zeros does the function $z^4 - 5z + 1$ have in the annulus $1 < |z| < 2$?

6. How many roots does the equation $f(z) = z$ have in the open disk $|z| < 1$ if $f(z)$ is analytic and of modulus less than 1 in the closed disk $|z| \leq 1$?

7. Prove that the equation

$$z + e^{-z} = \lambda \qquad (\lambda > 1)$$

has one and only one root z_0 in the right half-plane, where z_0 is real.

8. Use Liouville's theorem to give an alternative proof of the fundamental theorem of algebra.

9. Show that the converse of Corollary 12.25 is false.

10. Suppose $f(z)$ is analytic at a point z_0, and let $w_0 = f(z_0)$. Prove that there is a neighborhood K of z_0 and a corresponding neighborhood K^* of w_0 such that $f(z) - w$ has at least one zero in K for every $w \in K^*$.

11. Use the preceding problem to prove the following generalization of Theorem 9.12: If $f(z) \not\equiv$ constant is analytic in a domain G and if E is the image of G under the mapping $w = f(z)$, then E is also a domain (in the w-plane). Use this to give another proof of the maximum modulus principle.

12. Use a double integral to prove that

$$I = \int_0^\infty e^{-x^2}\,dx = \frac{\sqrt{\pi}}{2}.$$

13. Prove that $R^\alpha \ln R \longrightarrow 0$ as $R \longrightarrow \infty$ if $\alpha < 0$, while $r^\alpha \ln r \longrightarrow 0$ as $r \longrightarrow 0$ if $\alpha > 0$.

14. Use integration along the contour shown in Figure 35 to evaluate the following integrals:

a) $\displaystyle\int_{-\infty}^\infty \frac{x}{(x^2 + 4x + 13)^2}\,dx$; b) $\displaystyle\int_0^\infty \frac{x^2}{(x^2 + a^2)^2}\,dx \quad (a > 0)$;

c) $\displaystyle\int_{-\infty}^\infty \frac{dx}{(x^2 + a^2)(x^2 + b^2)} \quad (a > 0, b > 0)$; d) $\displaystyle\int_0^\infty \frac{x^2 + 1}{x^4 + 1}\,dx$.

15. Prove that

$$\int_{-\infty}^\infty \frac{dx}{(1 + x^2)^{n+1}} = \frac{(2n)!}{(n!)^2 2^{2n}}\pi \qquad (n = 1, 2, \ldots).$$

16. Use integration along the contour shown in Figure 36 to show that

$$\int_0^\infty \frac{\sin^2 x}{x^2}\,dx = \frac{\pi}{2}.$$

17. Verify the formula

$$\int_0^\infty \frac{x^{2m}}{1 + x^{2n}}\,dx = \frac{\pi}{n}\,\frac{1}{\sin \dfrac{2m + 1}{2n}\pi}, \tag{32}$$

where m and n are positive integers such that $m < n$.

18. Verify the formula

$$\int_{-\infty}^{\infty} \frac{x^{2m} - x^{2m'}}{1 - x^{2n}}\, dx = \frac{\pi}{n}\left(\cot \frac{2m+1}{2n}\pi - \cot \frac{2m'+1}{2n}\pi\right), \tag{33}$$

where m, m' and n are positive integers such that $m, m' < n$.

19. Show that (32) is a special case of (33).

20. Evaluate the following integrals:

a) $\displaystyle\int_0^{\infty} \frac{\cos x}{(x^2 + a^2)(x^2 + b^2)}\, dx \quad (a > 0, b > 0)$; b) $\displaystyle\int_0^{\infty} \frac{x \sin x}{(x^2 + a^2)^2}\, dx \quad (a > 0)$;

c) $\displaystyle\int_0^{\infty} \frac{\cos ax}{(x^2 + b^2)^2}\, dx \quad (a > 0, b > 0)$.

21. Suppose $f(x)$ becomes infinite at a point c where $a < c < b$. Then it will be recalled from calculus that the improper integral

$$\int_a^b f(x)\, dx \tag{34}$$

is defined as the limit

$$\lim_{\substack{\delta \to 0 \\ \epsilon \to 0}} \left\{\int_a^{c-\delta} f(x)\, dx + \int_{c+\epsilon}^b f(x)\, dx\right\} \qquad (\delta, \epsilon > 0)$$

as δ and ϵ approach zero *independently*.† The limit

$$\lim_{\epsilon \to 0} \left\{\int_a^{c-\epsilon} f(x)\, dx + \int_{c+\epsilon}^b f(x)\, dx\right\},$$

corresponding to the choice $\delta = \epsilon$, is called (*Cauchy*) *principal value* of the integral (34) and is denoted by

$$\text{P.V.} \int_a^b f(x)\, dx. \tag{35}$$

If (34) exists (i.e., converges), then obviously (35) also exists and equals (34), but (35) may well exist even though (34) fails to exist (i.e., diverges). For example, the integral

$$\int_{-1}^{1} \frac{dx}{x}$$

diverges, but its principal value exists and equals 0.

Calculate the principal value of the divergent integral

$$\int_{-\infty}^{\infty} \frac{\sin x}{(x^2 + 4)(x - 1)}\, dx.$$

22. In Example 12.42 we found that

$$\int_0^{\infty} \frac{x^{a-1}}{1 + x}\, dx = \frac{\pi}{\sin a\pi} \qquad (0 < a < 1).$$

Show that the same result can be obtained formally by applying the residue

†An extra limiting process is required if $a = -\infty$ or $b = +\infty$ (give the details).

theorem to the contour C^* which we get by setting $\epsilon = 0$ from the outset in Figure 39, thereby closing the "gap" in the contour C, although the new contour C^* is no longer a Jordan curve (draw a figure) and the function

$$f(z) = \frac{z^{a-1}}{1+z}$$

is no longer single-valued on C^* (why not?).

Comment. Direct use of "degenerate" contours like C^* often facilitates the evaluation of integrals involving multiple-valued functions by eliminating extra steps (like the passage to the limit as $\epsilon \to 0$ in Example 12.42).

23. Evaluate the following integrals:

a) $\displaystyle\int_0^\infty \frac{\ln^2 x}{1+x^2}\,dx$; b) $\displaystyle\int_0^\infty \frac{\ln x}{(x+a)^2+b^2}\,dx \quad (a > 0, b > 0)$;

c) $\displaystyle\int_0^\infty \frac{x^2 \ln x}{(1+x^2)^2}\,dx$; d) $\displaystyle\int_0^\infty \frac{x^a}{1+x^2} \quad (-1 < a < 1)$;

e) P.V. $\displaystyle\int_0^\infty \frac{x^{a-1}}{1-x}\,dx \quad (0 < a < 1)$.

CHAPTER THIRTEEN

Further Theory

13.1. More on Harmonic Functions

13.11. Resuming our study of harmonic functions, we begin by finding Fourier series expansions for a pair of conjugate harmonic functions:

THEOREM. *Let K be the disk $|z - z_0| < R$, and let $u = u(z)$ be a function harmonic in K with harmonic conjugate $v = v(z)$. Then u and v have expansions of the form*

$$u = u(z_0 + re^{i\varphi}) = a_0 + \sum_{n=1}^{\infty} (a_n \cos n\varphi - b_n \sin n\varphi)r^n, \tag{1}$$

$$v = v(z_0 + re^{i\varphi}) = b_0 + \sum_{n=1}^{\infty} (b_n \cos n\varphi + a_n \sin n\varphi)r^n, \tag{1'}$$

valid for all $0 \leq r < R$, $0 \leq \varphi \leq 2\pi$, where the convergence is uniform in every closed disk $0 \leq r \leq R' < R$, $0 \leq \varphi \leq 2\pi$.

Proof. As in Sec. 5.85, let $f(z)$ be the function (unique to within a purely imaginary constant) which is analytic in K and has u as its real part. Then by Theorem 10.13, $f(z)$ has a Taylor expansion

$$f(z) = \sum_{n=0}^{\infty} c_n(z - z_0)^n$$

in K, where the convergence is uniform in every closed disk $|z - z_0| \leq R' < R$ (by Theorem 7.17). To obtain (1), we substitute

$$c_n = a_n + ib_n, \qquad z - z_0 = re^{i\varphi}$$

into (2) and then take the real part of the resulting expansion

$$f(z) = \sum_{n=0}^{\infty} (a_n + ib_n)r^n e^{in\varphi}. \tag{2'}$$

To get (1′), we take the imaginary part of (2′). ∎

13.12. Remark. The expressions $u(z_0 + re^{i\varphi})$ and $v(z_0 + re^{i\varphi})$ figuring in (1) and (1′) will henceforth be written simply as $u(r, \varphi)$ and $v(r, \varphi)$. Moreover, as in Sec. 5.8, $u(x, y)$ and $v(x, y)$ will continue to mean the values of $u = u(z)$ and $v = v(z)$ at the point $z = x + iy$. In this sense, we have

$$u(z) = u(x, y) = u(r, \varphi), \qquad v(z) = v(x, y) = v(r, \varphi)$$

if $z = x + iy = z_0 + re^{i\varphi}$ (the slight abuse of notation being amply justified by its convenience), so that (1) and (1′) can be written as

$$u(r, \varphi) = a_0 + \sum_{n=1}^{\infty} (a_n \cos n\varphi - b_n \sin n\varphi)r^n, \tag{3}$$

$$v(r, \varphi) = b_0 + \sum_{n=1}^{\infty} (b_n \cos n\varphi + a_n \sin n\varphi)r^n. \tag{3'}$$

13.13. Example. The function

$$F(z) = \frac{Re^{i\theta} + (z - z_0)}{Re^{i\theta} - (z - z_0)}$$

is analytic in the disk $|z - z_0| < R$ (but in no larger disk). It follows from Theorem 5.82 that the real and imaginary parts of $F(z)$ are a pair of conjugate harmonic functions in the same disk. Writing $z - z_0 = re^{i\varphi}$ $(r < R)$, we have

$$\begin{aligned} F(z) = U(r, \varphi) + iV(r, \varphi) &= \frac{Re^{i\theta} + re^{i\varphi}}{Re^{i\theta} - re^{i\varphi}} = \frac{Re^{i\theta} + re^{i\varphi}}{Re^{i\theta} - re^{i\varphi}}\,\frac{Re^{-i\theta} - re^{-i\varphi}}{Re^{-i\theta} - re^{-i\varphi}} \\ &= \frac{R^2 - r^2 + 2iRr \sin(\varphi - \theta)}{R^2 + r^2 - 2Rr \cos(\varphi - \theta)} \end{aligned} \tag{4}$$

on the one hand and

$$\frac{Re^{i\theta} + re^{i\varphi}}{Re^{i\theta} - re^{i\varphi}} = \frac{1 + \dfrac{r}{R}e^{i(\varphi-\theta)}}{1 - \dfrac{r}{R}e^{i(\varphi-\theta)}} = -1 + \frac{2}{1 - \dfrac{r}{R}e^{i(\varphi-\theta)}} \tag{4'}$$

on the other, where the right-hand side of (4′) can be recognized at once as the sum of the convergent series

$$-1 + 2\left[1 + \frac{r}{R}e^{i(\varphi-\theta)} + \frac{r^2}{R^2}e^{2i(\varphi-\theta)} + \cdots\right]. \tag{5}$$

Equating real and imaginary parts of (4) and (5), we immediately get

$$U(r, \varphi) = \frac{R^2 - r^2}{R^2 + r^2 - 2Rr\cos(\varphi - \theta)}$$
$$= 1 + 2\sum_{n=1}^{\infty}\left(\frac{r}{R}\right)^n \cos n(\varphi - \theta), \tag{6}$$

$$V(r, \varphi) = \frac{2Rr\sin(\varphi - \theta)}{R^2 + r^2 - 2Rr\cos(\varphi - \theta)}$$
$$= 2\sum_{n=1}^{\infty}\left(\frac{r}{R}\right)^n \sin n(\varphi - \theta). \tag{6'}$$

These expansions are of the form (3) and (3′), with

$$a_0 = 1, \quad b_0 = 0, \quad a_n = \frac{2\cos n\theta}{R^n}, \quad b_n = -\frac{2\sin n\theta}{R^n} \qquad (n = 1, 2, \ldots).$$

Thus the series (6) and (6′) are uniformly convergent in every closed disk $0 \leq r \leq R' < R$, $0 \leq \varphi \leq 2\pi$, by Theorem 13.11. Multiplying (6) and (6′) by any function

$$\frac{1}{2\pi}u(R, \theta)$$

and using Theorem 6.38 and Chap. 6, Prob. 12 to integrate the resulting series term by term with respect to θ from 0 to 2π,† we find that

$$\frac{1}{2\pi}\int_0^{2\pi} u(R, \theta)\frac{R^2 - r^2}{R^2 + r^2 - 2Rr\cos(\varphi - \theta)}\,d\theta$$
$$= \frac{1}{2\pi}\int_0^{2\pi} u(R, \theta)\,d\theta + \frac{1}{\pi}\sum_{n=1}^{\infty}\int_0^{2\pi} u(R, \theta)\left(\frac{r}{R}\right)^n \cos n(\varphi - \theta)\,d\theta, \tag{7}$$

$$\frac{1}{2\pi}\int_0^{2\pi} u(R, \theta)\frac{2Rr\sin(\varphi - \theta)}{R^2 + r^2 - 2Rr\cos(\varphi - \theta)}\,d\theta$$
$$= \frac{1}{\pi}\sum_{n=1}^{\infty}\int_0^{2\pi} u(R, \theta)\left(\frac{r}{R}\right)^n \sin n(\varphi - \theta)\,d\theta. \tag{7'}$$

The last two formulas will be needed in a moment.

13.14. THEOREM. *Let $u = u(r, \varphi)$ be a function harmonic in the disk $0 \leq r < \rho$, $0 \leq \varphi \leq 2\pi$, with harmonic conjugate $v = v(r, \varphi)$. Then u and v both satisfy* ***Poisson's integral formula***

$$u(r, \varphi) = \frac{1}{2\pi}\int_0^{2\pi} u(R, \theta)\frac{R^2 - r^2}{R^2 + r^2 - 2Rr\cos(\varphi - \theta)}\,d\theta, \tag{8}$$

$$v(r, \varphi) = \frac{1}{2\pi}\int_0^{2\pi} v(R, \theta)\frac{R^2 - r^2}{R^2 + r^2 - 2Rr\cos(\varphi - \theta)}\,d\theta \tag{8'}$$

---

†We assume that $u(R, \theta)$ is continuous (and hence bounded) in the interval $0 \leq \theta \leq 2\pi$ for every fixed R. Note that (6) and (6′) are uniformly convergent in the interval $0 \leq \theta \leq 2\pi$ for fixed r $(< R)$ and φ.

for all $0 \leq r < R < \rho, 0 \leq \varphi \leq 2\pi$. *Moreover* v *is related to* u *by the formula*

$$v(r, \varphi) = b_0 + \frac{1}{2\pi}\int_0^{2\pi} u(R, \theta)\frac{2Rr\sin(\varphi - \theta)}{R^2 + r^2 - 2Rr\cos(\varphi - \theta)}\,d\theta, \tag{9}$$

where b_0 *is an arbitrary real constant.*

Proof. Replacing r by R ($< \rho$), φ by θ, and n by m in (3), we get

$$u(R, \theta) = a_0 + \sum_{m=1}^{\infty} (a_m \cos m\theta - b_m \sin m\theta)R^m, \tag{10}$$

where the convergence is uniform in every closed disk $0 \leq R \leq \rho' < \rho$, $0 \leq \theta \leq 2\pi$. Hence we can multiply (10) first by $\cos n\theta$ ($n = 0, 1, 2, \ldots$) and then by $\sin n\theta$ ($n = 1, 2, \ldots$), afterwards integrating term by term with respect to θ from 0 to 2π. This gives

$$a_0 = \frac{1}{2\pi}\int_0^{2\pi} u(R, \theta)\,d\theta, \qquad a_n = \frac{1}{\pi R^n}\int_0^{2\pi} u(R, \theta)\cos n\theta\,d\theta \quad (n = 1, 2, \ldots), \tag{11}$$

$$-b_n = \frac{1}{\pi R^n}\int_0^{2\pi} u(R, \theta)\sin n\theta\,d\theta \qquad (n = 1, 2, \ldots) \tag{11'}$$

(see Prob. 2). Substituting (11) and (11′) into (3) and (3′), we find that

$$u(r, \varphi) = \frac{1}{2\pi}\int_0^{2\pi} u(R, \theta)\,d\theta + \frac{1}{\pi}\sum_{n=1}^{\infty}\int_0^{2\pi} u(R, \theta)\left(\frac{r}{R}\right)^n \cos n(\varphi - \theta)\,d\theta, \tag{12}$$

$$v(r, \varphi) = b_0 + \frac{1}{\pi}\sum_{n=1}^{\infty}\int_0^{2\pi} u(R, \theta)\left(\frac{r}{R}\right)^n \sin n(\varphi - \theta)\,d\theta. \tag{12'}$$

Comparing (12) and (12′) with (7) and (7′), we immediately get (8) and (9). Moreover, since (8) holds for an arbitrary function harmonic in the given disk, we can replace u by v in (8), obtaining (8′). ∎

Setting $u \equiv 1$ in (8), we get the useful formula

$$\frac{1}{2\pi}\int_0^{2\pi} \frac{R^2 - r^2}{R^2 + r^2 - 2Rr\cos(\varphi - \theta)}\,d\theta = 1. \tag{13}$$

13.15. COROLLARY. *Let* $f(z)$ *be a function analytic in the disk* $|z - z_0| < \rho$, *with* $u(r, \varphi)$ *as its real part. Then*

$$f(z) = ib_0 + \frac{1}{2\pi}\int_0^{2\pi} u(R, \theta)\frac{Re^{i\theta} + (z - z_0)}{Re^{i\theta} - (z - z_0)}\,d\theta \tag{14}$$

(b_0 *real*, $|z - z_0| < R < \rho$), *a representation known as* ***Schwarz's formula.***

Proof. Multiply (9) by i and add the result to (8), afterwards using Example 13.13.

13.16. COROLLARY. *Let $u = u(z) = u(r, \varphi)$ be a function harmonic in the disk $|z - z_0| < \rho$. Then*

$$u(z_0) = \frac{1}{2\pi}\int_0^{2\pi} u(R, \theta)\, d\theta$$

($R < \rho$), i.e., the value of the function u at the point z_0 is the average of its values on the circle $|z - z_0| = R$.

Proof. Set $r = 0$ in formula (8). Alternatively, take the real part of formula (26), p. 146. ∎

13.2. The Dirichlet Problem

13.21. Definition. Let G be a *Jordan domain*, i.e., a domain whose boundary is a closed Jordan curve C, and let $h(z)$ be a continuous real function defined on C. Consider the problem of finding a function $u(z)$ harmonic in G such that

$$\lim_{\substack{z \to z_0 \\ z \in G}} u(z) = h(z_0) \tag{15}$$

for every $z_0 \in C$. This is the *Dirichlet problem* for G, of great importance both in complex analysis and mathematical physics.

13.22. Remark. If (15) holds, we say that "$u(z)$ takes the boundary values $h(z)$ on C." The function equal to $u(z)$ in G and $h(z)$ on C is then automatically continuous in $\bar{G}$, as well as harmonic in G.

13.23. We begin by solving the Dirichlet problem for a disk, specifically the disk of unit radius centered at the origin. Thus let G be the disk $|z| < 1$ and C the circle $|z| = 1$. If $h(z)$ coincides with the values taken on C by some function $u(z)$ which is harmonic in a disk $|z| < \rho$ of radius larger than 1, then since

$$u(re^{i\varphi}) = \frac{1 - r^2}{2\pi}\int_0^{2\pi} \frac{u(e^{i\theta})}{1 + r^2 - 2r\cos(\varphi - \theta)}\, d\theta$$

($r < 1$) by Theorem 13.14, the Dirichlet problem has the obvious solution

$$u(re^{i\varphi}) = \frac{1 - r^2}{2\pi}\int_0^{2\pi} \frac{h(e^{i\theta})}{1 + r^2 - 2r\cos(\varphi - \theta)}\, d\theta, \tag{16}$$

where the solution (16) is unique by Corollary 10.33d. As we now show, the solution of the Dirichlet problem for G is given by the very same formula (16) even when $h(e^{i\theta})$ is an *arbitrary* continuous function.

THEOREM. *Let G be the unit disk $|z| < 1$ and C the unit circle $|z| = 1$, and let $h(z) = h(e^{i\theta})$ be a continuous real function on C. Then the function* (16) *is the unique solution of the Dirichlet problem for G taking the boundary values $h(e^{i\theta})$ on C.*

Proof. Clearly (16) is the real part of the function

$$f(z) = \frac{1}{2\pi}\int_0^{2\pi} h(e^{i\theta})\frac{e^{i\theta}+z}{e^{i\theta}-z}\,d\theta,$$

obtained by setting $u(R, \theta) = h(e^{i\theta})$, $R = 1$, $z_0 = 0$, $b_0 = 0$ in Schwarz's formula (14). But $f(z)$ is analytic in G by Chap. 5, Prob. 28, since the integrand is obviously analytic in G for every θ in the interval $[0, 2\pi]$ and continuous in both variables z and θ for all $z \in G$, $\theta \in [0, 2\pi]$. It follows from Theorem 5.82 that $u(z)$ is harmonic in G.

The nub of the proof is to show that $u(re^{i\varphi}) \to h(e^{i\varphi_0})$ as $r \to 1$, $\varphi \to \varphi_0$ $(0 < r < 1,\ \varphi_0$ fixed). First we observe that

$$u(re^{i\varphi}) - h(\varphi) = \frac{1-r^2}{2\pi}\int_0^{2\pi}\frac{h(\theta)-h(\varphi)}{1+r^2-2r\cos(\varphi-\theta)}\,d\theta,$$

by (13) and (16), where for simplicity we write $h(\varphi)$ instead of $h(e^{i\varphi})$. But then

$$u(re^{i\varphi}) - h(\varphi) = \frac{1-r^2}{2\pi}\int_{-\pi}^{\pi}\frac{h(\varphi+\alpha)-h(\varphi)}{1+r^2-2r\cos\alpha}\,d\alpha,$$

since the integrand is periodic with period 2π (see Prob. 7). Let δ be any number such that $0 < \delta < \pi$, and let

$$M = \max_{0\le\varphi\le 2\pi}|h(\varphi)|, \qquad \omega(\delta, \varphi) = \max_{|\alpha|\le\delta}|h(\varphi+\alpha)-h(\varphi)|.$$

Then

$$\begin{aligned}\left|\frac{1-r^2}{2\pi}\int_{-\delta}^{\delta}\frac{h(\varphi+\alpha)-h(\varphi)}{1+r^2-2r\cos\alpha}\,d\alpha\right| &\le \omega(\delta,\varphi)\frac{1-r^2}{2\pi}\int_{-\delta}^{\delta}\frac{d\alpha}{1+r^2-2r\cos\alpha}\\ &< \omega(\delta,\varphi)\frac{1-r^2}{2\pi}\int_{-\pi}^{\pi}\frac{d\alpha}{1+r^2-2r\cos\alpha} = \omega(\delta,\varphi),\end{aligned}$$

where we again use (13), while†

$$\begin{aligned}\left|\frac{1-r^2}{2\pi}\int_{\delta\le|\alpha|\le\pi}\frac{h(\varphi+\alpha)-h(\varphi)}{1+r^2-2r\cos\alpha}\,d\alpha\right| &\le 2M\frac{1-r^2}{2\pi}\int_{\delta\le|\alpha|\le\pi}\frac{d\alpha}{1+r^2-2r\cos\alpha}\\ &\le 2M\frac{2(\pi-\delta)}{2\pi}\,\frac{1-r^2}{1+r^2-2r\cos\delta}\\ &< 2M\frac{\pi-\delta}{\pi}\,\frac{1-r^2}{2r-2r\cos\delta} < \frac{M}{r}\,\frac{1-r^2}{1-\cos\delta}.\end{aligned}$$

†The expression

$$\int_{\delta\le|\alpha|\le\pi} g(\varphi)\,d\varphi$$

is shorthand for

$$\int_{-\pi}^{-\delta} g(\varphi)\,d\varphi + \int_{\delta}^{\pi} g(\varphi)\,d\varphi.$$

It follows that

$$|u(re^{i\varphi}) - h(\varphi)| = \left|\frac{1-r^2}{2\pi}\int_{-\pi}^{\pi}\frac{h(\varphi+\alpha)-h(\varphi)}{1+r^2-2r\cos\alpha}\,d\alpha\right|$$
$$\leq \left|\frac{1-r^2}{2\pi}\int_{-\delta}^{\delta}\frac{h(\varphi+\alpha)-h(\varphi)}{1+r^2-2r\cos\alpha}\,d\alpha\right|$$
$$+\left|\frac{1-r^2}{2\pi}\int_{\delta\leq|\alpha|\leq\pi}\frac{h(\varphi+\alpha)-h(\varphi)}{1+r^2-2r\cos\alpha}\,d\alpha\right|$$
$$\leq \omega(\delta,\varphi)+\frac{M}{r}\frac{1-r^2}{1-\cos\delta},$$

and hence

$$|u(re^{i\varphi}) - h(\varphi_0)| \leq \omega(\delta,\varphi)+\frac{M}{r}\frac{1-r^2}{1-\cos\delta}+|h(\varphi)-h(\varphi_0)|, \tag{17}$$

where both $\omega(\delta,\varphi)\to 0$ as $\delta\to 0$ and $|h(\varphi)-h(\varphi_0)|\to 0$ as $\varphi\to\varphi_0$, by the continuity of $h(z)$ on C. Now let

$$\delta = \sqrt[4]{1-r^2}.$$

Then $\delta\to 0$ as $r\to 1$, while

$$\frac{M}{r}\frac{1-r^2}{1-\cos\delta} = \frac{2M}{r}\sqrt{1-r^2}\left(1+\frac{1}{12}\sqrt{1-r^2}+\cdots\right)\to 0$$

as $r\to 1$. Hence the right-hand side of (17) approaches zero as $r\to 1$, $\varphi\to\varphi_0$ or equivalently as $z = re^{i\varphi}\to e^{i\varphi_0}$, i.e.,

$$\lim_{z\to e^{i\varphi_0}} u(z) = h(\varphi_0),$$

as required. To complete the proof, we note that the uniqueness again follows from Corollary 10.33d. ∎

13.24. Once having established Theorem 13.23, we can now easily solve the Dirichlet problem for a half-plane:

THEOREM. *Let G be the upper half-plane* $\operatorname{Im} z > 0$ *and C the real axis, and let $h(z) = h(x)$ be a continuous real function on C. Then the function*

$$u(z) = \frac{y}{\pi}\int_{-\infty}^{\infty}\frac{h(\xi)}{(\xi-x)^2+y^2}\,d\xi \tag{18}$$

($z = x+iy$) is the unique solution of the Dirichlet problem for G taking the boundary values $h(x)$ on C.

Proof. By Example 8.29a, the function

$$w = f(z) = \frac{z-\zeta}{z-\bar{\zeta}} \qquad (\operatorname{Im}\zeta > 0) \tag{19}$$

maps the upper half-plane $\operatorname{Im} z > 0$ onto the disk $|w| < 1$, while carrying the point $z = \zeta$ into the point $w = 0$ and the real axis $-\infty < x < \infty$ into the circle $|w| = 1$. Let $z = \varphi(w)$ be the inverse of the fractional linear transformation (19). Then the function $h^*(w) = h(\varphi(w))$ is continuous on the circle $|w| = 1$ (why?). Let $u^*(w)$ be the unique solution of the Dirichlet problem for the disk $|w| < 1$ and the boundary values $h^*(w)$ given by Theorem 13.23, and let $u(z) = u^*(f(z))$. Then $u(z)$ is harmonic in G, being a harmonic function of an analytic function (see Chap. 5, Prob. 26), and takes the required boundary values $h^*(f(x)) = h(x)$ on C.

To get an explicit formula for $u(z)$, we choose $u = u^*$, $h = h^*$, $r = 0$ in (16), obtaining

$$u^*(0) = \frac{1}{2\pi}\int_0^{2\pi} h^*(e^{i\theta})\,d\theta. \tag{16'}$$

The point on the circle $|w| = 1$ corresponding to the point x on the real axis C is just

$$e^{i\theta} = \frac{x-\zeta}{x-\bar{\zeta}} \qquad (x \text{ real}),$$

so that

$$ie^{i\theta}\,d\theta = \frac{\zeta-\bar{\zeta}}{(x-\bar{\zeta})^2}\,dx,$$

and hence

$$d\theta = \frac{1}{i}\frac{x-\bar{\zeta}}{x-\zeta}\frac{\zeta-\bar{\zeta}}{(x-\bar{\zeta})^2}\,dx = \frac{2\eta}{|x-\zeta|^2}\,dx = \frac{2\eta}{(x-\xi)^2+\eta^2}\,dx,$$

where $\zeta = \xi + i\eta$. Therefore, expressing (16′) in terms of the variables ξ, η and x, we find that

$$u(\zeta) = u^*(0) = \frac{\eta}{\pi}\int_{-\infty}^{\infty}\frac{h(x)}{(x-\xi)^2+\eta^2}\,dx.$$

To get (18), we now merely replace x, ξ, η and ζ by ξ, x, y and z. ∎

13.3. More on Conformal Mapping

13.31. Let $w = f(z)$ be a univalent (i.e., one-to-one analytic) function in a domain G, which it maps onto a domain G^* (recall Theorem 9.12). Then $f(z)$ is called a *conformal mapping* of G onto G^*, the conformality of $f(z)$ at every point of G being guaranteed by Corollary 12.25; by the same token, G^* is called a *conformal image* of G. Note that by our definition, a function analytic in G and conformal at every point of G need not be a conformal mapping of G, since it may fail to be one-to-one in G (cf. Chap. 12, Prob. 9). The term "conformal mapping" is also used in a looser sense, meaning the branch of complex analysis involving problems in which conformality plays a key role.

13.32. THEOREM. *If G^* is a conformal image of G, then G is a conformal image of G^*. Moreover if G^* is a conformal image of G and G^{**} is a conformal image of G^*, then G^{**} is a conformal image of G.*†

Proof. Let $w = f(z)$ be a conformal mapping of G onto G^*, with inverse $z = \varphi(w)$, and let $\zeta = g(w)$ be a conformal mapping of G^* onto G^{**}. Then $z = \varphi(w)$ is a conformal mapping of G^* onto G, since $\varphi(w)$ is univalent in G^* by Theorem 9.13, while $\zeta = g(f(z))$ is a conformal mapping of G onto G^{**}, since a univalent function of a univalent function is itself univalent (why?). ∎

13.33. THEOREM. *If G^* and G^{**} are both conformal images of G, then G^{**} is a conformal image of G^* and conversely.*

Proof. Let $w = f(z)$ be a conformal mapping of G onto G^*, with inverse $z = \varphi(w)$, and let $\zeta = g(z)$ be a conformal mapping of G onto G^{**}, with inverse $z = \psi(\zeta)$. Then $\zeta = g(\varphi(w))$ is a conformal mapping of G^* onto G^{**}, while $w = f(\psi(\zeta))$ is a conformal mapping of G^{**} onto G^* ∎

13.34. Let G be a disk or a half-plane, and let z_0 be any point of G. Then, according to Sec. 8.29, G can be mapped conformally onto the unit disk $|w| < 1$ by a fractional linear transformation satisfying the conditions

$$f(z_0) = 0, \qquad f'(z_0) > 0. \tag{20}$$

This simple fact is susceptible to the following far-reaching generalization, which will not be proved here‡:

THEOREM (*Riemann*). *Let G be any simply connected domain in the extended plane whose boundary contains more than one point, and let z_0 be any point of G. Then there exists a unique univalent function $w = f(z)$ which maps G conformally onto the disk $|w| < 1$ and satisfies the conditions* (20).

The reason for the stipulation that the boundary of G contain more than one point is clear. Let Π_{z_0} be the domain equal to the whole extended plane minus the single point z_0, and suppose $w = f(z)$ is a conformal mapping of Π_{z_0} onto the unit disk $|w| < 1$. Then the function

$$g(z) = f\left(\frac{1}{z} + z_0\right)$$

is a conformal mapping of the whole finite plane Π_∞ onto the same disk. But no such function can exist, since if $|g(z)| < 1$ for all finite z, then $g(z)$ is

†Here, of course, G, G^* and G^{**} are all domains.

‡For the proof of the existence of $f(z)$, see e.g., A. I. Markushevich, *op. cit.*, Volume III, Theorem 1.2. The uniqueness part of the proof is elementary, and is indicated in Probs. 10–11.

a bounded entire function and hence constant, by Liouville's theorem, so that $g(z)$ obviously fails to be univalent in Π_∞!

13.35. Actually, the most general function univalent in Π_{z_0} is not hard to find:

THEOREM. *If $f(z)$ is univalent in Π_{z_0}, then $f(z)$ is a fractional linear transformation.*

Proof. Clearly z_0 cannot be a removable singular point of $f(z)$, since otherwise $f(z)$ can be made analytic in the whole extended plane (see Sec. 11.22) and hence bounded in the whole finite plane. But then $f(z)$ would be constant, by Liouville's theorem, which is impossible. Therefore $f(z)$ has either an essential singular point or a pole at z_0. Let $z_1 \neq z_0$, $A = f(z_1)$, and let K be any neighborhood of z_1 contained in Π_{z_0} and hence not containing z_0. Then $f(z)$ maps K into a domain K^* in the w-plane containing A and hence containing some neighborhood $|w - A| < \epsilon$ (recall Theorem 9.12). But then the function $w = f(z)$, being univalent, cannot take any value within ϵ of A in any deleted neighborhood of z_0 which does not intersect K, contrary to Theorem 11.26. Therefore z_0 cannot be an essential singular point, so that z_0 must be a pole. This pole must be simple, since otherwise, by Theorem 11.24b, the function

$$\varphi(z) = \begin{cases} \dfrac{1}{f(z)} & \text{if } z \neq z_0, \\ 0 & \text{if } z = z_0, \end{cases}$$

which is clearly univalent in a neighborhood of z_0, would have a multiple zero at z_0, so that $\varphi'(z_0) = 0$ contrary to Theorem 12.24. Thus the principal part of the Laurent expansion of $f(z)$ at z_0 is of the form

$$\frac{a}{z - z_0}$$

if z_0 is finite. But then

$$f(z) - \frac{a}{z - z_0}$$

is a bounded entire function (why?), and hence equals a constant, say b, again by Liouville's theorem, i.e., $f(z)$ is just the fractional linear transformation

$$f(z) = \frac{a}{z - z_0} + b. \tag{21}$$

If $z_0 = \infty$, then the principal part of $f(z)$ at z_0 is of the form az (see Chap. 11, Prob. 16), and we get the entire linear transformation

$$f(z) = az + b \tag{21'}$$

instead of (21). Note that in either case, $f(z)$ maps Π_{z_0} onto the whole finite plane Π_∞. ∎

13.36. The following proposition is a straightforward generalization of Riemann's theorem:

THEOREM. *Let G and G^* be any two simply connected domains in the extended plane, each with a boundary containing more than one point. Let z_0 be any point of G and w_0 any point of G^*. Then there exists a unique univalent function $w = f(z)$ which maps G conformally onto G^* and satisfies the conditions*

$$f(z_0) = w_0, \qquad f'(z_0) > 0. \tag{20'}$$

Proof. Let K be the disk $|\zeta| < 1$, let $\zeta = g(z)$ be the conformal mapping of G onto K with inverse $z = \varphi(\zeta)$ such that

$$g(z_0) = 0, \qquad g'(z_0) > 0$$

(whose existence is guaranteed by Riemann's theorem), and let $\zeta = h(w)$ be the conformal mapping of G^* onto K with inverse $w = \psi(\zeta)$ such that

$$h(w_0) = 0, \qquad h'(w_0) > 0.$$

Then

$$w = f(z) = \psi(g(z))$$

is a conformal mapping of G onto G^* such that

$$f(z_0) = \psi(g(z_0)) = \psi(0) = w_0,$$
$$f'(z_0) = \psi'(g(z_0))g'(z_0) = \psi'(0)g'(z_0) = \frac{g'(z_0)}{h'(z_0)} > 0$$

(cf. Theorem 9.13). ∎

13.37. The following important proposition, which we cite without proof, concerns the "boundary behavior" of conformal mappings of Jordan domains:

THEOREM. *Let $w = f(z)$ be a conformal mapping of a Jordan domain G with boundary C onto another Jordan domain G^* with boundary C^*,† and define $f(z)$ on C by setting*

$$f(z_0) = \lim_{\substack{z \to z_0 \\ z \in G}} f(z)$$

for all $z_0 \in C$. Then $f(z)$ is continuous in $\bar{G}$ and maps C onto C^*. This mapping of C onto C^* is one-to-one and "direction-preserving," i.e., as the point z traverses C, the image point $w = f(z)$ traverses C^* in the same direction.

---

†The existence of such a mapping follows from Theorem 13.36.

13.38. We are now in a position to solve the Dirichlet problem for an arbitrary Jordan domain:

THEOREM. *Let G be any Jordan domain with boundary C, and let $h(z)$ be a continuous real function on C. Then there exists a unique function harmonic in G taking the boundary values $h(z)$ on C.*

Proof. Let $w = f(z)$ be a conformal mapping of G onto the unit disk $|w| < 1$, with inverse $z = \varphi(w)$, and use Theorem 13.37 to define $\varphi(w)$ on the circle $|w| = 1$. Then the function $h^*(w) = h(\varphi(w))$ is continuous on the circle $|w| = 1$ (why?). Let $u^*(w)$ be the unique solution of the Dirichlet problem for the disk $|w| < 1$ and the boundary values $h^*(w)$ given by Theorem 13.23, and let $u(z) = u^*(f(z))$. Then $u(z)$ is harmonic in G (Chap. 5, Prob. 26) and takes the required boundary values $h^*(f(z)) = h(z)$ on C. ∎

13.4. Analytic Continuation

13.41. Given a domain G, let E be a subset of G with a limit point in G, and let $f(z)$ be a function defined in E. Suppose there exists an analytic function $\varphi(z)$ defined in G such that $f(z) = \varphi(z)$ for all $z \in E$. Then $\varphi(z)$ is called the *analytic continuation* of $f(z)$ from E into G. Note that the uniqueness of $\varphi(z)$ is guaranteed by Theorem 10.23.

13.42. Examples

a. Let E be the real axis, and let $f(x) = e^x$. Then

$$f(x) = \sum_{n=0}^{\infty} \frac{x^n}{n!} \qquad (-\infty < x < \infty), \tag{22}$$

so that the function

$$\varphi(z) = \sum_{n=0}^{\infty} \frac{z^n}{n!} \qquad (|z| < \infty) \tag{22'}$$

obtained by replacing x by z in (22) is the unique analytic continuation of $f(x)$ into the whole finite plane. This is of course entirely in keeping with the considerations of Sec. 8.11, where we set $\varphi(z) = e^z$ by definition.

b. Let E be the unit disk $|z| < 1$, let

$$f(z) = \sum_{n=0}^{\infty} z^n,$$

and let

$$\varphi(z) = \frac{1}{1 - z} \qquad (z \neq 1).$$

Then $f(z) = \varphi(z)$ for all $z \in E$, so that $\varphi(z)$ is the analytic continuation of $f(z)$ into the domain equal to the whole finite plane minus the single point $z = 1$.

c. Let E be the unit disk $|z| < 1$, and let

$$f(z) = \sum_{n=1}^{\infty} z^{n!} = z + z^2 + z^6 + z^{24} + \cdots. \tag{23}$$

Then the unit circle $|z| = 1$ is the "natural boundary" of the series (23), i.e., there is no analytic continuation of $f(z)$ into a larger domain G containing E. In fact, if there were such a continuation, then clearly G would contain some arc γ of the unit circle, and hence the limit

$$\lim_{r \to 1-} f(re^{2\pi i\alpha}) \tag{24}$$

would certainly be finite for all $e^{2\pi i\alpha} \in \gamma$. But this is impossible since γ must contain points $e^{2\pi i\alpha}$ with rational α and, as shown in Prob. 17, the limit (24) is infinite for all such points.

13.43. Next we consider a variant of analytic continuation involving "overlapping" domains. Let $\{G, f(z)\}$ be a set consisting of a domain G and a (single-valued) analytic function defined in G. Then $\{G, f(z)\}$ is called an *element*, with *domain* G. Two elements $\{G_1, f_1(z)\}$ and $\{G_2, f_2(z)\}$ are said to be *equal* if $G_1 = G_2$ and $f_1(z) \equiv f_2(z)$. Each of two elements $\{G_1, f_1(z)\}$ and $\{G_2, f_2(z)\}$ is said to be a *direct analytic continuation* of the other if $D = G_1 \cap G_2$ is a domain and if $f_1(z) = f_2(z)$ for all $z \in D$.† Note that in this case the function

$$\varphi(z) = \begin{cases} f_1(z) & \text{for all } z \in G_1, \\ f_2(z) & \text{for all } z \in G_2 \end{cases} \tag{25}$$

is the analytic continuation of both $f_1(z)$ and $f_2(z)$ into the domain $G = G_1 \cup G_2$, as defined in Sec. 13.41.

13.44. Examples

a. Let G_k be the domain‡

$$\frac{(k-1)\pi}{2} < \arg z < \frac{(k+1)\pi}{2} \qquad (k = 0, \pm 1, \pm 2, \ldots), \tag{26}$$

and let $f_k(z)$ be the function

$$f_k(z) = \ln|z| + i\theta_k,$$

where θ_k is the value of $\arg z$ satisfying the condition (26). Then the elements

†Given two sets A and B, by the *intersection* of A and B, denoted by $A \cap B$, we mean the set of all points belonging to both A and B (as in Sec. 10.23), while by the *union* of A and B, denoted by $A \cup B$, we mean the set of all points belonging to at least one of the sets A and B. Two sets A and B are said to be *disjoint* if their intersection $A \cap B$ is "empty," i.e., if they have no points in common.

‡Thus G_0 is the right half-plane, G_1 the upper half-plane, G_2 the left half-plane, G_3 the lower half-plane, G_4 the right half-plane again, etc.

$\{G_k, f_k(z)\}$ and $\{G_l, f_l(z)\}$ are direct analytic continuations of each other if and only if l takes one of the values $k-1$, k, $k+1$.

b. Let G_1 be the unit disk $|z| < 1$, and let

$$f_1(z) = \sum_{n=0}^{\infty} z^n = \frac{1}{1-z}, \tag{27}$$

as in Example 13.42b. Given any point $z_0 \in G_1$, let

$$\sum_{n=0}^{\infty} \frac{f_1^{(n)}(z_0)}{n!}(z-z_0)^n \tag{28}$$

be the Taylor expansion of $f_1(z)$ at z_0. Since

$$f_1^{(n)}(z_0) = \frac{n!}{(1-z_0)^{n+1}},$$

the radius of convergence of the series (28) is just

$$R = \lim_{n\to\infty} \sqrt[n]{|1-z_0|^{n+1}} = |1-z_0|$$

(by the Cauchy–Hadamard theorem), namely the distance between the points 1 and z_0. This is only natural (cf. Sec. 10.16), since $z = 1$ is the unique singular point of the function $1/(1-z)$. Let G_2 be the disk $|z - z_0| < R$, and let $f_2(z)$ denote the sum of the series (28). Then the elements $\{G_1, f_1(z)\}$ and $\{G_2, f_2(z)\}$ are obviously direct analytic continuations of each other, by the very meaning of (28). If the point $z_0 \in G_1$ lies on the positive real axis, then $|1 - z_0| = 1 - |z_0|$, so that G_1 contains G_2 and hence $G = G_1 \cup G_2 = G_1$, as in Figure 41a. In this case, the function (25) does not serve to continue the original function $f_1(z)$ from G_1 into a larger domain. On the other hand, for any other $z_0 \in G_1$ we have $|1 - z_0| > 1 - |z_0|$, so that G_2 "breaks through" the circle of convergence $|z| = 1$ of the original series (27), as in Figure 41b. In this case, the function (25) represents the analytic continuation of $f_1(z)$ from G_1 into the larger domain $G = G_1 \cup G_2$.

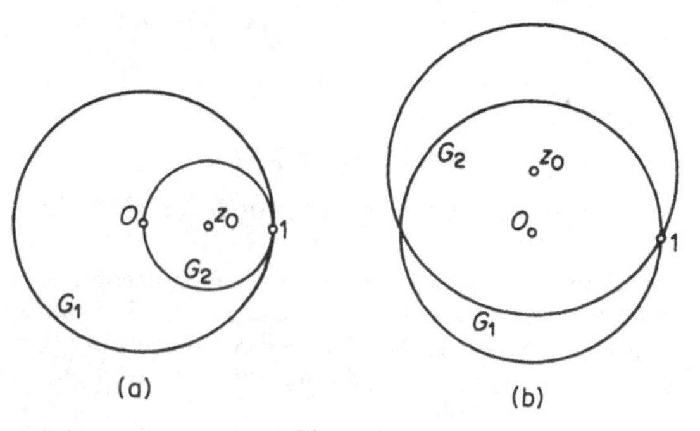

Figure 41

13.45. A set of elements

$$\{G_1, f_1(z)\}, \{G_2, f_2(z)\}, \ldots, \{G_n, f_n(z)\}$$

such that $\{G_{k+1}, f_{k+1}(z)\}$ is a direct analytic continuation of $\{G_k, f_k(z)\}$ for every $k = 1, 2, \ldots, n - 1$ is called a *chain of elements* joining $\{G_1, f_1(z)\}$ and $\{G_n, f_n(z)\}$.† Each of a pair of elements joined by a chain of elements is called an *analytic continuation* of the other. Clearly two elements which are direct analytic continuations of each other are automatically analytic continuations of each other, but two elements which are analytic continuations of each other are in general not direct analytic continuations of each other.

A finite or infinite set of elements F is said to be *connected* if every pair of elements in F can be joined by a chain of elements all belonging to F. In particular, if F is connected, then every element of F is clearly an analytic continuation of every other element of F. Such a connected set of elements F is called a *general analytic function*, and is said to be "generated" by any of its elements. The union of the domains of all the elements of a general analytic function F is itself a domain (why?), called the *domain* of F. Given a general analytic function F, let z_0 be any point in the domain of F. Then by a *value* of F at z_0, denoted by $F(z_0)$, we mean any value $f(z_0)$ where $\{G, f(z)\} \in F$, $z_0 \in G$. Note that F is in general *multiple-valued*, since F may well contain two elements $\{G, f(z)\}$ and $\{G^*, f^*(z)\}$ such that $f(z_0) \neq f^*(z_0)$ at some point $z_0 \in G \cap G^*$, provided of course that $\{G, f(z)\}$ and $\{G^*, f^*(z)\}$ are not *direct* analytic continuations of each other.

13.46. Examples

a. Let G_k and $f_k(z)$ be the same as in Example 13.44a. Then the set of all elements

$$\{G_k, f_k(z)\} \qquad (k = 0, \pm 1, \pm 2, \ldots)$$

is a general analytic function F, since any pair of elements $\{G_k, f_k(z)\}$ and $\{G_l, f_l(z)\}$ are joined by a chain of elements in F (which one?) and hence are analytic continuations of each other (as already noted, the continuations are direct if and only if l takes one of the values $k - 1, k, k + 1$). The domain of F is the whole finite plane minus the single point $z = 0$. If $z_0 \in G_{k_0}$, then $z_0 \in G_{k_0+4k} = G_{k_0}$, but $f_{k_0+4k}(z_0) = f_{k_0}(z_0) + 2\pi k i$. Thus F takes infinitely many values differing by integral multiples of $2\pi i$ at every point $z_0 \in D$. Obviously $F(z)$ should be identified with the multiple-valued function $\ln z$, since every G_k is a domain of univalence of $\ln z$ and every $f_k(z)$ is a single-valued branch of $\ln z$.

†A chain of "circular elements," i.e., elements with disks as their domains, is shown in Figure 31, in connection with the proof of Theorem 10.23.

b. Let G_1 and $f_1(z)$ be the same as in Example 13.44b, and let F be the set of all circular elements (i.e., elements with disks as their domains) which are analytic continuations of the element $\{G_1, f_1(z)\}$. Then F is a general analytic function with the same domain D as in the preceding example, but this time $F(z)$ should be identified with the single-valued function $1/(1 - z)$. Thus a general analytic function F can be either single-valued or multiple-valued if its domain D is multiply connected, as shown by these two examples. However, if D is simply connected, it can be shown† that F is necessarily single-valued, a result known as the *monodromy theorem.*

13.5. The Symmetry Principle

13.51. Next we consider an important kind of analytic continuation involving "adjacent" domains:

THEOREM. *Let G_1 and G_2 be two disjoint domains whose boundaries share a common piecewise smooth Jordan arc γ. Suppose $f_1(z)$ is analytic in G_1 and continuous in $G_1 \cup \gamma$, while $f_2(z)$ is analytic in G_2 and continuous in $G_2 \cup \gamma$, and suppose the set $D = G_1 \cup \gamma \cup G_2$ is a domain.‡ Moreover, suppose $f_1(z)$ and $f_2(z)$ coincide on γ. Then the function*

$$\varphi(z) = \begin{cases} f_1(z) & \text{if } z \in G_1, \\ f_1(z) = f_2(z) & \text{if } z \in \gamma, \\ f_2(z) & \text{if } z \in G_2 \end{cases} \tag{29}$$

is analytic in D, a fact summarized by calling $f_2(z)$ the analytic continuation of $f_1(z)$ from G_1 into G_2 ***across the arc*** *γ.*

Proof. Let C be any piecewise smooth closed Jordan curve contained in D, traversed in the positive direction. If C does not intersect γ, then either C is contained in G_1 or C is contained in G_2, so that

$$\int_C \varphi(z)\, dz = 0 \tag{30}$$

by Cauchy's integral theorem, where $\varphi(z)$ is the function (29). On the other hand, if C intersects γ, we divide C into two arcs C_1 and C_2 with end points

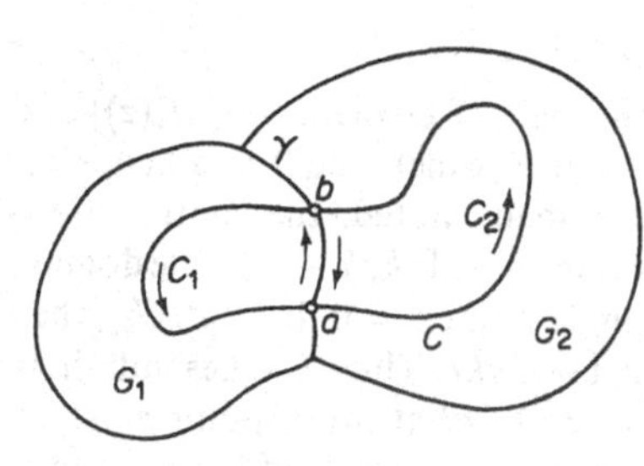

Figure 42

†A. I. Markushevich, *op. cit.*, Volume III, Theorem 8.5.

‡This is not an automatic consequence of the meaning of G_1, γ and G_2, but holds, for example, if G_1 and G_2 are Jordan domains (*ibid.*, Sec. 6).

$a \in \gamma$, $b \in \gamma$, as in Figure 42. It then follows from the *generalized* Cauchy integral theorem (see Sec. 5.42) that

$$\int_{C_1+\widehat{ab}} f_1(z)\,dz = \int_{C_2+\widehat{ba}} f_2(z)\,dz = 0,$$

and hence

$$\begin{aligned} 0 &= \int_{C_1} f_1(z)\,dz + \int_{\widehat{ab}} f_1(z)\,dz + \int_{C_2} f_2(z)\,dz + \int_{\widehat{ba}} f_2(z)\,dz \\ &= \int_{C_1} f_1(z)\,dz + \int_{C_2} f_2(z)\,dz, \end{aligned}$$

since the integrals over $\widehat{ab}$ and $\widehat{ba}$ cancel each other out because of Theorem 5.21 and the fact that $f_1(z)$ and $f_2(z)$ coincide on γ. But the sum on the right is just

$$\int_{C_1} \varphi(z)\,dz + \int_{C_2} \varphi(z)\,dz = \int_{C} \varphi(z)\,dz,$$

so that (30) again holds. Thus the integral of $\varphi(z)$ along any piecewise smooth closed Jordan curve C contained in D vanishes. But then $\varphi(z)$ is analytic in D, by Morera's theorem (Theorem 5.73).† ∎

13.52. We are now in a position to prove one of the most useful tools of complex analysis:

THEOREM ***(Symmetry principle).*** *Let G_1 be a domain whose boundary contains a circular arc or line segment γ, and let G_2 be the domain symmetric to G_1 with respect to γ,‡ where it is assumed that G_1 and G_2 are disjoint and that the set $G_1 \cup \gamma \cup G_2$ is a domain. Given a function $f_1(z)$ univalent in G_1 and continuous in $G_1 \cup \gamma$, suppose $f_1(z)$ maps G_1 conformally onto a domain G_1^* while carrying γ into a curve γ^* (part of the boundary of G_1^*), which is itself a circular arc or line segment. Then $f_1(z)$ has an analytic continuation from G_1 into G_2 across γ, and this analytic continuation $f_2(z)$ maps G_2 conformally onto the domain G_2^* symmetric to G_1^* with respect to γ^*. Moreover, if $\varphi(z)$ is the function* (29), *then $\varphi(z)$ is a conformal mapping of the domain $G_1 \cup \gamma \cup G_2$ onto the domain $G_1^* \cup \gamma^* \cup G_2^*$.*

Proof. The various domains and arcs figuring in the statement of the theorem are shown schematically in Figure 43. Let

$$\zeta = \frac{az+b}{cz+d} = l_1(z), \qquad \omega = \frac{\alpha w+\beta}{\gamma w+\delta} = l_2(z) \tag{31}$$

be fractional linear transformations carrying γ and γ^* into segments δ and

---

†Note that $\varphi(z)$ is the analytic continuation of $f_1(z)$ from the domain G_1 (or, for that matter, from the arc γ) into the domain D, in the sense of Sec. 13.41.

‡Thus G_2 is the set of all points symmetric to the points of G_1 with respect to γ (why is G_2 a domain?).

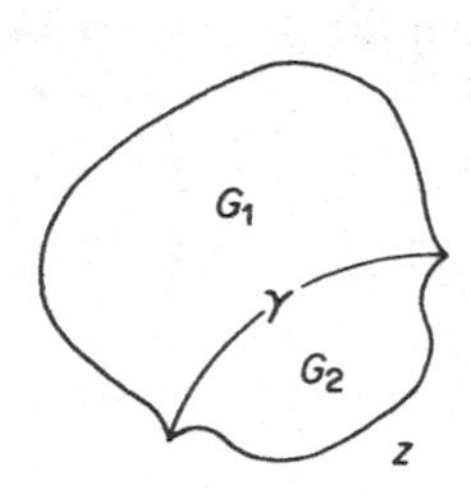

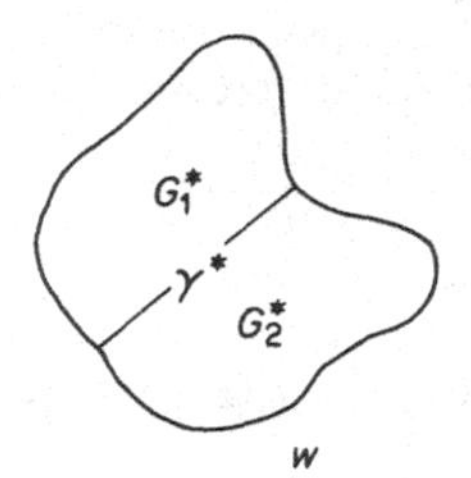

Figure 43

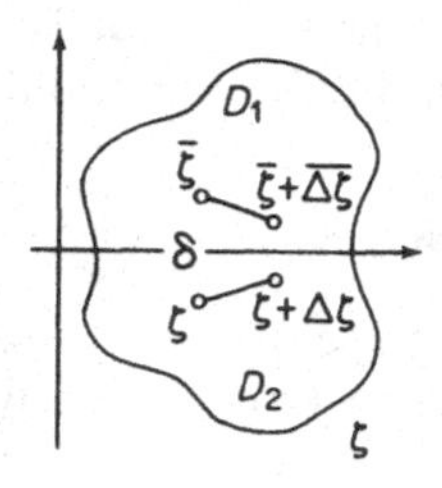

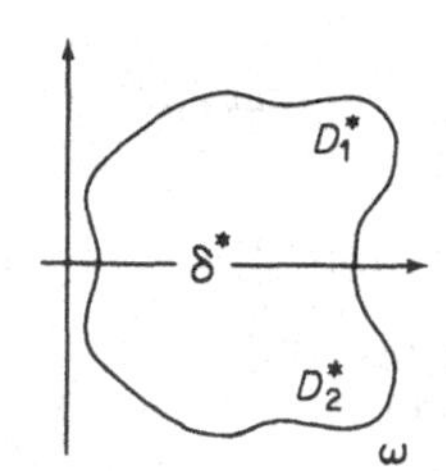

Figure 44

δ^* of the real axes in the ζ- and ω-planes, while carrying G_1 and G_1^* into domains D_1 and D_1^* respectively, as shown in Figure 44 (the existence of such transformations is guaranteed by Corollary 8.26b). If $z = \lambda_1(\zeta)$ is the inverse of $\zeta = l_1(z)$ then the function $w = l_2(f_1(\lambda_1(\zeta))) = g_1(\zeta)$ maps D_1 conformally onto D_1^*. Let D_2 be the domain symmetric to D_1 with respect to δ, and construct the function

$$\omega = g_2(\zeta) = \overline{g_1(\bar{\zeta})} \tag{32}$$

defined in D_2. Then $g_1(\zeta)$ and $g_2(\zeta)$ coincide on δ. In fact, if $\zeta \in \gamma$, then $\bar{\zeta} = \zeta$ and $\overline{g_1(\zeta)} = g_1(\zeta)$, since $g_1(\zeta) \in \delta^*$, so that

$$g_2(\zeta) = \overline{g_1(\bar{\zeta})} = \overline{g_1(\zeta)} = g_1(\zeta)$$

for all $\zeta \in \gamma$. Moreover $g_2(\zeta)$ is analytic in D_2. To see this, we observe that if ζ and $\zeta + \Delta\zeta$ are arbitrary points of D_2, then

$$\frac{g_2(\zeta + \Delta\zeta) - g_2(\zeta)}{\Delta\zeta} = \frac{\overline{g_1(\bar{\zeta} + \overline{\Delta\zeta}) - g_1(\bar{\zeta})}}{\Delta\zeta} = \overline{\left(\frac{g_1(\bar{\zeta} + \overline{\Delta\zeta}) - g_1(\bar{\zeta})}{\overline{\Delta\zeta}}\right)},$$

where $\bar{\zeta}$ and $\bar{\zeta} + \overline{\Delta\zeta}$ are points of D_1. It follows that the derivative

$$g_2'(\zeta) = \lim_{\Delta\zeta \to 0} \frac{g_2(\zeta + \Delta\zeta) - g_2(\zeta)}{\Delta\zeta} = \overline{\lim_{\overline{\Delta\zeta} \to 0} \left(\frac{g_1(\bar{\zeta} + \overline{\Delta\zeta}) - g_1(\bar{\zeta})}{\overline{\Delta\zeta}}\right)} = \overline{g_1'(\bar{\zeta})}$$

exists, since $g_1(\zeta)$ is analytic at $\bar{\zeta}$. The fact that $g_2(\zeta)$ is the analytic continuation of $g_1(\zeta)$ from D_1 into D_2 across δ is now an immediate consequence of

Theorem 13.51. Moreover, by its very construction, the function $g_2(\zeta)$ obviously maps D_2 conformally onto the domain D_2^* symmetric to D_1^* with respect to δ^*, and, by the same token, the function

$$\psi(\zeta) = \begin{cases} g_1(\zeta) & \text{if } \zeta \in D_1, \\ g_1(\zeta) = g_2(\zeta) & \text{if } \zeta \in \gamma, \\ g_2(\zeta) & \text{if } \zeta \in D_2 \end{cases}$$

maps the domain $D_1 \cup \delta \cup D_2$ conformally onto the domain $D_1^* \cup \delta^* \cup D_2^*$.

We now return to the original variables z and w by using the fractional linear transformations $z = \lambda_1(\zeta)$, $w = \lambda_2(\omega)$, which are the inverses of (31). The domain D_2 is then mapped onto the domain G_2 symmetric to G_1 with respect to γ, while D_2^* is mapped onto the domain G_2^* symmetric to G_1^* with respect to γ^* (recall Theorem 8.28). Let $f_2(z) = \lambda_2(g_2(l_1(z)))$. Then $f_2(z)$ is obviously the analytic continuation of $f_1(z)$ from G_1 into G_2 across γ. It is equally obvious that $f_2(z)$ maps G_2 conformally onto G_2^* and that the function (29) involving $f_2(z)$ and the original function $f_1(z)$ maps $G_1 \cup \gamma \cup G_2$ conformally onto $G_1^* \cup \gamma^* \cup G_2^*$. ∎

13.53. Finally we use the symmetry principle to prove an important principle of conformal mapping:

THEOREM. *Given two Jordan domains G and G^*, with boundaries C and C^* respectively, let z_1, z_2, z_3 be three distinct points of C and let w_1, w_2, w_3 be three distinct points of C^* arranged in the same order as z_1, z_2, z_3.*† *Then there exists a unique function $w = f(z)$ mapping G conformally onto G^* such that*

$$f(z_k) = w_k \qquad (k = 1, 2, 3).$$

Proof. By Theorem 13.37, $f(z)$ is continuous in $\bar{G}$ and maps C onto C^*, where the mapping of C onto C^* is one-to-one and direction-preserving. This explains the stipulation about the order of the boundary points.

Assuming first that G and G^* are the unit disks $|z| < 1$ and $|w| < 1$, suppose that among the infinitely many functions mapping G conformally onto G^* there are two functions $f(z)$ and $g(z)$ such that

$$f(z_k) = w_k, \qquad g(z_k) = w_k \qquad (k = 1, 2, 3).$$

By the symmetry principle, the functions $f(z)$ and $g(z)$ can be continued across the circle $|z| = 1$ into the domain symmetric to $|z| < 1$ with respect to $|z| = 1$, i.e., the domain $|z| > 1$. Therefore $f(z)$ and $g(z)$ are univalent in the whole extended plane except at the point z_0 symmetric to the point of the disk $|z| < 1$ mapped into the origin $w = 0$. It follows from Theorem 13.35 that $f(z)$ and $g(z)$ are fractional linear transformations. But a fractional

†I.e., encountered in the same order as the points z_1, z_2, z_3 when the two curves C and C^* are traversed in the same direction.

linear transformation is uniquely specified by its values at three distinct points (see Theorem 8.25), and hence $f(z) \equiv g(z)$.

Turning to the case of arbitrary Jordan domains G and G^*, let $\zeta = \sigma(z)$ and $\omega = \tau(w)$ be two functions mapping G and G^* conformally onto the unit disks $|\zeta| < 1$ and $|\omega| < 1$, with inverses $z = \varphi(\zeta)$ and $w = \psi(\omega)$ respectively, and let

$$\sigma(z_k) = \zeta_k, \qquad \tau(w_k) = \omega_k \qquad (k = 1, 2, 3).$$

Moreover, let $\omega = F(\zeta)$ be the unique conformal mapping of $|\zeta| < 1$ onto $|\omega| < 1$ such that

$$F(\zeta_k) = \omega_k \qquad (k = 1, 2, 3)$$

(the uniqueness of $F(\zeta)$ was just proved). Then obviously

$$w = f(z) = \psi(F(\sigma(z)))$$

maps G conformally onto G^* and carries the points z_1, z_2, z_3 into the points w_1, w_2, w_3. If there were another conformal mapping $w = g(z)$ of G onto G^* carrying z_1, z_2, z_3 into w_1, w_2, w_3, then there would be another function

$$\omega = G(\zeta) = \tau(g(\varphi(\zeta)))$$

mapping $|\zeta| < 1$ conformally onto $|\omega| < 1$ and carrying $\zeta_1, \zeta_2, \zeta_3$ into $\omega_1, \omega_2, \omega_3$. But this is impossible, as just shown, and hence $f(z) \equiv g(z)$. ∎

COMMENTS

13.1. We write capital letters U, V, F in Example 13.13 to avoid confusion between the special harmonic and analytic functions figuring in this example and the more general harmonic and analytic functions u, v, f figuring in Theorem 13.14 and Corollary 13.15. Note that u determines v only to within an arbitrary real constant in (9) and $f(z)$ only to within an arbitrary purely imaginary constant in (14). This is in keeping with Theorem 5.84 and Sec. 5.85.

13.2. We also allow C to be a closed Jordan curve passing through the point at infinity, i.e., the image under stereographic projection of a closed Jordan curve Γ on the Riemann sphere Σ passing through the pole of Σ (where Γ is defined in the obvious way). Thus G can be a half-plane (Sec. 13.24), a strip, a wedge, etc. It can be shown that Theorems 13.23 and 13.24 remain valid even if $h(z)$ has a finite number of jump discontinuities on C, provided that (15) is required to hold only at the points where $h(z)$ is continuous.

13.3. Concerning the first assertion in the proof of Theorem 13.35, we note that if $f(z)$ is analytic in the whole extended plane, then $f(z)$ is analytic at infinity and hence bounded in some domain $|z| > R$ (Chap. 4, Prob. 29).

But $f(z)$ is continuous and hence bounded in every closed disk $|z| \leq R$ (Chap. 3, Prob. 12a). It follows that $f(z)$ is bounded in the whole extended plane, and hence certainly in the whole finite plane. It is because of Theorems 13.37 and 13.38 that the Dirichlet problem was posed for an arbitrary Jordan domain in Sec. 13.21. For the proof of Theorem 13.37, see A. I. Markushevich, *op. cit.*, Volume II, p. 119 and Volume III, Corollary to Theorem 2.24 (see also the comment concerning the argument principle in Volume III, p. 319). Note the complete analogy between the proof of Theorem 13.38 and the first part of the proof of Theorem 13.24.

13.4. In its simplest terms, the problem of analytic continuation is the following: Starting from a given function $f(z)$ defined in some "initial" set E, find a function $\varphi(z)$ defined in a domain G containing E such that $\varphi(z)$ is analytic and coincides with $f(z)$ in E. This process of "continuation" or "extension" of $f(z)$ has the effect of making the "original" function $f(z)$ a "part" of a "larger" analytic function $\varphi(z)$, and is obviously possible only if $f(z)$ is "properly behaved" in E. For example, if E is the real line, $f(z)$ must be an infinitely differentiable function of the real variable $z = x$, as in Example 13.42a, while if E is a domain, $f(z)$ must be analytic in E from the very outset. Things are particularly nice in the case where E is a domain and the larger domain G is the union of $E = G_1$ and an overlapping domain G_2. This is essentially the situation in Sec. 13.43. More generally, G can be the union of a whole "chain" of overlapping domains, as in Sec. 13.45. This is not only a good way to make G as large as possible (cf. Prob. 16), but it also even allows the "master" function $\varphi(z)$ to be multiple-valued (!), with $f(z)$ merely one of its single-valued analytic branches, as in Example 13.46a.

13.5. Remarkably enough, analytic continuation is often possible even when the domains $E = G_1$ and G_2 (in the language of the preceding comment) are disjoint, provided the boundaries of G_1 and G_2 share a common arc γ. This is the idea behind Theorem 13.51. The proof of Theorem 13.51 is particularly elegant, involving as it does both the generalized Cauchy theorem and Morera's theorem. The proof of Theorem 13.53 rests on a number of carefully stockpiled ingredients, including Theorem 13.35 which may have seemed a digression at the time of its first appearance. Suppose that in the symmetry principle both γ and γ^* are already segments of the real axes in the z- and w-planes, so that the preliminary fractional linear transformations (31) are unnecessary. Then, by the same argument as that involving (32), the function $f_2(z) = \overline{f_1(\bar{z})}$ effects the analytic continuation of $f_1(z)$ from G_1 to G_2 across γ. Put somewhat differently, this leads at once to the following proposition, known as the *reflection principle*: Let $f(z)$ be analytic in a domain G that includes a segment δ of the real axis and is symmetric with respect to the real axis. Then

$$f(\bar{z}) = \overline{f(z)} \tag{33}$$

if and only if $f(z)$ is real for all $z \in \delta$ (for all real z if δ is the whole real axis). For example, the functions $z^2 + 1$, e^z, $\cos z$ satisfy (33), but not the functions $z^2 + i$, e^{iz}, $i \cos z$.

PROBLEMS

1. Prove that two given series of the form (3) and (3′) converge to a pair of conjugate harmonic functions in the disk $0 \le r \le R$, $0 \le \varphi \le 2\pi$, where
$$R = \frac{1}{\overline{\lim\limits_{n \to \infty}} \sqrt[n]{|a_n + ib_n|}},$$
but in no larger disk.

2. Let m and n be positive integers. Prove that
$$\int_0^{2\pi} \cos m\theta \cos n\theta \, d\theta = \int_0^{2\pi} \sin m\theta \sin n\theta \, d\theta = \begin{cases} \pi & \text{if } m = n, \\ 0 & \text{otherwise,} \end{cases}$$
while
$$\int_0^{2\pi} \cos m\theta \sin n\theta \, d\theta = 0.$$

3. Prove that if $f(z)$ is analytic and nonvanishing inside and on the circle $|z - z_0| = R$, then
$$\ln |f(z_0)| = \frac{1}{2\pi} \int_0^{2\pi} \ln |f(z_0 + Re^{i\theta})| \, d\theta.$$

4. Prove that
$$\int_0^{2\pi} \ln [\cosh^2 (\sin \theta) - \sin^2 (\cos \theta)] \, d\theta = 0.$$

5. Give an alternative proof of Poisson's integral formula, starting from Cauchy's integral formula.

6. Prove the following alternative version of Schwarz's formula (14):
$$f(z) = \frac{1}{\pi i} \int_{|\zeta - z_0| = R} \frac{u(\zeta)}{\zeta - z} \, d\zeta - \overline{f(z_0)}. \tag{14′}$$

7. Given a continuous real function $f(x)$ defined for all real x, suppose $f(x)$ is periodic with period $\omega > 0$, i.e., suppose $f(x + \omega) = f(x)$ for all x. Prove that
$$\int_a^{a+\omega} f(x) \, dx = \int_0^{\omega} f(x) \, dx$$
for arbitrary real a.

8. Solve the Dirichlet problem for the exterior of the unit circle $|z| = 1$. Show that the value of the solution at infinity equals the average of the boundary values on the circle.

9. Let G be the unit disk $|z| < 1$ and C the unit circle $|z| = 1$. Use (14′) to find the unique function $u(z)$ harmonic in G such that
$$\lim_{z \to e^{i\varphi_0}} \frac{\partial u(z)}{\partial r} = h(e^{i\varphi_0}) \qquad (0 \le \varphi_0 \le 2\pi),$$

where $\partial u/\partial r$ is the radial derivative of u, and $h(z)$ is a given function continuous on C, thereby solving the so-called *Neumann problem* for G.

10. Suppose there exist two functions $w = f(z)$ and $w = g(z)$, with inverses $z = \varphi(w)$ and $z = \psi(w)$ respectively, mapping a simply connected domain G onto the unit disk $|w| < 1$, while satisfying the conditions $f(z_0) = 0$, $f'(z_0) > 0$ and $g(z_0) = 0$, $g'(z_0) > 0$ $(z_0 \in G)$. Prove that the functions $F(w) = f(\psi(w))$ and $G(w) = g(\varphi(w))$ map the unit disk onto itself, while satisfying the conditions $F(0) = 0$, $F'(0) > 0$ and $G(0) = 0$, $G'(0) > 0$.

11. Use the preceding problem and Schwarz's lemma (Chap. 10, Prob. 28) to prove the uniqueness part of Riemann's theorem.

12. Given a piecewise smooth closed Jordan curve C with interior I, suppose $f(z)$ is analytic in $\bar{I}$ and one-to-one on C, and let C^* be the image of C under the mapping $w = f(z)$. Prove that $f(z)$ is univalent in the domain I, which it maps conformally onto the interior of C^*.

13. Prove the following generalization of Liouville's theorem: If $w = f(z)$ is an entire function and if $f(z)$ fails to take any values belonging to some curve γ in the w-plane, then $f(z)$ is a constant.

14. Let G_1 and G_2 be the disks $|z| < 1$ and $|z - 2| < 1$, respectively, and let

$$f_1(z) = \sum_{n=1}^{\infty} \frac{z^n}{n}, \qquad f_2(z) = i\pi + \sum_{n=1}^{\infty} (-1)^n \frac{(z-2)^n}{n}.$$

Prove that the elements $\{G_1, f_1(z)\}$ and $\{G_2, f_2(z)\}$ are analytic continuations of each other.

15. Let F be a general analytic function made up of elements $\{G_1, f_1(z)\}$, $\{G_2, f_2(z)\}, \ldots$ Prove that the set F' of all elements obtained by differentiating the elements of F, i.e., consisting of the elements $\{G_1, f'_1(z)\}, \{G_2, f'_2(z)\}, \ldots$, is itself a general analytic function (called the *derivative* of F).

16. A general analytic function which contains all analytic continuations of all its elements is called a *complete analytic function*, and the domain of such a function is often called its *domain of existence*. Find the domain of existence of the complete analytic function generated by the element $\{G_1, f_1(z)\}$, where G_1 and $f_1(z)$ are the same as in Prob. 14.

17. Prove that the limit (24) is infinite if α is rational.

18. Prove that the unit circle $|z| = 1$ is the natural boundary of the series

$$\sum_{n=1}^{\infty} z^{2^n}.$$

19. Interpret the proof of Theorem 10.23 from the standpoint of analytic continuation.

20. Discuss Riemann surfaces from the standpoint of analytic continuation and complete analytic functions (Prob. 16).

21. Prove that the annulus $r_1 < |z| < r_2$ can be mapped conformally onto the annulus $\rho_1 < |w| < \rho_2$ if and only if $\rho_2/\rho_1 = r_2/r_1$.

CHAPTER FOURTEEN

Mapping of Polygonal Domains

14.1. The Schwarz–Christoffel Transformation

14.11. By a *polygonal domain* Δ we mean a domain whose boundary consists entirely of line segments, possibly of infinite length or traversed more than once. The simplest polygonal domain is a *bounded polygon*, i.e., the interior of a closed polygonal Jordan curve (cf. Chap. 3, Prob. 1). First we pose the problem of finding a conformal mapping of the upper half-plane $\text{Im } z > 0$ onto such a bounded polygon. Conformal mappings of the upper half-plane onto more general polygonal domains will be found later (see Sec. 14.16). Mappings of this kind play an important role in a variety of physical applications, from such diverse fields as fluid dynamics, electrostatics and heat conduction.

14.12. We begin by examining the behavior of a conformal mapping near the "corners" of a polygonal domain Δ. Suppose the boundary of Δ has a "corner" at the point w_0, i.e., suppose two line segments meet at w_0 to form an angle $\alpha\pi$ $(0 < \alpha \leq 2)$. Let $w = f(z)$ be a conformal mapping of the upper half-plane Π_+ onto Δ, and suppose $f(z)$ carries the point z_0 of the real axis into the point w_0. Then $f(z)$ maps the "upper half-disk" K shown in Figure 45a onto the "sector-like" domain K' shown in Figure 45b. Therefore the function

$$\omega = \omega(z) = [f(z) - w_0]^{1/\alpha} \tag{1}$$

maps K onto the "distorted half-disk" K^* shown in Figure 45c. Moreover, $\omega(z)$ carries a segment δ of the real axis through the point $z = 0$ into a segment δ^* of some line through the point $\omega = 0$. Hence, by the symmetry

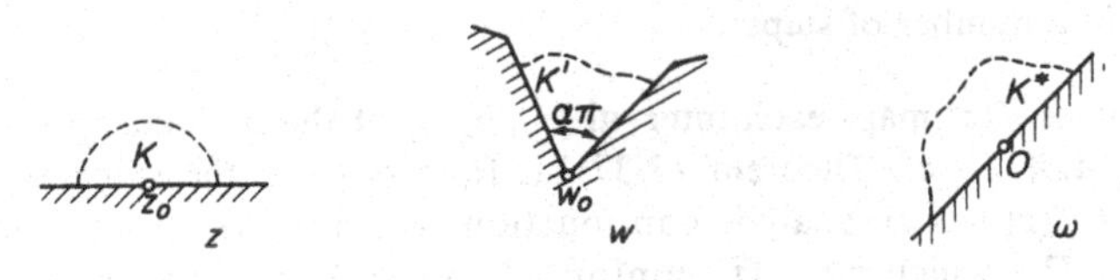

Figure 45

principle, $\omega(z)$ has an analytic continuation across δ, so that $\omega(z)$ is analytic in a "full" neighborhood of z_0 with a corresponding Taylor expansion

$$\omega(z) = c_1(z - z_0) + c_2(z - z_0)^2 + \cdots. \tag{2}$$

There is no constant in (2) since $\omega(z_0) = 0$, but $c_1 = \omega'(z_0) \neq 0$ by the conformality of $\omega(z)$. Using (1) to return to the function $f(z)$, we find that

$$f(z) = w_0 + (z - z_0)^\alpha[c_1 + c_2(z - z_0) + \cdots]^\alpha.$$

Choosing a fixed single-valued branch of the multiple-valued function†

$$[c_1 + c_2(z - z_0) + \cdots]^\alpha$$

and then expanding this branch in Taylor series, we finally find that

$$f(z) = w_0 + (z - z_0)^\alpha[c_1' + c_2'(z - z_0) + \cdots] \qquad (c_1' \neq 0). \tag{3}$$

14.13. Now let Δ be a bounded polygon with vertices $A_1, A_2, \ldots, A_n$ and interior angles $\alpha_1\pi, \alpha_2\pi, \ldots, \alpha_n\pi$, as shown in Figure 46. By Theorem 13.53, there exists a unique function $w = f(z)$ mapping the upper half-plane Π_+ onto Δ and carrying three given finite points a_1, a_2, a_3 of the real axis into three given points of the boundary of Δ. Let the latter points be the vertices A_1, A_2, A_3, and let $a_4, \ldots, a_n$ be the points of the real axis (again assumed to be finite) which $f(z)$ maps into the other vertices $A_4, \ldots, A_n$. The actual determination of $f(z)$ is based on the following argument, which

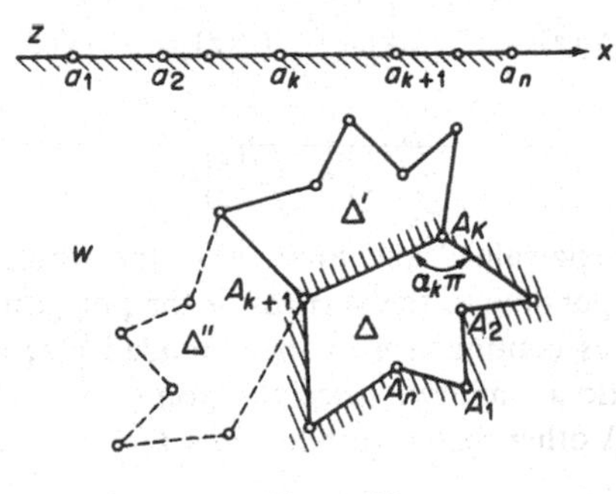

Figure 46

†This is possible since $c_1 + c_2(z - z_0) + \cdots$ is nonvanishing in some neighborhood of z_0.

we give in a number of steps:

a. Since $f(z)$ maps each interval $[a_k, a_{k+1}]$ of the real axis into the line segment A_kA_{k+1} (cf. Theorem 13.37), it follows from the symmetry principle that $f(z)$ has an analytic continuation across $[a_k, a_{k+1}]$ into the lower half-plane Π_-, which maps Π_- conformally onto the polygon Δ' symmetric to Δ with respect to the line segment $[a_k, a_{k+1}]$. This analytic continuation can then be continued in turn across any segment $[a_l, a_{l+1}]$ into the upper half-plane Π_+, with the new continuation mapping Π_+ conformally onto the polygon Δ'' symmetric to Δ' with respect to the segment A_lA_{l+1}, and so on (see the figure). Carrying out all possible analytic continuations of this type, we get in general an infinite-valued analytic function $F(z)$ with the original function $f(z)$ as one of its single-valued analytic branches in Π_+.

b. Let $f_1(z)$ and $f_2(z)$ be two such single-valued branches of $F(z)$ in the upper half-plane Π_+. Then $f_1(z)$ and $f_2(z)$ map Π_+ conformally onto two polygons Δ_1 and Δ_2 differing only by an even number of reflections in line segments. But every pair of reflections in two line segments reduces to a rotation and a shift (see Prob. 1), and hence

$$f_2(z) = e^{i\theta}f_1(z) + c, \tag{4}$$

where θ and c are constants (θ real). The same is true of any two branches of $F(z)$ in the lower half-plane Π_-. Let

$$g(z) = \frac{f''(z)}{f'(z)} = \frac{d \ln f'(z)}{dz}.$$

Then $g(z)$ is analytic in the upper half-plane, since $f'(z)$ is nonvanishing, being the derivative of a univalent function (cf. Theorem 12.24). Clearly $g(z)$ is the same for all possible continuations of $f(z)$ both into Π_- and Π_+, since (4) implies

$$f_2'(z) = e^{i\theta}f_1'(z), \qquad f_2''(z) = e^{i\theta}f_1''(z)$$

and hence

$$\frac{f_2''(z)}{f_2'(z)} = \frac{f_1''(z)}{f_1'(z)}.$$

Therefore $g(z)$ is single-valued and analytic in the whole z-plane except at the points a_k corresponding to the vertices of the polygon Δ. Moreover $f(z)$ and $g(z)$, regarded as continued into the extended plane minus the points a_k, are both analytic at infinity, since the point $z = \infty$ is mapped into a boundary point of Δ other than a vertex. Thus $f(z)$ has a Laurent expansion at ∞ of the form

$$f(z) = c_0 + \frac{c_{-m}}{z^m} + \frac{c_{-m-1}}{z^{m+1}} + \cdots \qquad (c_{-m} \neq 0)$$

(cf. Chap. 11, Prob. 16), and correspondingly

$$g(z) = \frac{f''(z)}{f'(z)} = \frac{m(m+1)\frac{c_{-m}}{z^{m+2}} + \cdots}{-\frac{mc_{-m}}{z^{m+1}} + \cdots}$$

$$= \frac{1}{z}\frac{m(m+1)c_{-m} + \cdots}{-mc_{-m} + \cdots} = -\frac{m+1}{z} + \cdots,$$

which implies

$$g(\infty) = 0. \tag{5}$$

c. To analyze the behavior of $g(z)$ at a_k, we apply formula (3) to the point a_k and the corresponding vertex A_k, obtaining

$$f(z) = A_k + (z - a_k)^{\alpha_k}[c'_1 + c'_2(z - a_k) + \cdots].$$

Hence the Laurent expansion of $g(z)$ at a_k is just

$$g(z) = \frac{f''(z)}{f'(z)} = \frac{\alpha_k(\alpha_k - 1)c'_1(z - a_k)^{\alpha_k - 2} + \cdots}{\alpha_k c'_1(z - a_k)^{\alpha_k - 1} + \cdots}$$

$$= \frac{1}{z - a_k}\frac{(\alpha_k - 1)\alpha_k c'_1 + \cdots}{\alpha_k c'_1 + \cdots} = \frac{\alpha_k - 1}{z - a_k} + c''_1 + c''_2(z - a_k) + \cdots,$$

i.e., the point a_k is a simple pole of $g(z)$ with residue $\alpha_k - 1$, where $\alpha_k\pi$ is the interior angle of the polygon Δ at the vertex A_k. Therefore $g(z)$ has precisely n singular points in the extended plane, namely simple poles at the points $a_1, a_2, \ldots, a_n$. It follows that the function

$$G(z) = g(z) - \frac{\alpha_1 - 1}{z - a_1} - \frac{\alpha_2 - 1}{z - a_2} - \cdots - \frac{\alpha_n - 1}{z - a_n}$$

is a bounded entire function, being analytic in the whole extended plane (see the comment to Sec. 13.3). Therefore $G(z) \equiv \text{const}$, by Liouville's theorem. But $G(\infty) = g(\infty) = 0$, by (5), and hence $G(z) \equiv 0$ or equivalently

$$g(z) = \frac{d \ln f'(z)}{dz} = \frac{\alpha_1 - 1}{z - a_1} - \frac{\alpha_2 - 1}{z - a_2} - \cdots - \frac{\alpha_n - 1}{z - a_n}. \tag{6}$$

d. Finally, integrating (6) twice along any path joining a fixed point $z_0 \in \Pi_+$ to a variable point $z \in \Pi_+$, we get first

$$\ln f'(z) = (\alpha_1 - 1)\ln(z - a_1) + (\alpha_2 - 1)\ln(z - a_2) + \cdots + (\alpha_n - 1)\ln(z - a_n) + \ln C$$

or equivalently

$$f'(z) = C(z - a_1)^{\alpha_1 - 1}(z - a_2)^{\alpha_2 - 1}\cdots(z - a_n)^{\alpha_n - 1}$$

and then

$$f(z) = C\int_{z_0}^{z}(z - a_1)^{\alpha_1 - 1}(z - a_2)^{\alpha_2 - 1}\cdots(z - a_n)^{\alpha_n - 1}\,dz + C_1, \tag{7}$$

where z_0, C and C_1 are constants. (For simplicity, we use the same symbol z for both the variable of integration and the upper limit of integration.) Formula (7) is the celebrated *Schwarz–Christoffel transformation*, giving the conformal mapping of the upper half-plane Π_+ onto the bounded polygon Δ. The constant z_0 can be fixed once and for all, e.g., by setting $z_0 = 0$, since changing z_0 amounts to changing C_1. Hence z_0 will no longer be regarded as an unknown parameter in (7).

14.14. Remark. According to Theorem 13.53, specification of three points a_1, a_2, a_3 of the real axis corresponding to three points A_1, A_2, A_3 of the polygon Δ automatically determines the remaining points $a_4, \ldots, a_n$ and the constants C, C_1. The actual determination of $a_4, \ldots, a_n$, C, C_1 is the chief difficulty associated with the use of the Schwarz–Christoffel transformation, but can always be accomplished by a little ingenuity, as in the examples considered in Sec. 14.2.

14.15. Next we remove the restriction that all the points a_k be finite. Suppose for example that $a_n = \infty$. Then to reduce this case to the one previously considered, we make a preliminary transformation†

$$\zeta = -\frac{1}{z} + a'_n, \tag{8}$$

where a'_n is any point of the real axis distinct from $a_1, a_2, \ldots, a_n$. This transformation maps the upper half-plane onto itself (why?) and carries the points $a_1, a_2, \ldots, a_n = \infty$ into finite points $a'_1, a'_2, \ldots, a'_n$. Applying (7) in the ζ-plane, we get

$$\begin{aligned} w &= C' \int_{\zeta_0}^{\zeta} (\zeta - a'_1)^{\alpha_1 - 1} (\zeta - a'_2)^{\alpha_2 - 1} \cdots (\zeta - a'_n)^{\alpha_n - 1} \, d\zeta + C_1 \\ &= C' \int_{z_0}^{z} \left(a'_n - a'_1 - \frac{1}{z}\right)^{\alpha_1 - 1} \left(a'_n - a'_2 - \frac{1}{z}\right)^{\alpha_2 - 1} \cdots \left(-\frac{1}{z}\right)^{\alpha_n - 1} \frac{dz}{z^2} + C_1 \\ &= C \int_{z_0}^{z} (z - b_1)^{\alpha_1 - 1} (z - b_2)^{\alpha_2 - 1} \cdots (z - b_{n-1})^{\alpha_{n-1} - 1} \frac{dz}{z^{\alpha_1 + \alpha_2 + \cdots + \alpha_n - n + 2}} + C_1 \end{aligned}$$

in terms of new constants

$$b_k = \frac{1}{a'_n - a'_k} \qquad (k = 1, 2, \ldots, n - 1)$$

(note that the new constant C absorbs C' and a number of other factors). But

$$\alpha_1 + \alpha_2 + \cdots + \alpha_n = n - 2, \tag{9}$$

†If one of the points a_k equals zero, then (8) should be replaced by

$$\zeta = -\frac{1}{z - a} + a'_n.$$

where a is any real number distinct from all the a_n.

since the sum of the interior angles of an n-sided polygon is just $(n-2)\pi$. Therefore

$$w = C\int_{z_0}^{z} (z-a_1)^{\alpha_1-1}(z-a_2)^{\alpha_2-1}\cdots(z-a_{n-1})^{\alpha_n-1}\,dz + C_1 \qquad (10)$$

where we change the b_k back to a_k for uniformity. Thus if one of the vertices of the polygon Δ corresponds to the point $z=\infty$, the corresponding factor in the Schwarz–Christoffel transformation simply drops out.

14.16. Finally we remove the restriction that all the vertices of the polygon Δ be finite. Suppose for example that $A_k=\infty$ while the other vertices of Δ are finite, and let A'_k, A''_k be arbitrary points, the first on the ray $A_{k-1}A_k$, the second on the ray A_kA_{k+1}. Joining A'_k and A''_k by a line segment, we get a new polygon Δ' which is bounded and has $n+1$ sides (see Figure 47). By (7), the function mapping the upper half-plane onto Δ' is just

$$w = C\int_{z_0}^{z} (z-a_1)^{\alpha_1-1}\cdots(z-a'_k)^{\alpha'_k-1}(z-a''_k)^{\alpha''_k-1}\cdots(z-a_n)^{\alpha_n-1}\,dz + C_1, \qquad (11)$$

where $\alpha'_k\pi$, $\alpha''_k\pi$ are the interior angles of Δ' at the vertices A'_k, A''_k and a'_k, a''_k are the points of the x-axis corresponding to these vertices. Now let the segment $A'_kA''_k$ approach infinity while remaining parallel to itself. Then the points a'_k and a''_k fuse into the single point a_k corresponding to the vertex A_k, while the factors in (11) containing a'_k and a''_k combine in the limit to give $(z-a_k)^{\alpha'_k+\alpha''_k-2}$. Let $\alpha_k\pi$ denote the negative of the angle between the rays $A_{k-1}A_k$ and A_kA_{k+1} at their finite point of intersection A^*_k (see the figure).† Then, examining the triangle $A'_kA''_kA^*_k$, we see that $\alpha'_k+\alpha''_k-\alpha_k=1$, i.e., $\alpha'_k+\alpha''_k-2=\alpha_k-1$, so that (11) takes the standard form

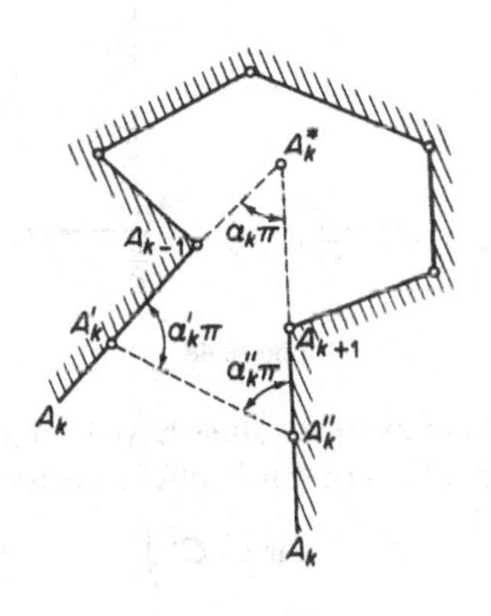

Figure 47

$$w = C\int_{z_0}^{z} (z-a_1)^{\alpha_1-1}\cdots(z-a_k)^{\alpha_k-1}\cdots(z-a_n)^{\alpha_n-1}\,dz + C_1. \qquad (12)$$

The same procedure applies to the case where several vertices of Δ lie at infinity. Thus the Schwarz–Christoffel transformation remains valid for polygons with one or several vertices at infinity, provided that the angle between two rays with a vertex at infinity is defined as the negative of the

†If $A_{k-1}A_k$ and A_kA_{k+1} are parallel, we choose $\alpha_k=0$.

angle between the rays at their finite point of intersection. Note that with our definition of an angle at infinity, formula (9) continues to hold. In fact, for the polygon Δ' with $n+1$ sides we have

$$\alpha_1 + \alpha_2 + \cdots + \alpha_{k-1} + \alpha_k' + \alpha_k'' + \alpha_{k+1} + \cdots + \alpha_n = n - 1. \qquad (9')$$

But $\alpha_k' + \alpha_k'' = \alpha_k + 1$, and hence (9′) reduces to

$$\alpha_1 + \alpha_2 + \cdots + \alpha_n + 1 = n - 1,$$

which is equivalent to (9).

14.2. Examples

14.21. Map the half-plane $\operatorname{Im} z > 0$ onto the half-strip $\pi/2 < \operatorname{Re} w < \pi/2$, $\operatorname{Im} w > 0$ (see Figure 48).

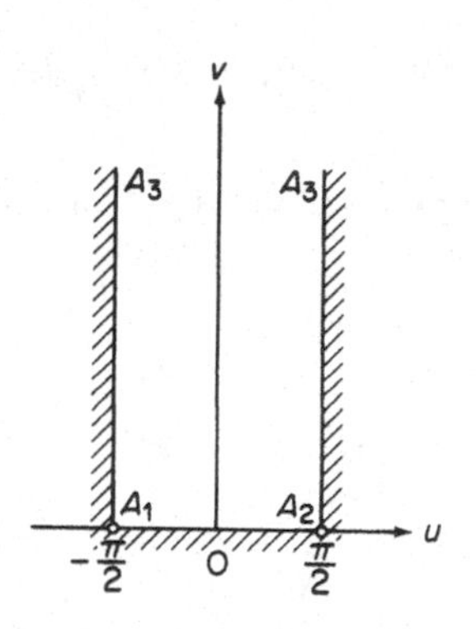

Figure 48

Solution. Regarding the half-strip as a "degenerate triangle" with a vertex at infinity, we exhibit the data of the problem as a table

k	A_k	α_k	a_k
1	$-\frac{\pi}{2}$	$\frac{1}{2}$	-1
2	$\frac{\pi}{2}$	$\frac{1}{2}$	1
3	∞	0	∞

where explicit values have been chosen for the numbers a_1, a_2, a_3.† Applying the Schwarz–Christoffel transformation in the form (10) with $z_0 = 0$, we get

$$w = C' \int_0^z (z+1)^{-1/2}(z-1)^{-1/2}\,dz + C_1$$

$$= C \int_0^z \frac{dz}{\sqrt{1-z^2}} + C_1 = C \arcsin z + C_1$$

(cf. Chap. 9, Prob. 14). To determine the constants C and C_1, we note that

$$-\frac{\pi}{2} = -C\frac{\pi}{2} + C_1,$$

$$\frac{\pi}{2} = \ \ C\frac{\pi}{2} + C_1,$$

†The particular values chosen make the subsequent calculation especially simple. The fact that $\alpha_3 = 0$ follows at once from the formula $\alpha_1 + \alpha_2 + \alpha_3 = 1$, valid for any triangle, or from the fact that the sides of the strip are parallel.

since the points a_1, a_2 go into the vertices A_1, A_2, and hence $C = 1$, $C_1 = 0$. Thus the required conformal mapping of the upper half-plane onto the indicated half-strip is given by the function $w = \arcsin z$, with inverse $z = \sin w$. This mapping has already figured in Chap. 9, Probs. 13–17.

14.22. Map the half-plane $\operatorname{Im} z > 0$ onto the rectangle shown in Figure 49.

Solution. Here the data of the problem are given by the table

k	A_k	α_k	a_k
1	K	$\frac{1}{2}$	1
2	$K + iK'$	$\frac{1}{2}$	$\lambda > 1$
3	$-K + iK'$	$\frac{1}{2}$	a_3
4	$-K$	$\frac{1}{2}$	a_4

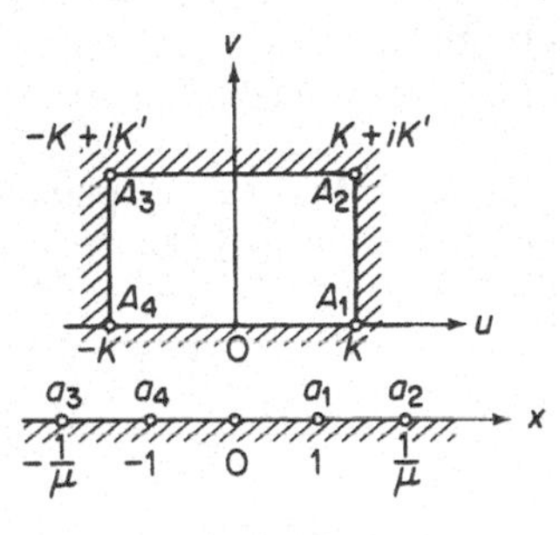

Figure 49

where the constants λ, a_3 and a_4 are yet to be specified. Suppose we require that the points $z = 0$ and $z = \infty$ go into the points $w = 0$ and $w = iK'$. Then the required mapping can be regarded as the analytic continuation across the imaginary axis (in accordance with the symmetry principle) of the mapping of the first quadrant of the z-plane onto the right half of the rectangle. It follows that $a_3 = -\lambda$ and $a_4 = -1$. Hence, by (7),

$$w = C' \int_0^z (z-1)^{-1/2}(z-\lambda)^{-1/2}(z+\lambda)^{-1/2}(z+1)^{-1/2}\,dz + C_1$$

$$= C' \int_0^z \frac{dz}{\sqrt{(z^2-1)(z^2-\lambda^2)}} + C_1 = C \int_0^z \frac{dz}{\sqrt{(1-z^2)(1-\mu^2 z^2)}}, \tag{13}$$

where $C_1 = 0$ since $w = 0$ for $z = 0$, and

$$\mu = \frac{1}{\lambda} \qquad (0 < \mu < 1)$$

is a new constant. Since A_1 is the image of a_1, we have

$$K = C \int_0^1 \frac{dx}{\sqrt{(1-x^2)(1-\mu^2 x^2)}}, \tag{14}$$

while

$$K + iK' = C \int_0^1 \frac{dx}{\sqrt{(1-x^2)(1-\mu^2x^2)}} + iC \int_1^{1/\mu} \frac{dx}{\sqrt{(x^2-1)(1-\mu^2x^2)}}, \tag{14'}$$

since A_2 is the image of a_2 (here we separate the integral from 0 to $\lambda = 1/\mu$

into two integrals). Comparing (14) and (14′), we find at once that

$$K' = C \int_1^{1/\mu} \frac{dx}{\sqrt{(x^2-1)(1-\mu^2x^2)}}. \tag{15}$$

Dividing (15) by (14), we get a formula relating the quantities μ and K'/K (C cancels out). Hence μ depends only on the ratio of the sides of the rectangle. As for the constant C, it depends on the actual size of the rectangle, and can be determined from either (14) or (15), once μ is known. The integrals in formulas (13)–(15) are called *elliptic integrals*, and cannot be expressed in terms of elementary functions.

14.23. Map the half-plane $\operatorname{Im} z > 0$ onto the polygonal domain shown in Figure 50, namely the strip $-h_1 < \operatorname{Im} w < h_2$ cut along the negative real axis.

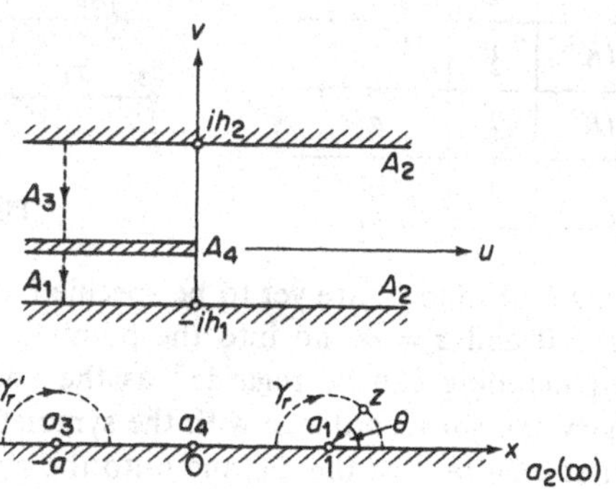

Figure 50

Solution. Regarding the strip as a "degenerate rectangle" with three vertices at infinity, we choose the values of a_k given in the table.

k	A_k	α_k	a_k
1	∞	0	1
2	∞	0	∞
3	∞	0	$-a<0$
4	0	2	0

It follows from the considerations of Secs. 14.15 and 14.16 that

$$w = C' \int_0^z (z-1)^{-1}(z+a)^{-1} z\, dz + C_1$$

$$= C' \int_0^z \frac{z}{(z-1)(z+a)}\, dz = C\left[\ln(1-z) + a\ln\left(1+\frac{z}{a}\right)\right], \tag{16}$$

where $C_1 = 0$ since $w = 0$ for $z = 0$. To determine the constants C and a, we argue as follows: Let the point z describe a little semicircle γ_r in the upper half-plane, with radius r and center $a_1 = 1$, as shown in the figure. Then the argument of the rotating vector $1 - z = re^{i\theta}$ changes from 0 to $-\pi$, and at the same time the image point w moves from the ray A_4A_1 to the ray A_1A_2, so that the corresponding increment of w equals

$$\Delta w = -ih_1 + \epsilon(r), \tag{17}$$

where $\epsilon(r) \longrightarrow 0$ as $r \longrightarrow 0$. This follows from the fact that as z describes γ_r, w describes a path that differs only slightly from a line segment perpendicular to A_4A_1 and A_1A_2 (justify this statement). But as z describes γ_r, the second term in brackets in the right-hand side of (16) changes only slightly, being continuous at $z = 1$, while the first term, namely $\ln(1 - z) = \ln r + i\theta$ changes by $-\pi i$, so that

$$\Delta w = -C\pi i + \eta(r), \tag{18}$$

where $\eta(r) \longrightarrow 0$ as $r \longrightarrow 0$. Equating (17) and (18), and then taking the limit as $r \longrightarrow 0$, we get

$$C = \frac{h_1}{\pi}.$$

Similarly, letting z describe a little semicircle γ'_r of radius r centered at $a_3 = -a$, we get

$$\Delta w = -ih_2 + \epsilon(r) \tag{17'}$$

instead of (17) and

$$\Delta w = -Ca\pi i + \eta(r) \tag{18'}$$

instead of (18), since this time the argument of the rotating vector $z + a = re^{i\theta}$ changes from π to 0, causing $\ln(z + a) = \ln r + i\theta$ to again change by $-\pi i$. It follows from (17') and (18') that

$$a = \frac{h_2}{C\pi} = \frac{h_2}{h_1}.$$

Thus, finally, the required conformal mapping of the upper half-plane onto the polygonal domain shown in Figure 50 is given by

$$w = \frac{h_1}{\pi}\ln(1 - z) + \frac{h_2}{\pi}\ln\left(1 + \frac{h_1}{h_2}z\right). \tag{19}$$

14.24. Map the strip $-\pi < \operatorname{Im} z < \pi$ onto the polygonal domain shown in Figure 51.

Solution. With a view to eventually applying the symmetry principle, we first map the upper half-plane $\operatorname{Im} z > 0$ onto the upper half of the indicated domain,

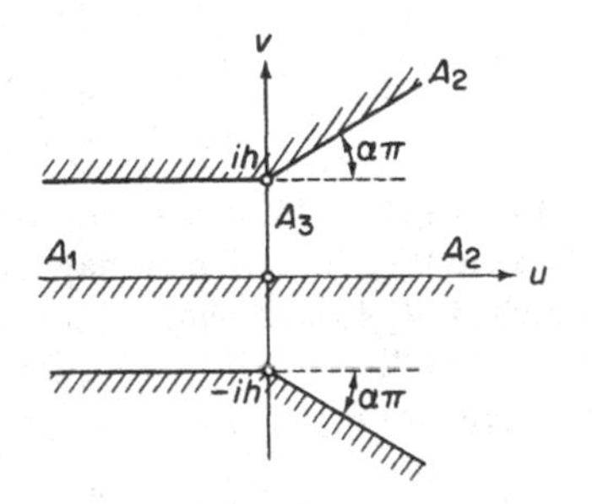

Figure 51

i.e., onto the "degenerate triangle" $A_1A_2A_3$ with two of its vertices at infinity, choosing the values of a_k given in the table.

k	A_k	α_k	a_k
1	∞	0	0
2	∞	$-\alpha$	∞
3	ih	$1+\alpha$	-1

The required mapping is just

$$w = C\int_{-1}^{z} \frac{(z+1)^{\alpha}}{z}\,dz + ih, \tag{20}$$

where we use the fact that A_3 is the image of $a_3 = -1$. To determine the constant C, let z describe a little semicircle γ_r of radius r centered at $z = 0$. Then the argument of the vector $z = re^{i\theta}$ varies from π to 0, and the corresponding increment of the function (20) is

$$\Delta w = C\int_{\gamma_r} \frac{dz}{z} + \epsilon(r) = -C\pi i + \epsilon(r), \tag{21}$$

where $\epsilon(r) \to 0$ as $r \to 0$. (Here we use the fact that $(z+1)^{\alpha}$ differs only slightly from 1 on γ_r.) On the other hand, the point w goes from the ray A_3A_1 to the ray A_1A_2 as z describes γ_r, and hence

$$\Delta w = -ih + \eta(r), \tag{22}$$

where $\eta(r) \to 0$ as $r \to 0$. Comparing (21) and (22), we get

$$C = \frac{h}{\pi},$$

so that (20) becomes

$$w = \frac{h}{\pi}\int_{-1}^{z} \frac{(z+1)^{\alpha}}{z}\,dz + ih. \tag{23}$$

We now replace z by e^z. Then (23) becomes the mapping

$$w = \frac{h}{\pi}\left\{\int_{\pi i}^{z} (e^z + 1)^{\alpha}\,dz + \pi i\right\} \tag{24}$$

of the strip $0 < \operatorname{Im} z < \pi$ onto the "triangle" $A_1A_2A_3$. But the lower edge of the strip $0 < \operatorname{Im} z < \pi$ is mapped onto the midline of the whole domain of which $A_1A_2A_3$ is the upper half, and hence, by the symmetry principle, (24) maps the "full" strip $-\pi < \operatorname{Im} z < \pi$ onto the whole domain. Note that (24) reduces to

$$w = \frac{h}{\pi}\{e^z + z + 1\} \tag{25}$$

if $\alpha = 1$.

COMMENTS

14.1. In Sec. 14.1 we used the powerful symmetry principle to *construct* a function mapping the upper half-plane onto a given polygon, thereby obtaining the Schwarz–Christoffel transformation (7). Conversely, suppose we are *given* formula (7), together with arbitrary complex numbers C, C_1 and real numbers a_k, α_k $(k = 1, \ldots, n)$ satisfying the conditions

$$-\infty < a_1 < a_2 < \cdots < a_n < \infty, \quad -2 \leq \alpha_k \leq 2,$$
$$\alpha_1 + \alpha_2 + \cdots + \alpha_n = n - 2.$$

Then it is not hard to show (cf. A. I. Markushevich, *op. cit.*, Volume II, Sec. 20) that (7) defines a conformal mapping of the upper half-plane onto some n-sided polygon with interior angles $\alpha_1\pi, \ldots, \alpha_n\pi$ (if $\alpha_k < 0$, the vertex corresponding to a_k lies at infinity). If the condition $\alpha_1 + \alpha_2 + \cdots + \alpha_n = n - 2$ fails to hold, it turns out that (7) maps the upper half-plane onto a polygon with $n + 1$ sides.

14.2. The theory of elliptic integrals and elliptic functions (hinted at in Prob. 16) constitutes an important chapter in complex analysis, which is the subject of an extensive literature. A very readable introduction to this topic can be found in A. I. Markushevich, *op. cit.*, Volume III, Part 2.

PROBLEMS

1. Let γ be the straight line through the point $z = a$ making the angle θ with the positive real axis. Show that reflection in γ is described by the transformation

$$w = e^{2i\theta}\overline{z - a} + a.$$

(θ real). Use this to show that every pair of reflections in two lines (or line segments) reduces to a rotation and a shift.

2. Let G be the whole z-plane minus two line segments, the first joining the points -1 and 1, the second joining the points $-2i$ and 0 (thus G is the exterior of a letter "T"). Using the symmetry principle, show that the function

$$w = \sqrt{\frac{\sqrt{z^2 - 1} + \sqrt{5}i}{i - \sqrt{z^2 - 1}}}$$

maps G onto the upper half-plane $\operatorname{Im} w > 0$.

3. Map the upper half-plane $\operatorname{Im} z > 0$ onto the upper half-plane $\operatorname{Im} w > 0$ minus the line segment joining the points 0 and ih.

4. Map the upper half-plane $\operatorname{Im} z > 0$ onto a rhombus in the w-plane with side length l and obtuse angle $\alpha\pi$.

5. Map the upper half-plane $\operatorname{Im} z > 0$ onto the following triangles in the w-plane with the indicated vertices A_1, A_2, A_3, in each case choosing $a_1 = 0$, $a_2 = 1$,

$a_3 = \infty \ (b > 0)$:

a) $0, b, b + \dfrac{ib}{\sqrt{3}}$; b) $0, b, b + ib$; c) $0, b, \dfrac{1 + i\sqrt{3}}{2}b$.

6. Map the upper half-plane $\operatorname{Im} z > 0$ onto the domain consisting of the upper half-plane $\operatorname{Im} w > 0$ cut along both the positive real axis and the ray

$$\operatorname{Re} w = 0, \qquad \operatorname{Im} w \geq h > 0. \tag{26}$$

7. Solve the preceding problem if (26) is replaced by

$$\operatorname{Re} w \leq 0, \qquad \operatorname{Im} w = h > 0. \tag{26'}$$

8. Prove that the Schwarz–Christoffel transformation (7) maps the unit disk $|z| < 1$ onto a bounded polygon Δ with interior angles $\alpha_1\pi, \alpha_2\pi, \ldots, \alpha_n\pi$, where the a_k are now the points of the unit circle $|z| = 1$ mapped into the vertices of Δ.

9. Onto what domain does the function

$$w = \int_0^z \frac{dz}{\sqrt{1 - z^4}}$$

map the unit disk $|z| < 1$? How about the function

$$w = \int_0^z \frac{dz}{(1 - z^n)^{2/n}}$$

(n a positive integer)?

10. Let Δ be a bounded polygon with vertices $A_1, A_2, \ldots, A_n$ and corresponding *exterior* angles $\alpha_1\pi, \alpha_2\pi, \ldots, \alpha_n\pi$, and let Δ' be the exterior of Δ (i.e., the exterior of the boundary of Δ). Show that the transformation

$$w = C \int_{z_0}^z (z - a_1)^{\alpha_1 - 1} \cdots (z - a_n)^{\alpha_n - 1} \frac{dz}{(z - a)^2 (z - \bar{a})^2} + C_1$$

maps the upper half-plane $\operatorname{Im} z > 0$ conformally onto Δ', while carrying the points $a_1, a_2, \ldots, a_n$ of the real axis into the vertices $A_1, A_2, \ldots, A_n$ and the point a of the upper half-plane into the point at infinity.

11. Show that the transformation

$$w = C \int_{z_0}^z (z - a_1)^{\alpha_1 - 1} \cdots (z - a_n)^{\alpha_n - 1} \frac{dz}{z^2} + C_1 \tag{27}$$

maps the unit disk $|z| < 1$ conformally onto the polygonal domain Δ' of the preceding problem, while carrying the points $a_1, a_2, \ldots, a_n$ of the unit circle $|z| = 1$ into the vertices $A_1, A_2, \ldots, A_n$ and the point $z = 0$ into the point at infinity.

12. Map the unit disk $|z| < 1$ onto the five-pointed star shown in Figure 52.

13. Map the unit disk $|z| < 1$ onto the domain exterior to a square.

14. Let C be the unit circle $|z| = 1$. Prove that the function

$$w = \frac{1}{2}\left(z + \frac{1}{z}\right)$$

Figure 52

maps both the interior and the exterior of C onto the w-plane cut along the segment joining the points $w = -1$ and $w = 1$.

15. Map the unit disk $|z| < 1$ onto the w-plane cut along the $2n$ line segments joining the origin to the points $w = 1, e^{\pi i/n}, \ldots, e^{(2n-1)\pi i/n}$.

16. Let $w = \operatorname{sn} z$ denote the inverse of the elliptic integral

$$z = \int_0^w \frac{dw}{\sqrt{(1 - w^2)(1 - \mu^2 w^2)}}.$$

Prove that the "elliptic function" $\operatorname{sn} z$ is odd, i.e., that $\operatorname{sn}(-z) \equiv -\operatorname{sn} z$. Where does $\operatorname{sn} z$ fail to be analytic? What happens if $\mu = 0$? Prove that

$$\operatorname{sn}(z + 4nK + 2n'K'i) \equiv \operatorname{sn} z$$

for all $n, n' = 0, \pm 1, \pm 2, \ldots$, where K and K' are given by (14) and (15) with $C = 1$.

Comment. Thus the function $\operatorname{sn} z$ is "doubly periodic," with periods $\omega = 4K$ and $\omega' = 2K'i$ (cf. Chap. 13, Prob. 7).

CHAPTER FIFTEEN

Some Physical Applications

15.1. Fluid Dynamics

15.11. Consider the motion of an incompressible fluid (i.e., a liquid or a gas) at velocities much less than the velocity of sound. By a *velocity field* or *flow* we mean a vector function giving the velocity of the fluid at every point of a given region and at every instant of time. Such a flow is said to be *stationary* if it is independent of time and *plane-parallel* if it is the same in all planes parallel to a given plane Π and has no components perpendicular to Π. Obviously in the latter case there is no loss of generality in assuming that Π is the xy-plane. Thus a stationary plane-parallel flow is characterized by a vector function of two spatial variables x and y, or equivalently by a complex function

$$w(x, y) = u(x, y) + iv(x, y),$$

where $u(x, y)$ is the x-component of the flow and $v(x, y)$ its y-component. All the flows considered below are assumed to be both stationary and plane-parallel.

15.12. Given a flow $w = u + iv$ defined in a domain G, let C be any piecewise smooth curve of length l contained in G, with parametric representation

$$z = z(s) \qquad (0 \leq s \leq l)$$

in terms of variable arc length along C. Then by Chap. 5, Prob. 6, the derivative $z'(s)$ exists and is of modulus 1 at all but a finite number of points of the interval $0 \leq s \leq l$. Let $\tau(s)$ be the unit tangent (vector) to C at the point $z(s)$, and let $n(s)$ be the unit (exterior) normal to C at $z(s)$, as in Figure 53 (where C is closed). Then, by elementary calculus, $\tau(s)$ has components dx/ds, dy/ds, so that

$$\tau(s) = z'(s) = \frac{dx}{ds} + i\frac{dy}{ds},$$

$$n(s) = \frac{1}{i}z'(s) = \frac{dy}{ds} - i\frac{dx}{ds}.$$

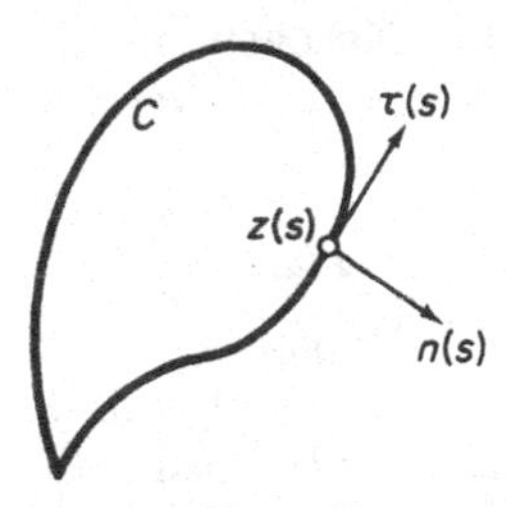

Figure 53

The component of the flow $w = u + iv$ tangent to C at the point $z(s)$, denoted by w_τ, is just the scalar product of the vectors $w = u + iv$ and $\tau(s)$, i.e.,

$$w_\tau = u\frac{dx}{ds} + v\frac{dy}{ds};$$

while the component of the flow normal to C at $z(s)$, denoted by w_n, is the scalar product of w and $n(s)$:

$$w_n = -v\frac{dx}{ds} + u\frac{dy}{ds}.$$

Suppose the flow $u + iv$ is continuously differentiable, i.e., suppose both u and v have continuous partial derivatives at every point of G, and let C be any piecewise smooth closed Jordan curve contained in G. Then the integral

$$\int_C w_\tau\, ds = \int_C (u + iv)_\tau\, ds = \int_C u\, dx + v\, dy \tag{1}$$

is called the *circulation* around C, while the integral

$$\int_C w_n\, ds = \int_C (u + iv)_n\, ds = \int_C -v\, dx + u\, dy \tag{2}$$

is called the *flux* through C.† If (1) vanishes for every curve C of the indicated type, we call the flow *irrotational* in G, while if (2) vanishes for every such C, we call the flow *solenoidal* in G. Since

$$\int_C \bar{w}\, dz = \int_C \overline{(u + iv)}(dx + i\, dy) = \int_C u\, dx + v\, dy + i\int_C -v\, dx + u\, dy,$$

we can write (1) and (2) in the alternative form

$$\int_C u\, dx + v\, dy = \operatorname{Re}\int_C \bar{w}\, dz, \tag{3}$$

$$\int_C -v\, dx + u\, dy = \operatorname{Im}\int_C \bar{w}\, dz. \tag{4}$$

†Concerning the physical meaning of these terms, see Probs. 1 and 2.

15.13. THEOREM. *A continuously differentiable flow $u + iv$ defined in a simply connected domain G is irrotational and solenoidal if and only if*

$$u + iv = \overline{f'(z)}, \tag{5}$$

where $f(z)$ is a function analytic in G, called the ***complex potential*** *of the flow.*

Proof. Suppose $u + iv$ is solenoidal and irrotational, so that the integrals (3) and (4) both vanish for every piecewise smooth closed Jordan curve C contained in G. Then both $u\,dx + v\,dy$ and $-v\,dx + u\,dy$ are exact differentials, i.e., there are real functions $\varphi = \varphi(x, y)$ and $\psi = \psi(x, y)$ such that

$$u\,dx + v\,dy = d\varphi,$$

$$-v\,dx + u\,dy = d\psi.$$

But then

$$u = \frac{\partial \varphi}{\partial x}, \qquad v = \frac{\partial \varphi}{\partial y}, \tag{6}$$

$$-v = \frac{\partial \psi}{\partial x}, \qquad u = \frac{\partial \psi}{\partial y}, \tag{7}$$

and in particular

$$\frac{\partial \varphi}{\partial x} = \frac{\partial \psi}{\partial y}, \qquad \frac{\partial \varphi}{\partial y} = -\frac{\partial \psi}{\partial x},$$

so that φ and ψ satisfy the Cauchy–Riemann equations in G. It follows that the function

$$f(z) = \varphi + i\psi$$

is analytic in G (see Sec. 4.23). Using (6) and (7), we immediately get

$$f'(z) = \frac{\partial \varphi}{\partial x} + i\frac{\partial \psi}{\partial x} = u - iv,$$

which is equivalent to (5).

Conversely, let $u + iv$ satisfy (5), where $f(z)$ is analytic in G, and let C be any piecewise smooth closed Jordan curve contained in G. Then

$$\int_C \bar{w}\,dz = \int_C f'(z)\,dz = 0$$

by Cauchy's integral theorem, since the derivative of an analytic function is itself analytic. But then the flow $u + iv$ is irrotational and solenoidal, because of (3) and (4). ∎

It follows from (5) that the functions u and $-v$ are conjugate harmonic functions in G. Also note that in terms of the complex potential $f(z)$, formulas (3) and (4) become

$$\int_C u\,dx + v\,dy = \operatorname{Re} \int_C f'(z)\,dz, \tag{3'}$$

$$\int_C -v\,dx + u\,dy = \operatorname{Im} \int_C f'(z)\,dz. \tag{4'}$$

15.14. The functions φ and ψ are called the (*velocity*) *potential* and the *stream function*, respectively, of the given flow, and correspondingly the curves

$$\varphi(x, y) = \text{const}, \qquad \psi(x, y) = \text{const} \tag{8}$$

("const" for "constant") are called *equipotentials* and *streamlines*. The mapping $\zeta = \xi + i\eta = f(z)$, where $f(z)$ is the complex potential, is conformal at every point of G except at points where the derivative $f'(z)$ vanishes (cf. Chap. 10, Prob. 24); these are also the points, known as *stagnation points*, where the velocity $u + iv$ vanishes. Clearly $\zeta = f(z)$ maps the curves (8) into the curves

$$\xi = \text{const}, \qquad \eta = \text{const}.$$

But the latter curves obviously form an orthogonal system, i.e., every curve $\xi = \text{const}$ is orthogonal to every curve $\eta = \text{const}$, and vice versa. Hence curves (8) also form an orthogonal system (except at stagnation points).

According to (8), the equipotentials are characterized by the condition

$$\frac{\partial \varphi}{\partial x}\,dx + \frac{\partial \varphi}{\partial y}\,dy = u\,dx + v\,dy = 0,$$

while the streamlines are characterized by the condition

$$\frac{\partial \psi}{\partial x}\,dx + \frac{\partial \psi}{\partial y}\,dy = -v\,dx + u\,dy = 0.$$

Thus at every point (x, y) of the flow (except at stagnation points), the velocity is normal to the equipotential through (x, y) and tangential to the streamline through (x, y). Incidentally, this proves once again that the equipotentials and streamlines form an orthonormal system. Moreover, the fact that the velocity is tangential to the streamlines shows that the streamlines are the actual trajectories of the moving fluid elements.

15.15. Any physical flow must satisfy the following condition: The surface of every object confining the flow, i.e., every curve making up the boundary Γ of the flow domain G,† must be part of a streamline $\psi(x, y) = \text{const}$, since the flow can have no component normal to such a surface. By the same token, if $f(z)$ is to be the complex potential of a flow, then $\psi(x, y) = \text{Im}\, f(z)$ must be constant on every curve making up Γ.

15.2. Examples

15.21. The entire linear function

$$f(z) = \alpha z \tag{9}$$

†Here we assume that Γ is made up of a finite number of piecewise smooth curves; Γ actually represents the projection onto the xy-plane of a set of cylindrical objects, regarded as very long in the z-direction (ideally, infinitely long).

can be regarded as the complex potential of a flow occupying the whole plane, with uniform velocity

$$\overline{f'(z)} = \bar{\alpha}$$

at every point. Writing $\alpha = a + ib$, we get the velocity potential and stream function

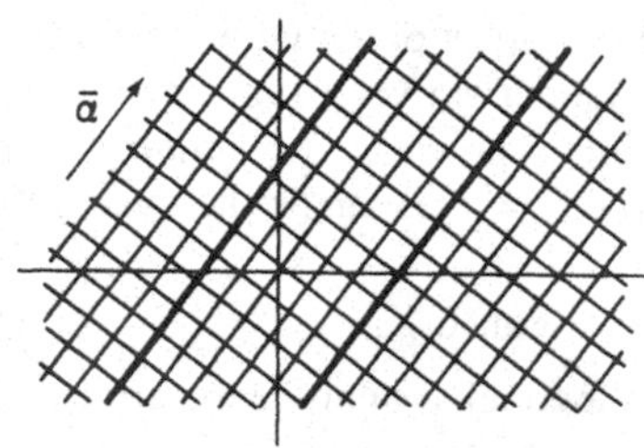

Figure 54

$$\varphi(x, y) = ax - by, \qquad \psi(x, y) = bx + ay.$$

The corresponding orthogonal system of equipotentials and streamlines

$$ax - by = \text{const}, \qquad bx + ay = \text{const}$$

is shown in Figure 54. The same function (9) represents the complex potential for uniform flow in a strip whose boundary consists of two lines parallel to the vector $\bar{\alpha}$ (like the two heavy lines shown in the figure).

15.22. The complex potential

$$f(z) = z^2 \tag{10}$$

also describes a flow occupying the whole plane, with nonuniform velocity

$$\overline{f'(z)} = 2\bar{z}.$$

This time the velocity potential and stream function are

$$\varphi(x, y) = x^2 - y^2, \qquad \psi(x, y) = 2xy,$$

while the corresponding orthogonal system of equipotentials and streamlines

$$x^2 - y^2 = \text{const}, \qquad 2xy = \text{const}$$

consists of two families of equilateral hyperbolas. The coordinates axes are themselves streamlines ($2xy = 0$) and intersect in a stagnation point at the origin. The same function (10) represents the complex potential of a flow in any quadrant of the xy-plane, with the sides of the quadrant regarded as the projections of the walls of a vessel containing the fluid (see Figure 55 for the case of flow in the first quadrant, a model of "flow around a corner").

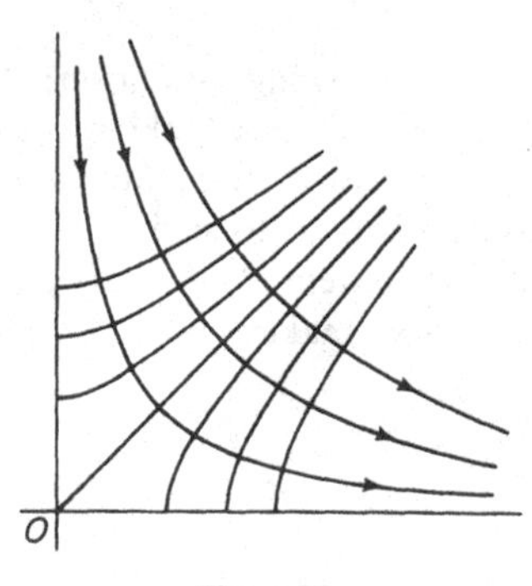

Figure 55

15.23. Let G be the domain $0 < |z| < \infty$, and let C be any piecewise smooth closed Jordan curve surrounding the origin and traversed in the positive direction. Then the flow with complex potential

$$f(z) = \frac{\kappa}{2\pi i} \ln z \qquad (\kappa \text{ real})$$

has circulation κ around C and zero flux through C. In fact, according to (3′) and (4′),

$$\int_C u\,dx + v\,dy = \text{Re}\frac{\kappa}{2\pi i}\int_C (\ln z)'\,dz = \text{Re}\frac{\kappa}{2\pi i}\int_C \frac{dz}{z} = \text{Re}\frac{\kappa}{2\pi i}2\pi i = \kappa,$$

$$\int_C -v\,dx + u\,dy = \text{Im}\frac{\kappa}{2\pi i}\int_C \frac{dz}{z} = 0.$$

On the other hand, the flow with complex potential

$$f(z) = \frac{\mu}{2\pi} \ln z \qquad (\mu \text{ real})$$

has zero circulation around C and flux μ through C, since now

$$\int_C u\,dx + v\,dy = \text{Re}\frac{\mu}{2\pi}\int_C \frac{dz}{z} = \text{Re}\frac{\mu}{2\pi}2\pi i = 0,$$

$$\int_C -v\,dx + u\,dy = \text{Im}\frac{\mu}{2\pi}\int_C \frac{dz}{z} = \text{Im}\, i\mu = \mu.$$

As an exercise, the reader should find the equipotentials and streamlines corresponding to both complex potentials. Note that both flows are irrotational and solenoidal in every simply connected subdomain of G, since

$$\int_C \frac{dz}{z} = 0$$

for every contour C which does not surround the origin.

15.24. Find the flow past a circular cylinder of radius R, assuming that the velocity at infinity equals $w_\infty = u_\infty + iv_\infty$† and that the flow is solenoidal in the flow domain G and irrotational in every simply connected subdomain of G.

Solution. According to (5), if $f(z)$ is the complex potential of the flow, then the derivative $f'(z)$ must be an analytic function in the domain $|z| > R$, taking the value $\bar{w}_\infty$ at infinity (see comment to Sec. 15.1). It follows that the Laurent expansion of $f'(z)$ at infinity is of the form

$$f'(z) = \bar{w}_\infty + \frac{c_1}{z} + \frac{c_2}{z^2} + \frac{c_3}{z^3} + \cdots \tag{11}$$

(cf. Chap. 11, Prob. 16), which implies

$$f(z) = \bar{w}_\infty z + c_1 \ln z - \frac{c_2}{z} - \frac{c_3}{2z^2} - \cdots, \tag{12}$$

---

†This is the natural idealization of the physical condition that the velocity has a given value w_∞ at a great distance from the cylinder.

where the constant of integration has been omitted since it has no effect on the velocity field.

To find the corresponding stream function $\psi(r, \theta) = \text{Im } f(z)$ in polar coordinates, we write

$$z = re^{i\theta}, \quad c_1 = a_1 + ib_1, \quad c_2 = a_2 + ib_2, \quad c_3 = a_3 + ib_3, \ldots,$$

obtaining the expansion

$$\psi(r, \theta) = a_1\theta + b_1 \ln r + \frac{a_2 + r^2 u_\infty}{r} \sin \theta - \frac{b_2 + r^2 v_\infty}{r} \cos \theta$$
$$+ \frac{a_3}{2r^2} \sin 2\theta - \frac{b_3}{2r^2} \cos 2\theta + \cdots. \tag{13}$$

Let Γ be the circle $|z| = R$. Then, since Γ must be one of the streamlines of the flow, the function $\psi(r, \theta)$ must be constant for $r = R$ and arbitrary θ (see Sec. 15.15). This condition will be satisfied if we make the following choice of coefficients in (13):

$$a_1 = 0, \quad a_2 + R^2 u_\infty = 0, \quad b_2 + R^2 v_\infty = 0, \quad a_3 = b_3 = \cdots = 0.$$

Then (11) and (12) reduce to

$$f'(z) = \bar{w}_\infty + \frac{ib_1}{z} - \frac{R^2 w_\infty}{z^2} \tag{14}$$

and

$$f(z) = ib_1 \ln z + \bar{w}_\infty z + \frac{R^2 w_\infty}{z}, \tag{15}$$

where b_1 is a real constant. To express b_1 in terms of the circulation κ around Γ (nonzero κ is allowed by the conditions of the problem), we note that by (3′),

$$\kappa = \text{Re} \int_\Gamma f'(z)\, dz = \text{Re} \int_{|z|=r} f'(z)\, dz$$

if $r > R$ (why?), and hence

$$\kappa = \text{Re} \int_{|z|=r} \left\{ \bar{w}_\infty + \frac{ib_1}{z} - \frac{R^2 w_\infty}{z^2} \right\} dz = -2\pi b_1. \tag{16}$$

Therefore

$$b_1 = -\frac{\kappa}{2\pi},$$

so that (14) and (15) take the form

$$f'(z) = \bar{w}_\infty + \frac{\kappa}{2\pi i z} - \frac{R^2 w_\infty}{z^2} \tag{17}$$

and

$$f(z) = \frac{\kappa}{2\pi i} \ln z + \bar{w}_\infty z + \frac{R^2 w_\infty}{z}, \tag{18}$$

where the first term in the right-hand side of (18) corresponds to a "purely circulatory flow" of the type considered in Example 15.23. Note that the coefficient c_1 of the second term in the expansion (11) equals $\kappa/2\pi i$. This holds quite generally, even in the case where Γ is an arbitrary piecewise smooth closed Jordan curve instead of the circle $|z| = R$.† In fact, let κ be the circulation around Γ, and let $|z| = r$ be any circle surrounding Γ. Then (16) is replaced by

$$\kappa = \operatorname{Re} \int_{|z|=r} \left\{ \bar{w}_\infty + \frac{c_1}{z} + \frac{c_2}{z^2} + \cdots \right\} dz = \operatorname{Re} 2\pi i c_1, \tag{16'}$$

while, on the other hand, the flux through Γ, equal to $\operatorname{Im} 2\pi i c_1$, must vanish since the flow is solenoidal in G. Thus c_1 is purely imaginary, and we again have

$$c_1 = \frac{\kappa}{2\pi i},$$

so that

$$f'(z) = \bar{w}_\infty + \frac{\kappa}{2\pi i z} + \frac{c_2}{z^2} + \cdots \tag{17'}$$

instead of (17).

Assuming for simplicity that $w_\infty = u_\infty > 0$ (this can always be achieved by making a preliminary rotation of the coordinate axes), we now look for the stagnation points of the flow, namely the points at which the velocity vanishes. Equating (17) to zero after setting $w_\infty = \bar{w}_\infty = u_\infty$, we get the quadratic equation

$$z^2 + \frac{\kappa}{2\pi i u_\infty} z - R^2 = 0,$$

with solutions

$$z_{1,2} = \frac{i\kappa}{4\pi u_\infty} \pm \sqrt{R^2 - \left(\frac{\kappa}{4\pi u_\infty}\right)^2}.$$

If $|\kappa| > 4\pi R u_\infty$, both stagnation points z_1 and z_2 are purely imaginary, but, because of the relation $z_1 z_2 = -R^2$, only one of these points lies outside the circle $|z| = R$, i.e., in the flow domain. The streamlines for this case are shown in Figure 56a. If $|\kappa| = 4\pi R u_\infty$, there is only one stagnation point, at one of the points in which the imaginary axis intersects the circle $|z| = R$ (see Figure 56b). If $|\kappa| < 4\pi R u_\infty$, there are two stagnation points z_1 and z_2 on the circle $|z| = R$ which are symmetric with respect to the imaginary axis (see Figure 56c). In the case of flow without circulation ($\kappa = 0$), there are obviously two stagnation points, at the points $\pm R$ in which the real axis intersects the circle $|z| = R$.

---

†We then say that the cylinder is of *cross section* Γ, even though the cylinder may well be solid.

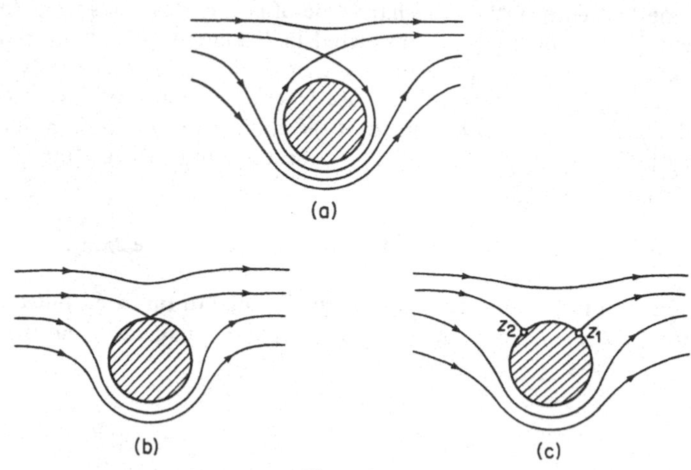

Figure 56

15.25. Under the same conditions as in the preceding example, find the flow past a cylinder of cross section Γ, where Γ is an arbitrary piecewise smooth closed Jordan curve.

Solution. Let $\zeta = g(z)$ be the unique conformal mapping of the exterior of Γ onto the exterior of the unit circle $|\zeta| = 1$ such that $g(\infty) = \infty$ while $g'(\infty)$ is a positive real number.† Then the Laurent expansion of $g(z)$ at infinity is of the form

$$\zeta = g(z) = cz + c_0 + \frac{c_1}{z} + \cdots \qquad (0 < c < \infty). \tag{19}$$

By (18), the complex potential for flow past the circle $|\zeta| = 1$ with velocity A at infinity and circulation κ around any circle $|\zeta| = r \geq 1$ is just

$$\Phi(\zeta) = \frac{\kappa}{2\pi i}\ln \zeta + \bar{A}\zeta + \frac{A}{\zeta} \tag{20}$$

(A will be suitably chosen in a moment). Replacing ζ by $g(z)$ in (20), we get a function

$$f(z) = \Phi(g(z)) = \frac{\kappa}{2\pi i}\ln g(z) + \bar{A}g(z) + \frac{A}{g(z)}, \tag{21}$$

†Here $g(z)$ fails to be analytic at infinity, because of the condition $g(\infty) = \infty$ corresponding to the presence of the term cz in (19). Hence, instead of defining $g'(\infty)$ as in Chap. 4, Prob. 30, we set

$$g'(\infty) = \lim_{z\to\infty} g'(z)$$

by definition, so that $g'(\infty) = c$, noting that $g(z)$ is conformal at infinity in this case (why?).

which has constant imaginary part on Γ as required,† and whose derivative is single-valued and analytic outside Γ. It follows that (21) is the complex potential for flow past Γ. To make the velocity of this flow equal to $w_\infty = u_\infty + iv_\infty$ at infinity, we note that

$$\bar{w}_\infty = f'(\infty) = \Phi'(\infty)g'(\infty) = \bar{A}c,$$

so that

$$A = \frac{w_\infty}{c} = \frac{w_\infty}{g'(\infty)}$$

is the proper choice of A. Thus, finally, the complex potential for flow past Γ with velocity w_∞ at infinity and circulation κ around Γ takes the form

$$f(z) = \frac{\kappa}{2\pi i}\ln g(z) + \frac{\bar{w}_\infty}{g'(\infty)}g(z) + \frac{w_\infty}{g'(\infty)g(z)}. \tag{21'}$$

15.26. Find the force exerted on a cylinder of cross section Γ by flow past the cylinder with velocity $w_\infty = u_\infty + iv_\infty$ at infinity.‡

Solution. If $P = P(x, y)$ is the pressure at the point (x, y) of the flow and if ρ is the density of the fluid (assumed to be constant), then *Bernoulli's law* states that the expression

$$P + \tfrac{1}{2}\rho\,|w|^2$$

is constant along streamlines and hence along the contour Γ. Thus

$$P = A - \tfrac{1}{2}\rho\,|w|^2$$

along Γ, where A is a positive constant. Since the pressure acting on an element $dz = dx + i\,dy$ of the contour Γ (assumed to be piecewise smooth) is directed along the interior normal to Γ, the force acting on dz due to the pressure is just

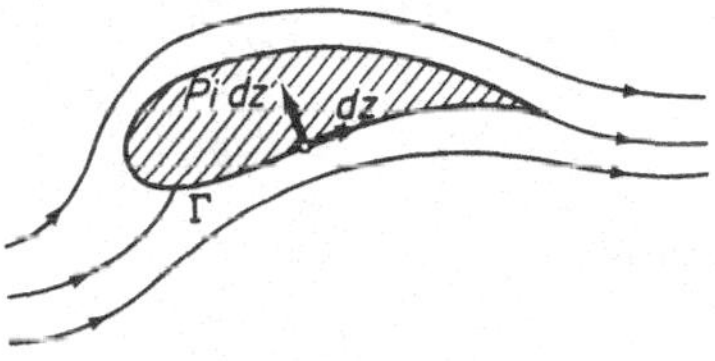

Figure 57

$$Pi\,dz = Ai\,dz - \tfrac{1}{2}\rho i\,|w|^2\,dz$$

(see Figure 57). Hence the total force $F = X + iY$ acting on Γ is given by the integral

$$X + iY = \int_\Gamma Pi\,dz = Ai\int_\Gamma dz - \tfrac{1}{2}\rho i\int_\Gamma |w|^2\,dz, \tag{22}$$

i.e.,

$$X + iY = -\tfrac{1}{2}\rho i\int_\Gamma |w|^2\,dz, \tag{23}$$

since the first integral in the right-hand side of (22) obviously vanishes. But

---

†Since $\Phi(\zeta)$ has constant imaginary part on the circle $|\zeta| = 1$, into which $\zeta = g(z)$ maps Γ (cf. Theorem 13.37).

‡More exactly, the force exerted on a unit length of the cylinder regarded as infinitely long in the z-direction.

the velocity w must be tangent to Γ, since Γ is a streamline, and hence

$$w = \overline{f'(z)} = |w|e^{i\theta},$$

where $f(z)$ is the complex potential of the flow and $\theta = \arg dz$, so that

$$|w| = \overline{f'(z)}e^{-i\theta}. \tag{24}$$

Substituting (24) into (23), we get

$$X + iY = -\tfrac{1}{2}\rho i \int_\Gamma [\overline{f'(z)}]^2 e^{-2i\theta}\, dz = -\tfrac{1}{2}\rho i \int_\Gamma [\overline{f'(z)}]^2\, \overline{dz}, \tag{25}$$

since

$$e^{-2i\theta}\, dz = e^{-i\theta}|dz| = \overline{dz}.$$

Taking the complex conjugate of (25), we finally get

$$X - iY = \tfrac{1}{2}\rho i \int_\Gamma [f'(z)]^2\, dz \tag{26}$$

or equivalently

$$X - iY = \tfrac{1}{2}\rho i \int_{|z|=r} [f'(z)]^2\, dz, \tag{26'}$$

where we replace Γ by any circle $|z| = r$ surrounding Γ (why is this justified?).

It is now a simple matter to express the force acting on Γ in terms of the circulation κ around Γ. In fact, substituting (17′) into (26′), we get

$$\begin{aligned} X - iY &= \frac{1}{2}\rho i \int_{|z|=r} \left\{\bar{w}_\infty + \frac{\kappa}{2\pi i z} + \frac{c_1}{z^2} + \cdots\right\}^2 dz \\ &= \frac{1}{2}\rho i \frac{2\kappa \bar{w}_\infty}{2\pi i} \int_C \frac{dz}{z} = \frac{1}{2}\rho i \frac{2\kappa \bar{w}_\infty}{2\pi i} 2\pi i = \rho\kappa\bar{w}_\infty i, \end{aligned}$$

and hence finally

$$X + iY = -\rho\kappa w_\infty i. \tag{27}$$

This is the celebrated *Kutta–Joukowski theorem*. In an aerodynamical context, (27) states that an airfoil at rest in a uniform wind of velocity w_∞ with circulation κ around the airfoil undergoes a force $\rho|\kappa w_\infty|$ perpendicular to the wind, where the direction of the force is obtained by rotating w_∞ through 90° in the direction opposite to the circulation.

15.3. Electrostatics

15.31. By an *electric field* we mean a vector field giving the force experienced by a unit positive charge at every point of a given region and at every instant of time. Just as in Sec. 15.11, such a field is said to be *stationary* if it is independent of time and *plane-parallel* if it is the same in all planes parallel to a given plane Π and has no components perpendicular to Π; in the latter case, there is again no loss of generality in assuming that Π is the xy-plane.

Thus a stationary plane-parallel electric field (concisely, an *electrostatic field in the plane*) is characterized by a vector function of two spatial variables x and y, or equivalently by a complex function

$$E(x, y) = E_x(x, y) + iE_y(x, y),$$

where $E_x(x, y)$ is the x-component of the field and $E_y(x, y)$ its y-component. Such a field is governed by the two *Maxwell equations*

$$\frac{\partial E_y}{\partial x} - \frac{\partial E_x}{\partial y} = 0, \tag{28}$$

$$\frac{\partial E_x}{\partial x} + \frac{\partial E_y}{\partial y} = 4\pi\rho, \tag{29}$$

where ρ is the surface charge density.

15.32. Now let G be any simply connected domain, and suppose $\rho \equiv 0$ in G (so that G is charge-free). Moreover, let C be any piecewise smooth closed Jordan curve contained in G. Then, by Green's theorem (see comment to Sec. 5.8), applied first to (28) and then to (29), we have

$$\int_C E_x\,dx + E_y\,dy = 0, \tag{30}$$

$$\int_C -E_y\,dx + E_x\,dy = 0, \tag{31}$$

so that the electrostatic field is both irrotational and solenoidal in G. Hence both $-E_x\,dx - E_y\,dy$ and $-E_y\,dx + E_x\,dy$ are exact differentials,† i.e., there are real functions $\varphi = \varphi(x, y)$ and $\psi = \psi(x, y)$ such that

$$-E_x\,dx - E_y\,dy = d\varphi,$$
$$-E_y\,dx + E_x\,dy = d\psi.$$

But then

$$E_x = -\frac{\partial\varphi}{\partial x}, \qquad E_y = -\frac{\partial\varphi}{\partial y}, \tag{32}$$

$$E_y = -\frac{\partial\psi}{\partial x}, \qquad E_x = \frac{\partial\psi}{\partial y}, \tag{33}$$

and in particular

$$\frac{\partial\psi}{\partial x} = \frac{\partial\varphi}{\partial y}, \qquad \frac{\partial\psi}{\partial y} = -\frac{\partial\varphi}{\partial x},$$

so that ψ and φ satisfy the Cauchy–Riemann equations in G. It follows that

---

†In keeping with historical convention, we introduce a minus sign in front of the expression $E_x\,dx + E_y\,dy$. Among other things, this causes the right-hand side of (34) to differ from that of (5) by the presence of the factor $-i$. The convention stems from the identification of φ with work done *against* the field rather than work done *by* the field.

the function

$$f(z) = \psi + i\varphi$$

is analytic in G. Using (32) and (33), we immediately get

$$f'(z) = \frac{\partial \psi}{\partial x} + i\frac{\partial \varphi}{\partial x} = -E_y - iE_x = -i(E_x - iE_y)$$

or equivalently

$$E_x + iE_y = -i\overline{f'(z)}. \tag{34}$$

Conversely, just as in the proof of Theorem 15.13, it is easy to see that if $E_x + iE_y$ satisfies (34), where $f(z)$ is analytic in G, then $E_x + iE_y$ is irrotational and solenoidal in G. It also follows from (34) that E_x and $-E_y$ are conjugate harmonic functions in G.

15.33. The functions ψ and φ are called the *stream function* and the (*electrostatic*) *potential*, respectively, of the given field, and correspondingly the curves

$$\psi(x, y) = \text{const}, \qquad \varphi(x, y) = \text{const}$$

are called *lines of force* and *equipotentials*. These curves form an orthogonal system for the same reason as in Sec. 15.14. Moreover, the electric field at (x, y) is tangential to the line of force through (x, y) and normal to the equipotential through (x, y). Thus the curves $\psi(x, y) = \text{const}$ are actually the curves along which the electric field acts (hence the term "lines of force").

Any conducting surface Γ situated in an electrostatic field must be part of an equipotential, i.e., the electric field can have no component tangential to Γ, since otherwise the field would produce motion of charge along Γ, contrary to the assumption that we are dealing with a stationary (i.e., time-independent) problem.†

15.34. Examples

a. The complex potential

$$f(z) = az \qquad (a > 0)$$

is defined in the whole plane, giving rise to an electric field

$$E_x + iE_y = -i\overline{f'(z)} = -ia$$

or equivalently

$$E_x = 0, \qquad E_y = -a.$$

†See e.g., J. A. Stratton, *Electromagnetic Theory*, McGraw-Hill Book Company, New York (1941), p. 164. By the same token, the lines of force cannot penetrate into the interior of a conductor.

Moreover

$$f(z) = a(x + iy) = ax + iay,$$

so that

$$\psi = ax, \qquad \varphi = ay.$$

Hence the lines of force are the vertical lines $x =$ const and the equipotentials are the horizontal lines $y =$ const, a fact which is obvious in advance from the symmetry of the problem. To get the field inside an infinite parallel-plate condenser of spacing $2h$, with upper plate at potential V and lower plate at potential $-V$, we need merely choose $a = V/h$, so that

$$\varphi = \frac{V}{h}y, \qquad E = -i\frac{V}{h}. \tag{35}$$

b. Find the electrostatic field near the edges of a parallel-plate condenser of spacing $2h$, with upper plate at potential V and lower plate at potential $-V$.

Solution. Here the full power of the complex variable method comes into play. Let $w = u + iv = f(z)$ be the function mapping the domain in the z-plane shown in Figure 58 (the exterior of a "semi-infinite condenser") onto

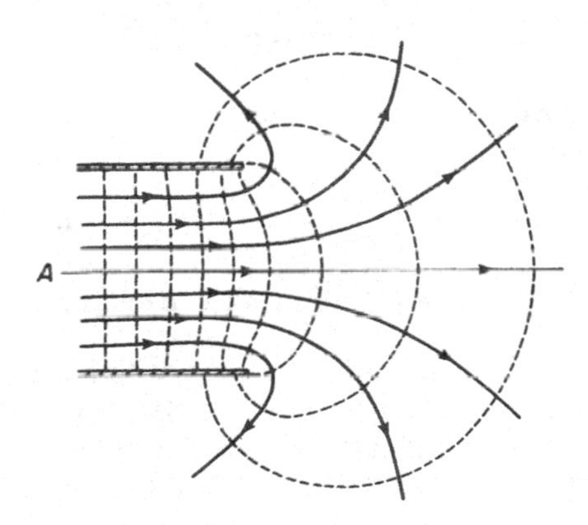

Figure 58

the strip $-V < v < V$ in the w-plane. This mapping has already been found (with the help of the Schwarz–Christoffel transformation) in Example 14.24, and is given by formula (25), p. 224 after interchanging the variables z and w, changing the width of the strip from 2π to $2V$, and dropping a superfluous additive constant:

$$z = \frac{h}{\pi}\left(e^{\pi w/V} + \frac{\pi w}{V}\right). \tag{36}$$

Taking real and imaginary parts of (36), we get two equations

$$x = \frac{h}{\pi}\left(e^{\pi u/V} \cos \frac{\pi v}{V} + \frac{\pi u}{V}\right),$$

$$y = \frac{h}{\pi}\left(e^{\pi u/V} \sin \frac{\pi v}{V} + \frac{\pi v}{V}\right).$$

The corresponding lines of force and equipotentials, obtained by setting first $u = \text{const}$ and then $v = \text{const}$ in these equations, are shown in the figure. The electric field E, given by (34), is just

$$E = -i\overline{\frac{dw}{dz}} = -i\frac{1}{\overline{\frac{dz}{dw}}} = -i\frac{V}{h}\frac{1}{1 + e^{\pi \bar{w}/V}}.$$

Deep inside the condenser, i.e., for z near the point A shown in the figure, $\bar{w}$ is near $-\infty$ and hence E is near $-iV/h$, in keeping with (35). However $\bar{w} \to \pm Vi$ near the edge of the condenser, causing the field E to become infinitely large. This cannot actually happen in the laboratory, of course, since (among other things) no physical condenser can have perfectly sharp edges.

COMMENTS

15.1. The functions φ and ψ figuring in the proof of Theorem 15.13 are given by the integrals

$$\varphi(x, y) = \int_{(x_0, y_0)}^{(x, y)} u\, dx + v\, dy + \text{const},$$

$$\psi(x, y) = \int_{(x_0, y_0)}^{(x, y)} -v\, dx + u\, dy + \text{const}$$

taken along any piecewise smooth curve in G with initial point $(x_0, y_0) \in G$ and final point $(x, y) \in G$ (see comment to Sec. 5.8), and hence are themselves continuously differentiable. Theorem 15.13 can be generalized by allowing the flow domain G to be multiply connected and the complex potential to be a multiple-valued analytic function (in the sense of Sec. 13.45), with single-valued analytic branches in every simply connected subdomain of G. Then the flow is still irrotational and solenoidal in every simply connected subdomain of G, but it may fail to be so in G itself (as in Examples 15.23–15.26). However, even in this more general case, the derivative $f'(z)$ is single-valued and analytic in G (see Chap. 9, Prob. 8).

15.2. To see how an airplane works, note that (27) predicts a *lifting force* $Y = \rho|\kappa|w_\infty > 0$ if $\kappa < 0$, $w_\infty > 0$. For the way in which the sharp trailing edge of an actual airfoil leads to circulation around the airfoil, see e.g., A. I. Markushevich, *op. cit.*, Volume II, p. 194ff.

15.3. The reader familiar with electromagnetic theory will recognize (28) and (29) as the two-dimensional versions of the Maxwell equations

$$\operatorname{curl} \vec{E} = 0, \qquad \operatorname{div} \vec{E} = 4\pi\rho,$$

themselves just special cases of a more general set of four Maxwell equations. In writing (28) and (29), we assume that ρ is continuous and that E is continuously differentiable. The generalization of the considerations of Sec. 15.32 to the case of a multiply connected domain G and a multiple-valued complex potential $f(z)$ is the same as in the comment to Sec. 15.1, except that unlike the velocity field, the electric field is *always* irrotational, because of (28).

PROBLEMS

1. Show that the flux of a vector field $w = u + iv$ through a contour C is just the net amount of fluid flowing outward through C per unit time. Hence show that if a flow is solenoidal in a domain G, then there are no "sources" or "sinks" in G, i.e., no points at which fluid enters or leaves G (cf. Prob. 3).

2. Suppose a fluid element in a flow $w = u + iv$ suddenly "freezes" and afterwards moves freely with respect to the rest of the fluid. Prove that the element rotates with angular velocity

$$\frac{1}{2}\left(\frac{\partial v}{\partial x} - \frac{\partial u}{\partial y}\right).$$

Hence show that if a flow is irrotational in a domain G, then the fluid elements undergo translational motion with accompanying distortion, but no rotation (hence the word "irrotational").

3. Sketch equipotentials and streamlines for the flow with complex potential

$$f(z) = \frac{\mu}{2\pi} \ln \frac{z - z_1}{z - z_2} \qquad (\mu \text{ real}). \tag{37}$$

Prove that the flux through every contour (i.e., piecewise smooth closed Jordan curve) surrounding z_1 but not z_2 is μ, while that through every contour surrounding z_2 but not z_1 is $-\mu$, a fact summarized by saying that the flow has a *source of strength* μ at z_1 and a source of strength $-\mu$ at z_2. A source of negative strength m is usually called a *sink of strength* $|m|$.

4. Sketch equipotentials and streamlines for the flow with complex potential

$$f(z) = \frac{\kappa}{2\pi i} \ln \frac{z - z_1}{z - z_2} \qquad (\kappa \text{ real}). \tag{37'}$$

Prove that the circulation around every contour surrounding z_1 but not z_2 is κ while that around every contour surrounding z_2 but not z_1 is $-\kappa$, a fact summarized by saying that the flow has a *vortex of strength* κ at z_1 and a vortex of strength $-\kappa$ at z_2.

5. Sketch equipotentials and streamlines for the flow whose complex potential is the sum of the potentials (37) and (37').

Comment. Each of the points z_1 and z_2 is now a *spiral vortex*, i.e., a point where a source and a vortex are combined.

6. Analyze the flow described by each of the following complex potentials, i.e., plot the velocity field, equipotentials and streamlines, and look for stagnation points, sources, vortices, etc:

   a) $f(z) = \ln(z^2 - a^2) \quad (a > 0)$; b) $f(z) = \ln \dfrac{z^2 - a^2}{z^2 + a^2} \quad (a > 0)$;

   c) $f(z) = \ln\left(1 + \dfrac{1}{z^2}\right)$; d) $f(z) = az + \dfrac{\kappa}{2\pi i} \ln z \quad (a > 0, \kappa > 0)$.

7. Let $|\kappa| < 4\pi R u_\infty$ in the problem of flow past a circular cylinder (Example 15.24), so that there are two stagnation points $z_1 = Re^{i\varphi_1}$ and $z_2 = Re^{i\varphi_2}$ on the circle $|z| = R$, as in Figure 56c. Prove that

$$\varphi_1 = \arc\sin \frac{\kappa}{4\pi R u_\infty}, \qquad \varphi_2 = \pi - \varphi_1,$$

   so that in particular $\kappa = 4\pi R u_\infty \sin \varphi_1$.

8. Find the complex potential for flow past an elliptic cylinder with semiaxes a and b, given that the flow has circulation around the cylinder and velocity $w_\infty = u_\infty + iv_\infty$ at infinity.

9. Prove that a cylinder of cross section Γ immersed in a flow described by a complex potential $f(z)$ experiences a torque (per unit length) about the origin equal to

$$T = -\tfrac{1}{2}\rho \operatorname{Re} \int_\Gamma z[f'(z)]^2\, dz,$$

   where ρ is the density of the fluid. Prove that T vanishes in the case of a circular cylinder.

10. Calculate the torque T acting on an elliptic cylinder with semiaxes a and b immersed in a flow with no circulation and velocity $|w_\infty|e^{i\alpha}$ at infinity.

11. Prove that the mapping

$$w = (z - 1)^\alpha \left(1 + \frac{\alpha z}{1 - \alpha}\right)^{1-\alpha}$$

   maps the upper half-plane $\operatorname{Im} z > 0$ onto the upper half-plane $\operatorname{Im} w > 0$ cut along the line segment joining the points 0 and $e^{i\alpha\pi}$ (see Figure 59).

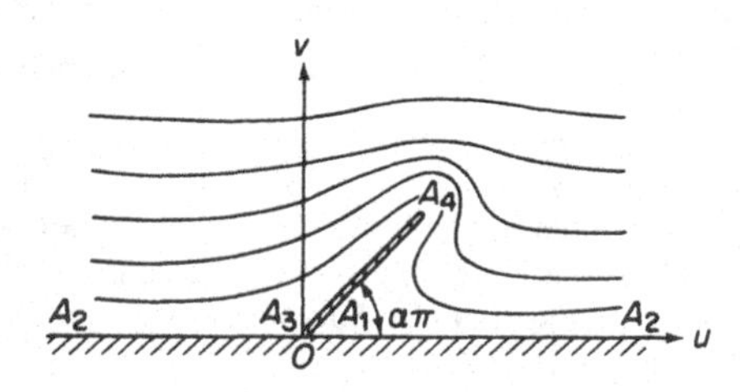

Figure 59

12. Use the result of the preceding problem to find the flow of a fluid in an infinitely deep basin of the form shown in Figure 59, where this time the segment representing the obstacle is of length h.

13. Investigate the electrostatic field corresponding to the complex potential

$$f(z) = 2qi \ln \frac{1}{z}.$$

14. Show that the field of the preceding problem is produced by an infinite charged line perpendicular to the xy-plane at the origin, carrying charge q per unit length.

15. Find the lines of force, equipotentials and electric field due to the complex potential $f(z) = 1/z^2$.

16. Find the lines of force and complex potential of the electric field with potential

$$\varphi = \arc\tan \frac{\tan \pi y}{\tanh \pi x}.$$

17. Suppose an electrostatic field has the circles $x^2 + y^2 = 2ax$ as equipotentials. Find the ratio between the magnitude of the field at the point $(2a, 0)$ and its magnitude at the point (a, a).

18. Show that the complex potential of the electrostatic field produced by the condenser shown in Figure 60, consisting of coplanar plates a distance $2a$

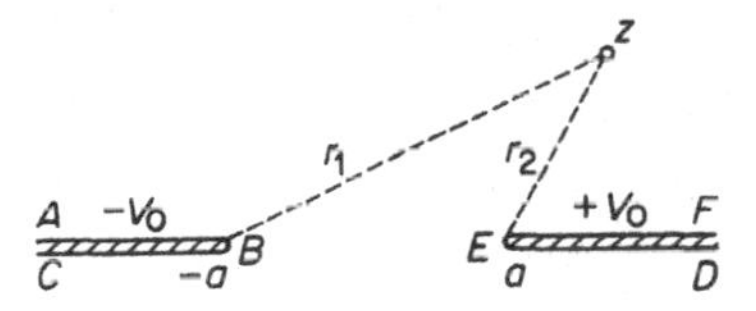

Figure 60

apart, one at potential V, the other at potential $-V$, is given by

$$f(z) = \frac{2V}{\pi} \ln (z + \sqrt{z^2 - a^2}).$$

Find the corresponding electric field.

19. Let $T = T(z)$ be the temperature at the point $z = x + iy$ in a stationary plane-parallel temperature field. Then it can be shown that T satisfies Laplace's equation

$$\frac{\partial^2 T}{\partial x^2} + \frac{\partial^2 T}{\partial y^2} = 0.$$

Prove that the temperature distribution in the semi-infinite slab of width a shown in Figure 61, whose sides are at temperature zero and whose bottom

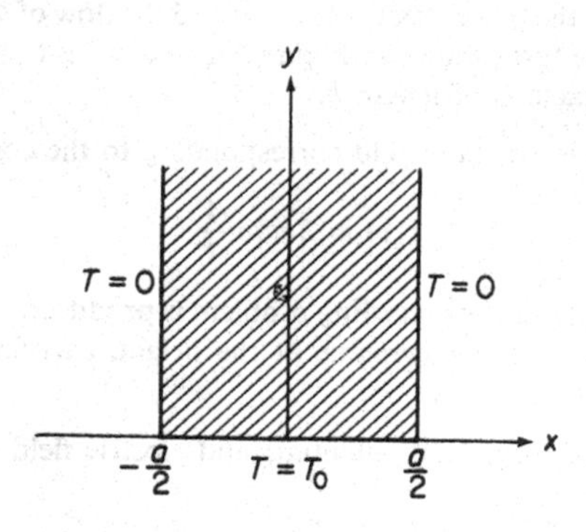

Figure 61

is at temperature T_0, is given by

$$T = \frac{2T_0}{\pi} \arctan \frac{\cos \dfrac{\pi x}{a}}{\sinh \dfrac{\pi y}{a}}$$

Selected Hints and Answers

Chapter 1

2. *Ans.* b) $0, 1, i, -1, -i$.

4. *Ans.* The parabola $r = 1/(1 + \cos\theta)$ and its interior.

5. *Ans.* b) The upper half-plane excluding the real axis; d) A hyperbola reducing to a pair of straight lines if $a = 0$; f) The right half-plane including the imaginary axis.

9. *Ans.* When the ratio $(z_1 - z_3)/(z_2 - z_3)$ is real.

10. *Ans.* $z = (\lambda_2 z_1 + \lambda_1 z_2)/(\lambda_1 + \lambda_2)$.

11. *Ans.* $z_4 = z_1 - z_2 + z_3$.

15. *Ans.* a) $2^{12}(1 + i)$; c) $(2 - \sqrt{3})^{12}$.

16. *Ans.* $\cos nx = \cos^n x - \binom{n}{2}\cos^{n-2} x \sin^2 x + \binom{n}{4}\cos^{n-4} x \sin^4 x - \cdots,$

$\sin nx = n\cos^{n-1} x \sin x - \binom{n}{3}\cos^{n-3} x \sin^3 x + \binom{n}{5}\cos^{n-5} x \sin^5 x - \cdots,$

where $\binom{n}{k}$ denotes the binomial coefficient $n!/k!(n-k)!$.

17. *Hint.* Use Prob. 16.

18. *Ans.* $\sqrt{1+i} = \pm\sqrt{2}\left(\cos\frac{\pi}{8} + i\sin\frac{\pi}{8}\right)$.

19. *Ans.* $x = \pm\sqrt{\frac{\sqrt{a^2+b^2}+a}{2}}$, $y = \pm\sqrt{\frac{\sqrt{a^2+b^2}-a}{2}}$, where the signs are taken to be the same if $b > 0$ and opposite if $b < 0$.

22. *Hint.* Multiply by $1 - \epsilon$.

25. *Hint.* The equation of any circle or straight line in the xy-plane can be written in the form

$$A(x^2 + y^2) + 2Bx + 2Cy + D = 0 \qquad (A, B, C, D \text{ real}),$$

where we have a circle if $A \neq 0$, $B^2 + C^2 - AD > 0$ (why?) or a straight line if $A = 0$ and at least one of the numbers B, C is nonzero. Now let $E = B + iC$.

Chapter 2

1. *Ans.* b) $0, 2, 1 \pm i$.
2. *Hint.* Cf. Example 2.23a.
3. *Ans.* b) Every point in the complex plane.
6. *Hint.* Consider the sequence $1, -1, 1, -1, \ldots$
7. *Hint.* Let n_0 be the smallest value of n for which there exists a neighborhood of α containing all the points $z_n, z_{n+1}, \ldots$ but not the origin, and let θ_n be the unique value of $\arg z_n$ satisfying the inequality $|\theta_n - \theta| < \pi/2$. Then $\theta_n \to \theta$ as $n \to \infty$ (why?).
8. *Hint.* Consider the sequence $z_n = -1 + (-1)^n \dfrac{i}{n}$.
12. *Ans.* b) 1.
13. *Hint.* There is no loss of generality in assuming that $\alpha = 0$.
16. *Hint.* Any "spherical cap" containing N cut off from Σ by a plane perpendicular to the diameter ON is a neighborhood of N.
17. *Hint.* Cf. Example 2.23a.
18. *Hint.* Cf. Theorem 2.25. Define a limit point at ∞ in the obvious way.
21. *Ans.* A circle through the pole N.
22. *Ans.* No.

Chapter 3

2. *Ans.* The domain $0 < |z| < 1$.
7. *Ans.* a_m/b_n if $m = n$, 0 if $m < n$, ∞ if $m > n$.
10. *Hint.* $||f(z)| - |f(z_0)|| \leq |f(z) - f(z_0)|$, formula (17), p. 9.
12. *Hint.* a) Use the Heine–Borel theorem to show that $\mathcal{E}$ can be covered by a finite number of neighborhoods and hence is bounded; b) Let η be a limit point of $\mathcal{E}$, let w_n be a sequence of distinct points of $\mathcal{E}$ converging to η (Chap. 2, Prob. 2), and let z_n be any point of E such that $f(z_n) = w_n$. The sequence z_n is bounded, and hence has a limit point ζ by the Bolzano–Weierstrass theorem. Use the continuity of $f(z)$ to show that $f(\zeta) = \eta$ and hence that $\eta \in \mathcal{E}$, so that $\mathcal{E}$ is closed (as well as bounded); c) The set $\mathcal{E}^*$ of all images of points $z \in E$ under the mapping $w = |f(z)|$ is bounded and closed, by b) and Prob. 8. Let m be the greatest lower bound and M the least upper bound of $\mathcal{E}^*$.

Show that m and M belong to $\mathcal{E}^*$, so that there exist points z_0, $z \in E$ such that $|f(z_0)| = m$, $|f(Z)| = M$.

13. *Ans.* Continuous but not uniformly continuous.

14. *Hint.* Use Prob. 6.

16. *Ans.* Only $f(z) = \dfrac{z \operatorname{Re} z}{|z|}$, $f(0) = 0$.

17. *Hint.* First use the fact that Γ is closed (cf. Prob. 11) to prove that the distance r_z between Γ and any point z not in Γ is positive, r_z being defined as the greatest lower bound of all numbers $|z - \zeta|$ with $\zeta \in \Gamma$. Given any point $z \in C$, let K_z be the open disk of radius $\frac{1}{2}r_z$ centered at z. Apply the Heine–Borel theorem to the bounded closed set C (cf. Prob. 11), thereby covering C with a finite number of disks $K_{z_1}, \ldots, K_{z_n}$. Let δ be the smallest radius of these disks. If z is any point of C, then z belongs to some disk K_{z_ν}. But the disk of radius r_{z_ν} centered at z_ν contains no points of Γ, and hence $|z - \zeta| \geq \frac{1}{2}r_{z_\nu} \geq \delta$ for all $\zeta \in \Gamma$.

Chapter 4

2. *Hint.* See formula (2).

4. *Hint.* In the first case $f'(z) = \dfrac{\partial u}{\partial x} + i\dfrac{\partial v}{\partial x}$, in the second case $f'(z) = -i\dfrac{\partial u}{\partial y} + \dfrac{\partial v}{\partial y}$ (cf. Example 4.13).

10. *Ans.* The arc of the parabola $v^2 = 4(1 - u)$ joining the points $(0, -2)$ and $(0, 2)$.

11. *Ans.* a) $\alpha = \pi/2$, $\mu = 2$; c) $\alpha = \pi/4$, $\mu = 2\sqrt{2}$.

15. *Ans.* $w = (z - z_0)^5$.

19. *Ans.* b) $\pi/2, 1, 2 + 2i$.

20. *Ans.* $w = (2 + i)z + 1 - 3i$.

21. *Ans.* $w = (1 + i)(1 - z)$.

26. *Hint.* The mapping $\eta = (cz + d)/(az + b)$ is conformal at $z = \delta$ and carries C, C^* into two curves L, L^* in the η-plane forming an angle of α radians with its vertex at the origin. But the mapping $w = 1/\eta$, equivalent to (11′), carries L, L^* into two curves Γ, Γ^* in the w-plane forming an angle of α radians with its vertex at infinity.

27. *Hint.* The mapping $\zeta = 1/z$ carries C, C^* into two curves L, L^* in the ζ-plane forming an angle of α radians with its vertex at the origin. But the mapping $w = (b\zeta + a)/(d\zeta + c)$, equivalent to (11′), is conformal at $\zeta = 0$ and carries L, L^* into two curves Γ, Γ^* in the w-plane forming an angle of α radians with its vertex at $A = a/c$.

Chapter 5

1. *Hint.* Cf. Sec. 4.31.

2. *Hint.* It follows from the mean value theorem that

$$\sum_{k=1}^{n} |\Delta z_k| = \sum_{k=1}^{n} \sqrt{[x(t_k) - x(t_{k-1})]^2 + [y(t_k) - y(t_{k-1})]^2}$$
$$= \sum_{k=1}^{n} \sqrt{[x'(\tau_k)]^2 + [y'(\tau_k^*)]^2}\, \Delta t_k \qquad (\Delta t_k = t_k - t_{k-1}),$$

where τ_k and τ_k^* are appropriate points in the interval $[t_{k-1}, t_k]$.

3. *Hint.* Suppose $\widehat{AB}$ is rectifiable, of length l. Inscribe arbitrary polygonal curves of lengths p' and p'' in $\widehat{AP}$ and $\widehat{PB}$ respectively. Together these polygonal curves make up a polygonal curve of length

$$p' + p'' = p \leq l \tag{i}$$

inscribed in C. In particular, $p' \leq l$, $p'' \leq l$, so that p' and p'' are bounded from above and hence have least upper bounds l' and l'', by the completeness of the real number system. Therefore $\widehat{AP}$ and $\widehat{PB}$ are rectifiable, of lengths l' and l''. Moreover

$$l' + l'' \leq l, \tag{ii}$$

after taking least upper bounds in (i).

Conversely suppose $\widehat{AP}$ and $\widehat{PB}$ are rectifiable, of lengths l' and l'', respectively. Inscribe an arbitrary polygonal curve L of length p in $\widehat{AB}$. If P is a vertex of L, then L decomposes at once into two polygonal curves, the first of length p' inscribed in $\widehat{AP}$, the second of length p'' inscribed in $\widehat{PB}$. Otherwise replace L by a new polygonal curve L^* of length p^* with the same vertices as L plus the extra vertex P. Adding an extra vertex cannot decrease the length of a polygonal curve inscribed in $\widehat{AB}$, and hence $p \leq p^*$. Thus, in any event,

$$p \leq p' + p'' \leq l' + l'', \tag{iii}$$

so that p is bounded from above and hence has a least upper bound l. Therefore $\widehat{AB}$ is rectifiable, of length l. Moreover

$$l \leq l' + l'' \tag{iv}$$

after taking least upper bounds in (iii). Now compare (ii) and (iv).

4. *Hint.* Let $s(t)$ be the length of the variable arc of C with initial point $z(a)$ and final point $z(t)$. (The existence of $s(t)$ follows from Probs. 2 and 3.) If $\Delta s = s(t + \Delta t) - s(t)$, then by Prob. 2,

$$\sqrt{m_x^2 + m_y^2} \leq \frac{\Delta s}{\Delta t} \leq \sqrt{M_x^2 + M_y^2},$$

where the quantities m_x, m_y, M_x, M_y now pertain to the interval $[t, t + \Delta t]$ rather than to the interval $[a, b]$. But $m_x, M_x \longrightarrow |x'(t)|$ as $\Delta t \longrightarrow 0$, while $m_y, M_y \longrightarrow |y'(t)|$ as $\Delta t \longrightarrow 0$, by the continuity of $x'(t)$ and $y'(t)$. Therefore

$$s'(t) = \lim_{\Delta t \to 0} \frac{\Delta s}{\Delta t} = \sqrt{[x'(t)]^2 + [y'(t)]^2}.$$

To get l, integrate $s'(t)$ from a to b.

5. *Hint.* Show that the difference between the sum

$$\sum_{k=1}^{n} f(\zeta_k)\Delta z_k \qquad (\Delta z_k = z_k - z_{k-1})$$

corresponding to an arbitrary "partition" $z_0, z_1, z_2, \ldots, z_{n-1}, z_n$ of C and the sum corresponding to the same partition plus the end points of the arcs $C_1, C_2, \ldots, C_n$ is of modulus no greater than $2M\lambda(n-1)$, where λ is the same as in Sec. 5.11 and $M = \max_{z \in C} |f(z)|$.

6. *Hint.* Let $z(t(s)) = \tilde{z}(s)$, and note that

$$s = \int_0^{t(s)} |z'(t)|\, dt.$$

7. *Hint.* Since C is closed, $z(a) = z(b)$.

8. *Hint.* Given any closed curve C in G, suppose (48) holds and let C, γ and $C_2 = \gamma^-$ be the same as in Prob. 7. Then

$$0 = \int_C f(z)\, dz = \int_{C_1} f(z)\, dz + \int_\gamma f(z)\, dz = \int_{C_1} f(z)\, dz - \int_{C_2} f(z)\, dz,$$

by Theorem 5.21. Conversely, given any two curves C_1 and C_2 in G with the same initial and final points, suppose (47) holds and form the closed curve $C = C_1 + C_2^-$. Then

$$0 = \int_{C_1} f(z)\, dz - \int_{C_2} f(z)\, dz = \int_{C_1} f(z)\, dz + \int_{C_2^-} f(z)\, dz = \int_C f(z)\, dz.$$

9. *Hint.* Note that

$$\int_{C_1} \operatorname{Re} z\, dz = 2 + i, \qquad \int_{C_2} \operatorname{Re} z\, dz = 2 + 2i, \qquad \int_C \operatorname{Re} z\, dz = -i.$$

10. *Ans.* a) 1; b) 2; c) 0.

11. *Hint.* Choosing first $\zeta_k = z_k$ and then $\zeta_k = z_{k-1}$, we get

$$\int_C z\, dz = \lim_{\lambda \to 0} \sum_{k=1}^{n} \zeta_k \Delta z_k = \tfrac{1}{2} \lim_{\lambda \to 0} \sum_{k=1}^{n} (z_k + z_{k-1})(z_k - z_{k-1})$$

$$= \tfrac{1}{2} \lim_{\lambda \to 0} \sum_{k=1}^{n} (z_k^2 - z_{k-1}^2) = \tfrac{1}{2}(z_n^2 - z_0^2) = \tfrac{1}{2}(Z^2 - z_0^2).$$

12. *Hint.* Use Theorem 5.23.

14. *Hint.* If $f(z) = u + iv$, then

$$\int_C f(z)\, dz = \int_C u\, dx - v\, dy + \int_C v\, dx + u\, dy$$

$$= \int_I \left(-\frac{\partial v}{\partial x} - \frac{\partial u}{\partial y}\right) dx\, dy + \int_I \left(\frac{\partial u}{\partial x} - \frac{\partial v}{\partial y}\right) dx\, dy,$$

where I is the interior of C. Now use the Cauchy–Riemann equations.

15. *Hint.* $\dfrac{1}{z^2+1} = \dfrac{1}{2i}\left(\dfrac{1}{z-i} - \dfrac{1}{z+i}\right).$

16. *Hint.* b) $-\pi$.

17. *Ans.* No.

18. *Hint.* $\dfrac{z^4 + z^2 + 1}{z(z^2 + 1)} = z + \dfrac{1}{z} - \dfrac{1}{2}\left(\dfrac{1}{z + i} + \dfrac{1}{z - i}\right)$.

19. *Ans.* $\dfrac{\pi}{4} + k\pi$ $(k = 0, \pm 1, \pm 2, \ldots)$, where C must not go through the points $\pm i$.

22. *Ans.* No; $f(u) = a + bu$.

26. *Hint.* If $U = U(w) = U(u + iv)$, then

$$U_{xx} = (u_x + iv_x)^2 U'' + (u_{xx} + iv_{xx})U',$$
$$U_{yy} = (u_y + iv_y)^2 U'' + (u_{yy} + iv_{yy})U',$$

and hence $U_{xx} + U_{yy} = 0$ (why?). Alternatively construct an analytic function $\varphi(w)$ with U as its real part, and then consider the composite function $\varphi(f(z))$.

27. *Hint.* The proof is virtually the same as that of Theorem 5.71.

28. *Hint.* If C is any piecewise smooth closed Jordan curve contained in G, then

$$\int_C F(z)\,dz = \int_C \left\{\int_\Gamma f(z, \zeta)\,d\zeta\right\} dz = \int_\Gamma \left\{\int_C f(z, \zeta)\,dz\right\} d\zeta$$

(why?). But

$$\int_C f(z, \zeta)\,dz = 0$$

for every $\zeta \in \Gamma$, by Cauchy's integral theorem, and hence

$$\int_C F(z)\,dz = 0.$$

The analyticity of $F(z)$ now follows from Morera's theorem.

Chapter 6

2. *Hint.* Cf. Chap. 2, Prob. 5.

4. *Hint.* Use Prob. 3.

5. *Ans.* No.

6. *Hint.* If $s_n = z_1 + \cdots + z_n$, then obviously $|s_n| \leq |z_1| + \cdots + |z_n|$ and hence certainly $|s_n| \leq |z_1| + \cdots + |z_n| + \cdots$. Now take the limit of the left-hand side as $n \to \infty$.

7. *Hint.* The series

$$1 - \tfrac{1}{2} + \tfrac{1}{3} - \tfrac{1}{4} + \tfrac{1}{5} - \tfrac{1}{6} + \cdots$$

converges, but not the two series

$$1 + \tfrac{1}{3} + \tfrac{1}{5} + \cdots, \qquad -\tfrac{1}{2} - \tfrac{1}{4} - \tfrac{1}{6} - \cdots.$$

9. *Ans.* a) Absolutely convergent; b) Divergent (oscillatory); c) Absolutely convergent.

10. *Hint.* $\dfrac{1}{\sqrt{1}}\dfrac{1}{\sqrt{n}} + \dfrac{1}{\sqrt{2}}\dfrac{1}{\sqrt{n-1}} + \cdots + \dfrac{1}{\sqrt{n}}\dfrac{1}{\sqrt{1}} > \dfrac{n}{\sqrt{n}\sqrt{n}} = 1.$

11. *Hint.* Cf. Example 6.34.

16. *Hint.* On the one hand $|1 + z + \cdots + z^{n-1}| < n$ if $|z| < 1$. On the other hand the sum $s(z)$ of the series is just $1/(1 - z)$ and hence $|s_n(z) - s(z)|$ becomes arbitrarily large for z sufficiently close to 1.

19. *Hint.* Consider the (real) series
$$\sum_{n=1}^{\infty} 2x[n^2 e^{-n^2x^2} - (n - 1)^2 e^{-(n-1)^2x^2}] \qquad (0 \le x \le 1)$$
with nth partial sum
$$s_n(x) = 2xn^2 e^{-n^2x^2}$$
and sum $s(x)$ identically equal to 0. The convergence cannot be uniform since $s_n(1/n) = 2n/e$. Clearly
$$\lim_{n\to\infty} \int_0^1 s_n(x)\, dx = \lim_{n\to\infty} (1 - e^{-n^2}) = 1 \neq 0 = \int_0^1 s(x)\, dx.$$
Incidentally, this example shows that a series can have a continuous sum without being uniformly convergent (Theorem 6.35 notwithstanding).

21. *Hint.* The series
$$z + \frac{z^2}{2^2} + \cdots + \frac{z^n}{n^2} + \cdots$$
is uniformly convergent in the disk $|z| < 1$, but the same is not true of the differentiated series.

22. *Hint.* The series
$$\sin x + \left(\frac{\sin 2x}{2} - \sin x\right) + \left(\frac{\sin 3x}{3} - \frac{\sin 2x}{2}\right) + \cdots$$
converges uniformly to zero on the real axis (why?). Examine the result of differentiating this series term by term.

24. *Hint.* First use Theorem 6.38 to integrate the series
$$s(z) = f_1(z) + \cdots + f_n(z) + \cdots$$
term by term.

Chapter 7

1. *Hint.* For example, the conditions $|z| < R$ and $|z| > R$ in Theorem 7.15 become $|z - a| < R$ and $|z - a| > R$, the circle of convergence becomes the circle $|z - a| = R$, and so on.

2. *Hint.* Use Theorem 6.36.

4. *Ans.* a) 1; b) 0; c) 1; d) $\frac{1}{4}$; e) 1 if $|a| \le 1$, $1/|a|$ if $|a| > 1$.

5. *Hint.* It follows from the expansion
$$e^n = 1 + \frac{n}{1!} + \cdots + \frac{n^{n-1}}{(n-1)!} + \frac{n^n}{n!}\left[1 + \frac{n}{n+1} + \frac{n^2}{(n+1)(n+2)} + \cdots\right]$$
that
$$e^n < n\frac{n^n}{n!} + \frac{n^n}{n!}\left[1 + \frac{n}{n+1} + \left(\frac{n}{n+1}\right)^2 + \cdots\right] = (2n + 1)\frac{n^n}{n!}.$$

On the other hand, it is obvious that

$$e^n > \frac{n^n}{n!}$$

and hence

$$\frac{1}{e^n} < \frac{n!}{n^n} < \frac{2n+1}{e^n}.$$

6. *Ans.* 1.

7. *Ans.* a) R; b) ∞; c) R^k; d) $\sqrt[k]{R}$.

8. *Ans.* a) $R \geq \min\{r, r'\}$; b) $R \geq rr'$; c) $R \leq r/r'$.

10. *Hint.* $\overline{\lim\limits_{n\to\infty}} \sqrt[n]{|(n+1)c_{n+1}|} = \overline{\lim\limits_{n\to\infty}} [(n+1)^{1/n} (\sqrt[n+1]{|c_{n+1}|})^{(n+1)/n}]$

$$= \overline{\lim_{n\to\infty}} \sqrt[n+1]{|c_{n+1}|} = \overline{\lim_{n\to\infty}} \sqrt[n]{|c_n|}.$$

11. *Hint.* First show that there is no loss of generality in assuming that $R = z_0 = 1$ and $s(z_0) = s(1) = 0$. Use Theorem 6.27 to multiply the given series by the geometric series

$$\frac{1}{1-z} = 1 + z + \cdots + z^n + \cdots,$$

obtaining

$$\frac{s(z)}{1-z} = s_0 + s_1 z + \cdots + s_n z^n + \cdots,$$

where $s_n = c_0 + c_1 + \cdots + c_n$ and the series on the right has radius of convergence 1 (why?). Given any $\epsilon > 0$, let m be a positive integer such that $|s_n| < \epsilon/2$ for all $n > m$. Then

$$|s(z)| = \left|(1-z)\sum_{n=0}^{m} s_n z^n + (1-z)\sum_{n=m+1}^{\infty} s_n z^n\right|$$

$$< (1-z)M + \frac{\epsilon}{2}(1-z)\sum_{n=m+1}^{\infty} z^n$$

if $0 < z < 1$, where $M = |s_0| + |s_1| + \cdots + |s_m|$. But

$$\sum_{n=m+1}^{\infty} z^n = \frac{z^{m+1}}{1-z},$$

and hence

$$|s(z)| < (1-z)M + \frac{\epsilon}{2} z^{m+1} < (1-z)M + \frac{\epsilon}{2}$$

if $0 < z < 1$. Therefore

$$|s(z)| < \frac{\epsilon}{2} + \frac{\epsilon}{2} = \epsilon$$

if $1 - z < \epsilon/2M$, i.e., if z is sufficiently close to 1. Thus

$$\lim_{z\to 1} s(z) = 0 \qquad (0 < z < 1),$$

as required.

12. *Hint.* Start from the familiar expansion

$$\ln(1 + x) = x - \frac{x^2}{2} + \frac{x^3}{3} - \cdots \qquad (-1 < x < 1),$$

noting that

$$\lim_{x \to 1-} \ln(1 + x) = \ln 2.$$

13. *Hint.* Each of the power series

$$\sum_{n=1}^{\infty} z_n \zeta^n, \qquad \sum_{n=1}^{\infty} z'_n \zeta^n$$

"generated" by the numerical series (19) has a radius of convergence ≥ 1 (why?). Therefore

$$\sum_{n=1}^{\infty} z_n \zeta^n \sum_{n=1}^{\infty} z'_n \zeta^n = \sum_{n=1}^{\infty} Z_n \zeta^n$$

for all $|\zeta| < 1$, where $Z_n = z_1 z'_n + z_2 z'_{n-1} + \cdots + z_n z'_1$, and hence

$$\lim_{\zeta \to 1} \sum_{n=1}^{\infty} z_n \zeta^n \lim_{\zeta \to 1} \sum_{n=1}^{\infty} z'_n \zeta^n = \lim_{\zeta \to 1} \sum_{n=1}^{\infty} Z_n \zeta^n \qquad (0 < \zeta < 1).$$

Now apply Abel's theorem to all three series.

14. *Hint.* Consider the series

$$\frac{1}{1 + z} = 1 - z + z^2 - \cdots.$$

15. *Hint.* Note that

$$\begin{aligned}
\sum_{n=0}^{m} c_n - s(z) &= \sum_{n=0}^{m} c_n(1 - z^n) - \sum_{n=m+1}^{\infty} c_n z^n \\
&\leq (1 - z) \sum_{n=0}^{m} |c_n|(1 + z + \cdots + z^{n-1}) + \sum_{n=m+1}^{\infty} |c_n| z^n \\
&< m(1 - z) \frac{1}{m} \sum_{n=0}^{m} n|c_n| + \sum_{n=m+1}^{\infty} n|c_n| \frac{z^n}{n}
\end{aligned}$$

if $0 < z < 1$. Now recall Chap. 2, Prob. 13.

Chapter 8

5. *Ans.* $e^z \longrightarrow \infty$ if $-\pi/2 < \alpha < \pi/2$, $e^z \longrightarrow 0$ if $\pi/2 < \alpha < 3\pi/2$, e^z approaches no limit if $\alpha = \pm\pi/2$.

6. *Hint.* By choosing $z = (1 + i)x$, show that the function fails to be continuous at the origin.

9. *Ans.* $z = \pm\frac{\pi i}{3} + 2k\pi i \qquad (k = 0, \pm 1, \pm 2, \ldots).$

10. *Ans.* The logarithmic spiral with polar equation $r = ce^{\theta/\alpha}$ where $c = e^{-\beta/\alpha}$.

13. *Ans.* $w = \dfrac{z - i}{iz - 1}.$

15. *Ans.* b) $w = \dfrac{iz + 2 + i}{z + 1}.$

17. *Ans.* b) $\frac{9}{2} + i$.

22. *Ans.* a) The half-disk $|w| < 1$, $\operatorname{Im} w < 0$; c) The domain bounded by the circle $|w - \frac{1}{2}| = \frac{1}{2}$ and the tangent line $\operatorname{Re} w = 1$.

23. *Hint.* If $c \neq 0$, $(a - d)^2 + 4bc = 0$, we have a single fixed point $z_0 = (a - d)/2c$ and the relation

$$\frac{1}{w - z_0} = \frac{1}{z - z_0} + k \qquad (k \neq 0)$$

instead of (31). If $c = 0$, see Chap. 4, Prob. 17.

25. *Ans.* a) $w = \dfrac{(3 + i)z - (1 + i)}{(1 - i)z + (1 + i)}$; c) $w = \dfrac{(2i - 1)z + 1}{z - 1}$.

26. *Ans.* a) $w = \left(\dfrac{z + 1}{z - 1}\right)^2$; b) $w = \left(\dfrac{z^n + 1}{z^n - 1}\right)^2$; c) $w = \left(\dfrac{z - a}{z - b}\right)^n$;

d) $w = \left(\dfrac{e^{-z} - 1}{e^{-z} + 1}\right)^2$.

27. *Hint.* If $f_1(z) = 1/z$, $f_2(z) = 1/(1 - z)$, then $f_1 \circ f_2(z) = 1 - z$ while $f_2 \circ f_1(z) = z/(z - 1)$.

Chapter 9

1. *Hint.* $f(z) = f(-2i - z)$.

2. *Hint.* $f(2\sqrt{3} + i) = f(-2\sqrt{3} + i)$.

3. *Hint.* $\dfrac{1}{2}\left(z' + \dfrac{1}{z'}\right) = \dfrac{1}{2}\left(z'' + \dfrac{1}{z''}\right)$ if and only if $z' = z''$ or $z'z'' = 1$.

4. *Ans.* Both domains are mapped onto the w-plane cut along the interval $[-1, 1]$ of the real axis.

5. *Ans.* b) $(2k + \frac{1}{4})\pi i \qquad (k = 0, \pm 1, \pm 2, \ldots)$.

6. *Ans.* b) $e^{-(2k+(1/2))\pi} \qquad (k = 0, \pm 1, \pm 2, \ldots)$.

7. *Ans.* When α is not an integer. The function is n-valued if α is a rational number of the form m/n, where m and $n > 0$ are relatively prime integers; the function is infinite-valued if α is irrational or if $\operatorname{Im} \alpha \neq 0$. The branch points are $z = 0$ and $z = \infty$, of order $n - 1$ in the first case and of infinite order in the second case.

9. *Ans.* b) Yes.

10. *Ans.* a) $2\pi/3$; c) π.

11. *Ans.* b) 0.

13. *Hint.* The mapping $w = \sin z$ can be regarded as the result of the following four consecutive mappings:

$$z_1 = iz, \quad z_2 = e^{z_1}, \quad z_3 = -iz_2 = \frac{e^{iz}}{i}, \quad w = \frac{1}{2}\left(z_3 + \frac{1}{z_3}\right).$$

The first and third mappings are univalent in any domain. The second is univalent in any domain containing no pair of points z_1', z_1'' such that $z_1' - z_1''$

$= 2k\pi i$ ($k = \pm 1, \pm 2, \ldots$), while the fourth is univalent in any domain containing no pair of points z'_3, z''_3 such that $z'_3 z''_3 = 1$ (cf. Prob. 3). Hence $w = \sin z$ is univalent in any domain containing no pair of points z', z'' such that either $z' - z'' = 2k\pi$ or $e^{i(z'+z'')} = -1$ (i.e., $z' + z'' = (2k + 1)\pi$, where $k = 0, \pm 1, \pm 2, \ldots$).

14. *Hint.* Show that $w = \sin z$ maps the family of parallel vertical lines $z = x = c$ ($-\pi/2 < c < \pi/2$) into a family of confocal hyperbolas filling up E.

19. *Ans.* b) $2k\pi - i \ln(\sqrt{2} - 1)$, $(2k + 1)\pi - i \ln(\sqrt{2} + 1)$, where $k = 0, \pm 1, \pm 2, \ldots$

20. *Hint.* Show that every time the curve "winds around" the origin, the values of $L(z)$ change by $2\pi i$ (the curve must not go through the origin).

21. *Ans.* $\arg f(z)$ and $\ln|f(z)|$ are harmonic, but not $|f(z)|$.

Chapter 10

2. *Ans.* $-\ln(1 - z)$.

4. *Ans.* $e^a \sum_{n=0}^{\infty} \frac{(z - a)^n}{n!} \qquad (|z| < \infty)$.

5. *Ans.* $\sum_{n=0}^{\infty} \cos\left(\frac{\pi}{4} + \frac{n\pi}{2}\right) \frac{\left(z - \frac{\pi}{4}\right)^n}{n!} \qquad (|z| < \infty)$.

6. *Hint.* Differentiate the geometric series

$$\frac{1}{1 - z} = \sum_{n=0}^{\infty} z^n \qquad (|z| < 1).$$

8. *Hint.* $\frac{1}{(z + 1)(z - 2)} = \frac{1}{3}\left(\frac{1}{z - 2} - \frac{1}{z + 1}\right)$.

9. *Hint.* $\frac{1}{1 + z + z^2} = \frac{1 - z}{1 - z^3}$.

10. *Hint.* Use Theorems 6.39 and 10.13, noting that uniform convergence in every closed disk $|z - z_0| \leq r < R$ is equivalent to uniform convergence in every bounded closed domain contained in the open disk $|z - z_0| < R$.

11. *Hint.* If $|z| \leq r < 1$, then

$$\left|\frac{z^k}{1 - z^k}\right| \leq \frac{r^k}{1 - r^k} \leq \frac{r^k}{1 - r}.$$

It follows from Theorem 6.36 that the series (32) is uniformly convergent in every closed disk $|z| \leq r < 1$.

12. *Ans.* $f(z) = \sum_{n=1}^{\infty} \tau(n) z^n$, where $\tau(n)$ is the number of positive integers which divide n (including 1 and n).†

13. *Hint.* Use Prob. 10 and Theorem 6.27.

†Thus $\tau(1) = 1$, $\tau(2) = 2$, $\tau(3) = 2$, $\tau(4) = 3$, $\tau(5) = 2, \ldots$

14. *Ans.* b) $\sigma + \tau z + (\tau - \frac{1}{2}\sigma)z^2 + (\frac{5}{6}\tau - \sigma)z^3 + \cdots$, where $\sigma = \sin 1$, $\tau = \cos 1$; d) $1 - \frac{1}{4}z^2 - \frac{1}{96}z^4 - \frac{19}{5760}z^6 - \cdots$.

16. *Hint.* Use Cauchy's inequalities.

17. *Ans.* b) No; d) Yes, $\dfrac{1}{1+z}$.

18. *Ans.* a) Yes; b) No.

19. *Hint.* Cf. Sec. 10.24.

21. *Ans.* 15.

24. *Hint.* The Taylor expansion of $f(z)$ at z_0 is of the form

$$f(z) = f(z_0) + \frac{f^{(m)}(z_0)}{m!}(z - z_0)^m + \cdots \qquad (m > 1).$$

Let $\Delta z = z - z_0$, $\Delta w = f(z) - f(z_0)$. Then, by the same reasoning as in Sec. 4.31,

$$\lim_{\Delta z \to 0} \arg \Delta w = m \lim_{\Delta z \to 0} \arg \Delta z + \arg f^{(m)}(z_0).$$

Therefore the mapping produces m-fold enlargement of angles between curves passing through z_0.

25. *Hint.* Use the Cauchy–Riemann equations.

26. *Hint.* Use Prob. 25.

27. *Hint.* Use Corollary 10.32d.

28. *Hint.* Apply the maximum modulus principle to the function $\varphi(z) = f(z)/z$.

30. *Hint.* Choose $f(z) = \cos z$ in formula (26).

Chapter 11

2. *Ans.* b) $\dfrac{1}{a-b}\left[\dfrac{1}{z-a} + \sum_{n=0}^{\infty} \dfrac{(z-a)^n}{(b-a)^{n+1}}\right]$; d) $\dfrac{1}{a-b}\sum_{n=1}^{\infty} \dfrac{a^n - b^n}{z^{n+1}}$.

3. *Ans.* $\sum_{n=0}^{\infty} c_n z^n + \sum_{n=1}^{\infty} c_{-n} z^n$, where $c_n = c_{-n} = \sum_{k=0}^{\infty} \dfrac{1}{k!(n+k)!}$.

4. *Ans.* b) $\dfrac{2}{z - 2\pi i}$; d) $\dfrac{e^{-b}}{2ib}\dfrac{1}{z - ib}$; f) $\dfrac{1}{\pi(z-n)}$.

5. *Ans.* No.

6. *Ans.* $-z - \dfrac{1}{2z} + \sum_{n=2}^{\infty} (-1)^n \dfrac{1 \cdot 3 \cdots (2n-3)}{n! 2^n} z^{-2n+1}$.

7. *Ans.* A pole at each zero of the denominator, of the same order as that of the zero.

10. *Ans.* b) Simple poles at $z = \dfrac{1 \pm i}{\sqrt{2}}, \dfrac{-1 \pm i}{\sqrt{2}}$; d) Simple pole at $z = 0$, poles of order 2 at $z = \pm 2i$; f) No finite singular points; h) Simple poles at $z = 2n\pi i$ $(n = \pm 1, \pm 2, \ldots)$.

11. *Ans.* a) Simple poles at $z = n\pi$ $(n = \pm 1, \pm 2, \ldots)$, removable singular point

at $z = 0$; c) Simple poles at $z = 1/n\pi$ ($n = \pm 1, \pm 2, \ldots$), limit point of poles at $z = 0$; e) Simple poles at $z = n\pi$ ($n = \pm 1, \pm 2, \ldots$); g) Essential singular points at $z = 1/n\pi$ ($n = \pm 1, \pm 2, \ldots$), limit point of essential singular points at $z = 0$.

14. *Hint.* Cf. Chap. 9, Prob. 15.

19. *Ans.* b) Removable singular point at ∞; d) Removable singular point (zero of order 5) at ∞; f) Essential singular point at ∞; h) Limit point of poles at ∞.

20. *Ans.* a) Limit point of poles at ∞; c) Simple pole at ∞; e) Limit point of poles at ∞; g) Essential singular point at ∞.

21. *Ans.* a) $\operatorname*{Res}_{z=0} f(z) = 1$, $\operatorname*{Res}_{z=\pm 1} f(z) = -\frac{1}{2}$; c) $\operatorname*{Res}_{z=-1} f(z) = 2 \sin 2$;
e) $\operatorname*{Res}_{z=n\pi} f(z) = (-1)^n \quad (n = 0, \pm 1, \pm 2, \ldots)$; g) $\operatorname*{Res}_{z=0} f(z) = 0$.

22. *Ans.* a) $c_{-1} f(z_0)$.

23. *Ans.* $2\pi i$.

24. *Ans.* $-\pi i/\sqrt{2}$.

26. *Ans.* $-2\pi i$.

27. *Ans.* $-4ni$.

28. *Ans.* $-2c_0 c_1$.

30. *Hint.* Use Prob. 29.

31. *Hint.* If $e^{ix} = z$, then

$$dx = \frac{dz}{iz}, \qquad \cos x = \frac{e^{ix} + e^{-ix}}{2} = \frac{1}{2}\left(z + \frac{1}{z}\right),$$

so that the integral becomes

$$i \int_{|z|=1} \frac{dz}{pz^2 - (p^2 + 1)z + p}.$$

The integrand has simple poles at $z = p$ and $z = 1/p$, but only the former lies inside the circle $|z| = 1$. Use formula (34) to calculate the residue of the integrand at $z = p$.

32. *Hint.* The same substitutions as in Prob. 31 reduce the integral to

$$-4i \int_{|z|=1} \frac{z\,dz}{(qz^2 + 2pz + q)^2}.$$

The integrand has poles of order 2 at the points

$$z_1 = \frac{1}{q}(-p + \sqrt{p^2 - q^2}), \qquad z_2 = \frac{1}{q}(-p - \sqrt{p^2 - q^2}),$$

but only z_1 lies inside the circle $|z| = 1$. Use formula (36) to calculate the residue of the integrand $z = z_1$.

33. *Hint.* If $e^{2i(x-a)} = z$, then

$$dx = \frac{dz}{2iz}, \qquad \cot(x - a) = i\frac{e^{i(x-a)} + e^{-i(x-a)}}{e^{i(x-a)} - e^{-i(x-a)}} = i\frac{z+1}{z-1}.$$

Chapter 12

3. *Ans.* $n_+ = 2, n_- = 1, \nu = 1$.

4. *Ans.* b) 0.

5. *Ans.* 3.

6. *Ans.* 1.

7. *Hint.* Consider the contour made up of the semicircle $|z| = R$, $\operatorname{Re} z \geq 0$ and the line segment joining the points $\pm iR$, choosing $f(z) = z - \lambda$, $g(z) = e^{-z}$ in Rouché's theorem.

8. *Hint.* If $P(z)$ has no zeros, then $1/P(z)$ is a bounded entire function and hence constant, which is absurd.

9. *Hint.* Let $f(z) = e^z$.

10. *Hint.* Cf. Theorem 12.24.

11. *Hint.* The fact that E is connected is almost obvious (recall the proof of Theorem 9.12). To prove that E is open, let w_0 be any point of E, and let z_0 be any point of G such that $f(z_0) = w_0$. By Prob. 10 there is a neighborhood K of z_0 and a corresponding neighborhood K^* of w_0 such that given any $w \in K^*$, there is at least one point $z \in K$ for which $f(z) = w$. But then K^* is contained in E, and hence w_0 is an interior point of E. To prove the maximum modulus principle, let z_0 be any point of G, and let K be a neighborhood of z_0 contained in G. The image of K under $w = f(z)$ is itself a domain, and hence K contains a point z whose image $w = f(z)$ is further from the origin of the w-plane than the point $w_0 = f(z_0)$, i.e., a point z such that $|f(z)| > |f(z_0)|$.

12. *Hint.* Note that

$$I^2 = \int_0^\infty e^{-x^2}\,dx \int_0^\infty e^{-y^2}\,dy = \int_0^\infty \int_0^\infty e^{-(x^2+y^2)}\,dx\,dy,$$

and hence

$$I^2 = \int_0^{\pi/2} d\theta \int_0^\infty e^{-r^2}\,r\,dr$$

after transforming to polar coordinates.

13. *Hint.* Use L'Hospital's rule.

14. *Ans.* a) $-\frac{\pi}{27}$; c) $\frac{\pi}{ab(a+b)}$.

16. *Hint.* Integrate the function $f(z) = \frac{e^{2iz} - 1}{z^2}$.

19. *Hint.* Replace n by $2n$ and set $m' = m + n$ $(m < n)$.

20. *Ans.* b) $\frac{\pi}{4a} e^{-a}$.

21. *Ans.* $\frac{\pi}{5}\left(\cos 1 - \frac{1}{e^2}\right)$.

22. *Hint.* Note that σ_1 is now the "upper edge" and σ_2 the "lower edge" of

the interval $[r, R]$ of the axis. Thus $z = x$ on σ_1, while $z = xe^{2\pi i}$ on σ_2, where now $0 \le \operatorname{Im} \ln z = \arg z \le 2\pi$ instead of (27).

23. *Ans.* a) $\dfrac{\pi^3}{8}$; c) $\dfrac{\pi}{4}$; e) $\pi \cot a\pi$.

Chapter 13

1. *Hint.* Use Theorem 7.24.
2. *Hint.* Use Euler's formula (Sec. 8.13a).
3. *Hint.* Use Corollary 13.16.
4. *Hint.* Choose $f(z) = \cos z$ in Prob. 3, recalling Chap. 8, Prob. 8a.
5. *Hint.* Let C be the circle of radius R centered at z_0, let z be any point inside C, and let

$$z^* = z_0 + \frac{R^2}{z - z_0}$$

be the point outside C obtained by inverting z in C (cf. Sec. 1.42). Then

$$f(z) = \frac{1}{2\pi i}\int_C \frac{f(\zeta)}{\zeta - z}\,d\zeta, \qquad 0 = \frac{1}{2\pi i}\int_C \frac{f(\zeta)}{\zeta - z^*}\,d\zeta,$$

and subtracting the second equation from the first, we get

$$f(z) = \frac{1}{2\pi i}\int_C f(\zeta)\left[\frac{1}{\zeta - z} - \frac{1}{\zeta - z^*}\right]d\zeta$$

or

$$\begin{aligned} f(re^{i\varphi}) &= \frac{1}{2\pi}\int_0^{2\pi} f(Re^{i\theta})\left[\frac{R}{R - re^{i(\varphi-\theta)}} + \frac{re^{i(\theta-\varphi)}}{R - re^{i(\theta-\varphi)}}\right]d\theta \\ &= \frac{1}{2\pi}\int_0^{2\pi} f(Re^{i\theta})\frac{R^2 - r^2}{R^2 + r^2 - 2Rr\cos(\varphi - \theta)}\,d\theta. \end{aligned}$$

6. *Hint.* $\dfrac{\zeta - z_0 + (z - z_0)}{(\zeta - z_0)(\zeta - z)} = \dfrac{2}{\zeta - z} - \dfrac{1}{\zeta - z_0}.$
7. *Hint.* If n is the unique integer such that $a \le n\omega < a + \omega$, then

$$\int_a^{a+\omega} f(x)\,dx = \int_a^{n\omega} f(x)\,dx + \int_{n\omega}^{a+\omega} f(x)\,dx = \int_a^{n\omega} f(x)\,dx + \int_{(n-1)\omega}^{a} f(x)\,dx.$$

9. *Ans.* $u(z) = -\dfrac{1}{\pi}\displaystyle\int_0^{2\pi} h(e^{i\theta}) \ln|e^{i\theta} - z|\,d\theta + \text{const.}$

11. *Hint.* By Schwarz's lemma, $|F(w)| \le |w|, |G(w)| \le |w|$, and hence $|f(z)| \le |g(z)|$, $|g(z)| \le |f(z)|$ or equivalently $|f(z)| = |g(z)|$. But then $|F(w)| = |w|$ and hence, by Schwarz's lemma again, $F(w) = e^{i\theta}w$, where $e^{i\theta} = 1$ since $F'(0) > 0$.
12. *Hint.* Use the argument principle (Sec. 12.13).
13. *Hint.* Consider the function $\varphi(f(z))$, where $\varphi(w)$ is the function mapping the domain with boundary γ onto the unit disk $|z| < 1$.

16. *Ans.* The finite plane minus the point $z = 1$.

17. *Hint.* Let $\alpha = p/q$ where p and $q > 0$ are integers, and let $z_0 = e^{2\pi i\alpha}$, $z = \rho z_0$ $(0 < \rho < 1)$. Then

$$f(z) = \sum_{n=1}^{q-1} z^{n!} + \sum_{n=q}^{\infty} \rho^{n!}.$$

If $M = 2q + N$, where N is an arbitrary positive integer, then

$$|f(z)| > \sum_{n=q}^{M} \rho^{n!} - \sum_{n=1}^{q-1} |z|^{n!} > (M - q + 1)\rho^{M!} - (q - 1),$$

where the right-hand side approaches $M - 2q + 2 = N + 2$ as $\rho \longrightarrow 1$.

18. *Hint.* There is obviously a singular point at $z = 1$; moreover there are singular points for $z^2 = 1$ since $f(z) = z^2 + f(z^2)$, singular points for $z^4 = 1$ since $f(z) = z^2 + z^4 + f(z^4)$, etc.

21. *Hint.* By repeated application of the symmetry principle, enlarge the annulus $r_1 < |z| < r_2$ to the whole extended plane minus the point $z = 0$ or the point $z = \infty$. Then use Theorem 13.35.

Chapter 14

2. *Hint.* First map the right half of G onto the upper half-plane by making the transformation $\omega = \sqrt{z^2 - 1}$. After applying the symmetry principle, make the further transformations $\omega_1 = \dfrac{\omega + \sqrt{5}\,i}{\omega - i}$, $w = \sqrt{-\omega_1}$.

3. *Ans.* $w = h\sqrt{z^2 - 1}$.

5. *Ans.* b) $w = C\displaystyle\int_0^z z^{-3/4}(1 - z)^{-1/2}\,dz$, where

$$C = \frac{b}{\displaystyle\int_0^1 x^{-3/4}(1 - x)^{-1/2}\,dx}.$$

6. *Ans.* $w = \dfrac{3h}{16}\sqrt{3z}\left(1 - \dfrac{1}{z}\right)^2$, where the points $0, 1, \infty, -3$ correspond to the vertices $\infty, 0, \infty, ih$.

10. *Hint.* The function (6) now contains extra terms $-\dfrac{2}{z - a} - \dfrac{2}{z - \bar{a}}$, corresponding to the simple poles at a and $\bar{a}$.

11. *Hint.* Make an auxiliary fractional linear transformation.

13. *Ans.* $w = C\displaystyle\int_{z_0}^{z} \frac{\sqrt{1 - z^4}}{z^2}\,dz + C_1$.

14. *Hint.* By (27) with $a_1 = -1$, $a_2 = 1$, $\alpha_1 = \alpha_2 = 2$,

$$w = C\int_{z_0}^{z} \frac{(z + 1)(z - 1)}{z^2}\,dz + C_1 = C\int_{z_0}^{z} \left(1 - \frac{1}{z^2}\right) dz + C_1$$

$$= C\left(z + \frac{1}{z}\right) + C_1'.$$

Since $z = \pm 1$ is mapped into $w = \pm 1$, we have $\pm 1 = \pm 2C + C_1'$ and hence $C = \frac{1}{2}$, $C_1' = 0$.

15. *Ans.* $w = [\frac{1}{2}(z^n + z^{-n})]^{1/n}$, where the points $z = 1, e^{\pi i/2n}, e^{\pi i/n}$ go into the points $w = 1, 0, e^{\pi i/n}$.

16. *Hint.* Use the symmetry principle.

Chapter 15

1. *Hint.* Think of the flow as taking place between parallel planes a unit distance apart, with the planes offering no resistance to the motion of the fluid, and let dS be the "elementary cylinder" with base ds, where ds is an element of arc along C. Then the volume of fluid flowing outward through dS in time dt is just $w_n\,ds\,dt$.

8. *Hint.* The function

$$\zeta = g(z) = \frac{1}{a+b}(z + \sqrt{z^2 - c^2}) \qquad (c^2 = a^2 - b^2)$$

(choose the positive square root) maps the exterior of the ellipse with semiaxes a and b onto the exterior of the circle $|\zeta| = 1$.

9. *Hint.* The contribution to T due to the pressure forces acting on an element $dz = dx + i\,dy$ of the contour Γ equals $P(x\,dx + y\,dy) = \text{Re}\,(Pz\,\overline{dz})$.

10. *Ans.* $T = -\frac{1}{2}\pi\rho c^2|w|^2 \sin 2\alpha$, where $c = \sqrt{a^2 - b^2}$.

11. *Hint.* Start from Example 14.23, using the auxiliary transformation

$$\omega = e^{\pi(w+ih_1)/H} \qquad (H = h_1 + h_2)$$

to map the domain shown in Figure 50 onto the domain shown in Figure 59.

14. *Hint.* Let L be the z-axis. Then, by Coulomb's law, the contribution to the electric field due to the elementary charge $q\,dz$ is just

$$|d\vec{E}| = q\frac{dz}{r^2 + z^2},$$

where $r^2 = x^2 + y^2$ (draw a figure). But the vector $\vec{E}$ lies in the xy-plane, and hence

$$|E_x + iE_y| = \cos\theta\,|d\vec{E}| = q\int_{-\infty}^{\infty}\frac{\cos\theta}{r^2 + z^2}\,dz = q\int_{-\pi/2}^{\pi/2}\frac{\cos\theta}{r}\,d\theta = \frac{2q}{r},$$

where θ is the angle between $d\vec{E}$ and the xy-plane.

17. *Ans.* $\frac{1}{2}$.

19. *Hint.* First use the function $w = \sin(\pi z/a)$ to map the slab onto the half-plane $\text{Im}\,w > 0$ (cf. Example 14.21), with the sides of the slab going into the intervals $(-\infty, -1]$, $[-1, 1]$, $[1, \infty)$ of the real axis. Then by Theorem 13.24,

$$T = \frac{T_0}{\pi}\int_{-1}^{1}\frac{v}{(\xi - u)^2 + v^2}\,d\xi = \frac{T_0}{\pi}\left(\arctan\frac{1-u}{v} - \arctan\frac{-1-u}{v}\right)$$

$$= \frac{T_0}{\pi}\left(\text{arc cot}\,\frac{u-1}{v} - \text{arc cot}\,\frac{u+1}{v}\right) = \frac{T_0}{\pi}\arg\frac{w-1}{w+1}$$

(recall the comment to Sec. 13.2).

Bibliography

AHLFORS, L. V., *Complex Analysis*, McGraw-Hill Book Company, New York (1953).

BIEBERBACH, L., *Conformal Mapping* (translated by F. Steinhardt), Chelsea Publishing Company, New York (1953).

CHURCHILL, R. V., *Complex Variables and Applications*, second edition, McGraw-Hill Book Company, New York (1960).

COPSON, E. T., *An Introduction to the Theory of Functions of a Complex Variable*, Oxford University Press, Inc., New York (1935).

HILLE, E., *Analytic Function Theory*, in two volumes, Ginn and Company, Boston (1959, 1962).

KNOPP, K., *Theory of Functions* (translated by F. Bagemihl), in two volumes, Dover Publications, Inc., New York (1945, 1947).

LEVINSON, N., and R. M. REDHEFFER, *Complex Variables*, Holden-Day, Inc., San Francisco (1970).

MACROBERT, T. M., *Functions of a Complex Variable*, fourth edition, The Macmillan Company, New York (1958).

MARKUSHEVICH, A. I., *Theory of Functions of a Complex Variable* (translated by R. A. Silverman), in three volumes, Prentice-Hall, Inc., Englewood Cliffs, N.J. (1965, 1967).

NEHARI, Z., *Conformal Mapping*, McGraw-Hill Book Company, New York (1952).

NEHARI, Z., *Introduction to Complex Analysis*, Allyn and Bacon, Inc., Boston (1962).

PHILLIPS, E. G., *Functions of a Complex Variable with Applications*, Oliver & Boyd Ltd., Edinburgh (1961).

SILVERMAN, R. A., *Introductory Complex Analysis*, Dover Publications, Inc., New York (1972).

TITCHMARSH, E. C., *The Theory of Functions*, second edition, Oxford University Press, Inc., New York (1939).

WHITTAKER, E. T., and G. N. WATSON, *A Course of Modern Analysis*, fourth edition, Cambridge University Press, New York (1963).

Index

H

I

J

K

L

M

A CATALOG OF SELECTED

DOVER BOOKS

IN SCIENCE AND MATHEMATICS

Astronomy

CHARIOTS FOR APOLLO: The NASA History of Manned Lunar Spacecraft to 1969, Courtney G. Brooks, James M. Grimwood, and Loyd S. Swenson, Jr. This illustrated history by a trio of experts is the definitive reference on the Apollo spacecraft and lunar modules. It traces the vehicles' design, development, and operation in space. More than 100 photographs and illustrations. 576pp. 6 3/4 x 9 1/4. 0-486-46756-2

EXPLORING THE MOON THROUGH BINOCULARS AND SMALL TELESCOPES, Ernest H. Cherrington, Jr. Informative, profusely illustrated guide to locating and identifying craters, rills, seas, mountains, other lunar features. Newly revised and updated with special section of new photos. Over 100 photos and diagrams. 240pp. 8 1/4 x 11. 0-486-24491-1

WHERE NO MAN HAS GONE BEFORE: A History of NASA's Apollo Lunar Expeditions, William David Compton. Introduction by Paul Dickson. This official NASA history traces behind-the-scenes conflicts and cooperation between scientists and engineers. The first half concerns preparations for the Moon landings, and the second half documents the flights that followed Apollo 11. 1989 edition. 432pp. 7 x 10. 0-486-47888-2

APOLLO EXPEDITIONS TO THE MOON: The NASA History, Edited by Edgar M. Cortright. Official NASA publication marks the 40th anniversary of the first lunar landing and features essays by project participants recalling engineering and administrative challenges. Accessible, jargon-free accounts, highlighted by numerous illustrations. 336pp. 8 3/8 x 10 7/8. 0-486-47175-6

ON MARS: Exploration of the Red Planet, 1958-1978--The NASA History, Edward Clinton Ezell and Linda Neuman Ezell. NASA's official history chronicles the start of our explorations of our planetary neighbor. It recounts cooperation among government, industry, and academia, and it features dozens of photos from Viking cameras. 560pp. 6 3/4 x 9 1/4. 0-486-46757-0

ARISTARCHUS OF SAMOS: The Ancient Copernicus, Sir Thomas Heath. Heath's history of astronomy ranges from Homer and Hesiod to Aristarchus and includes quotes from numerous thinkers, compilers, and scholasticists from Thales and Anaximander through Pythagoras, Plato, Aristotle, and Heraclides. 34 figures. 448pp. 5 3/8 x 8 1/2. 0-486-43886-4

AN INTRODUCTION TO CELESTIAL MECHANICS, Forest Ray Moulton. Classic text still unsurpassed in presentation of fundamental principles. Covers rectilinear motion, central forces, problems of two and three bodies, much more. Includes over 200 problems, some with answers. 437pp. 5 3/8 x 8 1/2. 0-486-64687-4

BEYOND THE ATMOSPHERE: Early Years of Space Science, Homer E. Newell. This exciting survey is the work of a top NASA administrator who chronicles technological advances, the relationship of space science to general science, and the space program's social, political, and economic contexts. 528pp. 6 3/4 x 9 1/4. 0-486-47464-X

STAR LORE: Myths, Legends, and Facts, William Tyler Olcott. Captivating retellings of the origins and histories of ancient star groups include Pegasus, Ursa Major, Pleiades, signs of the zodiac, and other constellations. "Classic." – *Sky & Telescope.* 58 illustrations. 544pp. 5 3/8 x 8 1/2. 0-486-43581-4

A COMPLETE MANUAL OF AMATEUR ASTRONOMY: Tools and Techniques for Astronomical Observations, P. Clay Sherrod with Thomas L. Koed. Concise, highly readable book discusses the selection, set-up, and maintenance of a telescope; amateur studies of the sun; lunar topography and occultations; and more. 124 figures. 26 halftones. 37 tables. 335pp. 6 1/2 x 9 1/4. 0-486-42820-6

Chemistry

MOLECULAR COLLISION THEORY, M. S. Child. This high-level monograph offers an analytical treatment of classical scattering by a central force, quantum scattering by a central force, elastic scattering phase shifts, and semi-classical elastic scattering. 1974 edition. 310pp. 5 3/8 x 8 1/2. 0-486-69437-2

HANDBOOK OF COMPUTATIONAL QUANTUM CHEMISTRY, David B. Cook. This comprehensive text provides upper-level undergraduates and graduate students with an accessible introduction to the implementation of quantum ideas in molecular modeling, exploring practical applications alongside theoretical explanations. 1998 edition. 832pp. 5 3/8 x 8 1/2. 0-486-44307-8

RADIOACTIVE SUBSTANCES, Marie Curie. The celebrated scientist's thesis, which directly preceded her 1903 Nobel Prize, discusses establishing atomic character of radioactivity; extraction from pitchblende of polonium and radium; isolation of pure radium chloride; more. 96pp. 5 3/8 x 8 1/2. 0-486-42550-9

CHEMICAL MAGIC, Leonard A. Ford. Classic guide provides intriguing entertainment while elucidating sound scientific principles, with more than 100 unusual stunts: cold fire, dust explosions, a nylon rope trick, a disappearing beaker, much more. 128pp. 5 3/8 x 8 1/2. 0-486-67628-5

ALCHEMY, E. J. Holmyard. Classic study by noted authority covers 2,000 years of alchemical history: religious, mystical overtones; apparatus; signs, symbols, and secret terms; advent of scientific method, much more. Illustrated. 320pp. 5 3/8 x 8 1/2. 0-486-26298-7

CHEMICAL KINETICS AND REACTION DYNAMICS, Paul L. Houston. This text teaches the principles underlying modern chemical kinetics in a clear, direct fashion, using several examples to enhance basic understanding. Solutions to selected problems. 2001 edition. 352pp. 8 3/8 x 11. 0-486-45334-0

PROBLEMS AND SOLUTIONS IN QUANTUM CHEMISTRY AND PHYSICS, Charles S. Johnson and Lee G. Pedersen. Unusually varied problems, with detailed solutions, cover of quantum mechanics, wave mechanics, angular momentum, molecular spectroscopy, scattering theory, more. 280 problems, plus 139 supplementary exercises. 430pp. 6 1/2 x 9 1/4. 0-486-65236-X

ELEMENTS OF CHEMISTRY, Antoine Lavoisier. Monumental classic by the founder of modern chemistry features first explicit statement of law of conservation of matter in chemical change, and more. Facsimile reprint of original (1790) Kerr translation. 539pp. 5 3/8 x 8 1/2. 0-486-64624-6

MAGNETISM AND TRANSITION METAL COMPLEXES, F. E. Mabbs and D. J. Machin. A detailed view of the calculation methods involved in the magnetic properties of transition metal complexes, this volume offers sufficient background for original work in the field. 1973 edition. 240pp. 5 3/8 x 8 1/2. 0-486-46284-6

GENERAL CHEMISTRY, Linus Pauling. Revised third edition of classic first-year text by Nobel laureate. Atomic and molecular structure, quantum mechanics, statistical mechanics, thermodynamics correlated with descriptive chemistry. Problems. 992pp. 5 3/8 x 8 1/2. 0-486-65622-5

ELECTROLYTE SOLUTIONS: Second Revised Edition, R. A. Robinson and R. H. Stokes. Classic text deals primarily with measurement, interpretation of conductance, chemical potential, and diffusion in electrolyte solutions. Detailed theoretical interpretations, plus extensive tables of thermodynamic and transport properties. 1970 edition. 590pp. 5 3/8 x 8 1/2. 0-486-42225-9

Engineering

FUNDAMENTALS OF ASTRODYNAMICS, Roger R. Bate, Donald D. Mueller, and Jerry E. White. Teaching text developed by U.S. Air Force Academy develops the basic two-body and n-body equations of motion; orbit determination; classical orbital elements, coordinate transformations; differential correction; more. 1971 edition. 455pp. 5 3/8 x 8 1/2. 0-486-60061-0

INTRODUCTION TO CONTINUUM MECHANICS FOR ENGINEERS: Revised Edition, Ray M. Bowen. This self-contained text introduces classical continuum models within a modern framework. Its numerous exercises illustrate the governing principles, linearizations, and other approximations that constitute classical continuum models. 2007 edition. 320pp. 6 1/8 x 9 1/4. 0-486-47460-7

ENGINEERING MECHANICS FOR STRUCTURES, Louis L. Bucciarelli. This text explores the mechanics of solids and statics as well as the strength of materials and elasticity theory. Its many design exercises encourage creative initiative and systems thinking. 2009 edition. 320pp. 6 1/8 x 9 1/4. 0-486-46855-0

FEEDBACK CONTROL THEORY, John C. Doyle, Bruce A. Francis and Allen R. Tannenbaum. This excellent introduction to feedback control system design offers a theoretical approach that captures the essential issues and can be applied to a wide range of practical problems. 1992 edition. 224pp. 6 1/2 x 9 1/4. 0-486-46933-6

THE FORCES OF MATTER, Michael Faraday. These lectures by a famous inventor offer an easy-to-understand introduction to the interactions of the universe's physical forces. Six essays explore gravitation, cohesion, chemical affinity, heat, magnetism, and electricity. 1993 edition. 96pp. 5 3/8 x 8 1/2. 0-486-47482-8

DYNAMICS, Lawrence E. Goodman and William H. Warner. Beginning engineering text introduces calculus of vectors, particle motion, dynamics of particle systems and plane rigid bodies, technical applications in plane motions, and more. Exercises and answers in every chapter. 619pp. 5 3/8 x 8 1/2. 0-486-42006-X

ADAPTIVE FILTERING PREDICTION AND CONTROL, Graham C. Goodwin and Kwai Sang Sin. This unified survey focuses on linear discrete-time systems and explores natural extensions to nonlinear systems. It emphasizes discrete-time systems, summarizing theoretical and practical aspects of a large class of adaptive algorithms. 1984 edition. 560pp. 6 1/2 x 9 1/4. 0-486-46932-8

INDUCTANCE CALCULATIONS, Frederick W. Grover. This authoritative reference enables the design of virtually every type of inductor. It features a single simple formula for each type of inductor, together with tables containing essential numerical factors. 1946 edition. 304pp. 5 3/8 x 8 1/2. 0-486-47440-2

THERMODYNAMICS: Foundations and Applications, Elias P. Gyftopoulos and Gian Paolo Beretta. Designed by two MIT professors, this authoritative text discusses basic concepts and applications in detail, emphasizing generality, definitions, and logical consistency. More than 300 solved problems cover realistic energy systems and processes. 800pp. 6 1/8 x 9 1/4. 0-486-43932-1

THE FINITE ELEMENT METHOD: Linear Static and Dynamic Finite Element Analysis, Thomas J. R. Hughes. Text for students without in-depth mathematical training, this text includes a comprehensive presentation and analysis of algorithms of time-dependent phenomena plus beam, plate, and shell theories. Solution guide available upon request. 672pp. 6 1/2 x 9 1/4. 0-486-41181-8

Mathematics–Bestsellers

HANDBOOK OF MATHEMATICAL FUNCTIONS: with Formulas, Graphs, and Mathematical Tables, Edited by Milton Abramowitz and Irene A. Stegun. A classic resource for working with special functions, standard trig, and exponential logarithmic definitions and extensions, it features 29 sets of tables, some to as high as 20 places. 1046pp. 8 x 10 1/2.
0-486-61272-4

ABSTRACT AND CONCRETE CATEGORIES: The Joy of Cats, Jiri Adamek, Horst Herrlich, and George E. Strecker. This up-to-date introductory treatment employs category theory to explore the theory of structures. Its unique approach stresses concrete categories and presents a systematic view of factorization structures. Numerous examples. 1990 edition, updated 2004. 528pp. 6 1/8 x 9 1/4.
0-486-46934-4

MATHEMATICS: Its Content, Methods and Meaning, A. D. Aleksandrov, A. N. Kolmogorov, and M. A. Lavrent'ev. Major survey offers comprehensive, coherent discussions of analytic geometry, algebra, differential equations, calculus of variations, functions of a complex variable, prime numbers, linear and non-Euclidean geometry, topology, functional analysis, more. 1963 edition. 1120pp. 5 3/8 x 8 1/2.
0-486-40916-3

INTRODUCTION TO VECTORS AND TENSORS: Second Edition--Two Volumes Bound as One, Ray M. Bowen and C.-C. Wang. Convenient single-volume compilation of two texts offers both introduction and in-depth survey. Geared toward engineering and science students rather than mathematicians, it focuses on physics and engineering applications. 1976 edition. 560pp. 6 1/2 x 9 1/4.
0-486-46914-X

AN INTRODUCTION TO ORTHOGONAL POLYNOMIALS, Theodore S. Chihara. Concise introduction covers general elementary theory, including the representation theorem and distribution functions, continued fractions and chain sequences, the recurrence formula, special functions, and some specific systems. 1978 edition. 272pp. 5 3/8 x 8 1/2.
0-486-47929-3

ADVANCED MATHEMATICS FOR ENGINEERS AND SCIENTISTS, Paul DuChateau. This primary text and supplemental reference focuses on linear algebra, calculus, and ordinary differential equations. Additional topics include partial differential equations and approximation methods. Includes solved problems. 1992 edition. 400pp. 7 1/2 x 9 1/4.
0-486-47930-7

PARTIAL DIFFERENTIAL EQUATIONS FOR SCIENTISTS AND ENGINEERS, Stanley J. Farlow. Practical text shows how to formulate and solve partial differential equations. Coverage of diffusion-type problems, hyperbolic-type problems, elliptic-type problems, numerical and approximate methods. Solution guide available upon request. 1982 edition. 414pp. 6 1/8 x 9 1/4.
0-486-67620-X

VARIATIONAL PRINCIPLES AND FREE-BOUNDARY PROBLEMS, Avner Friedman. Advanced graduate-level text examines variational methods in partial differential equations and illustrates their applications to free-boundary problems. Features detailed statements of standard theory of elliptic and parabolic operators. 1982 edition. 720pp. 6 1/8 x 9 1/4.
0-486-47853-X

LINEAR ANALYSIS AND REPRESENTATION THEORY, Steven A. Gaal. Unified treatment covers topics from the theory of operators and operator algebras on Hilbert spaces; integration and representation theory for topological groups; and the theory of Lie algebras, Lie groups, and transform groups. 1973 edition. 704pp. 6 1/8 x 9 1/4.
0-486-47851-3

Mathematics–Logic and Problem Solving

PERPLEXING PUZZLES AND TANTALIZING TEASERS, Martin Gardner. Ninety-three riddles, mazes, illusions, tricky questions, word and picture puzzles, and other challenges offer hours of entertainment for youngsters. Filled with rib-tickling drawings. Solutions. 224pp. 5 3/8 x 8 1/2. 0-486-25637-5

MY BEST MATHEMATICAL AND LOGIC PUZZLES, Martin Gardner. The noted expert selects 70 of his favorite "short" puzzles. Includes The Returning Explorer, The Mutilated Chessboard, Scrambled Box Tops, and dozens more. Complete solutions included. 96pp. 5 3/8 x 8 1/2. 0-486-28152-3

THE LADY OR THE TIGER?: and Other Logic Puzzles, Raymond M. Smullyan. Created by a renowned puzzle master, these whimsically themed challenges involve paradoxes about probability, time, and change; metapuzzles; and self-referentiality. Nineteen chapters advance in difficulty from relatively simple to highly complex. 1982 edition. 240pp. 5 3/8 x 8 1/2. 0-486-47027-X

SATAN, CANTOR AND INFINITY: Mind-Boggling Puzzles, Raymond M. Smullyan. A renowned mathematician tells stories of knights and knaves in an entertaining look at the logical precepts behind infinity, probability, time, and change. Requires a strong background in mathematics. Complete solutions. 288pp. 5 3/8 x 8 1/2. 0-486-47036-9

THE RED BOOK OF MATHEMATICAL PROBLEMS, Kenneth S. Williams and Kenneth Hardy. Handy compilation of 100 practice problems, hints and solutions indispensable for students preparing for the William Lowell Putnam and other mathematical competitions. Preface to the First Edition. Sources. 1988 edition. 192pp. 5 3/8 x 8 1/2. 0-486-69415-1

KING ARTHUR IN SEARCH OF HIS DOG AND OTHER CURIOUS PUZZLES, Raymond M. Smullyan. This fanciful, original collection for readers of all ages features arithmetic puzzles, logic problems related to crime detection, and logic and arithmetic puzzles involving King Arthur and his Dogs of the Round Table. 160pp. 5 3/8 x 8 1/2. 0-486-47435-6

UNDECIDABLE THEORIES: Studies in Logic and the Foundation of Mathematics, Alfred Tarski in collaboration with Andrzej Mostowski and Raphael M. Robinson. This well-known book by the famed logician consists of three treatises: "A General Method in Proofs of Undecidability," "Undecidability and Essential Undecidability in Mathematics," and "Undecidability of the Elementary Theory of Groups." 1953 edition. 112pp. 5 3/8 x 8 1/2. 0-486-47703-7

LOGIC FOR MATHEMATICIANS, J. Barkley Rosser. Examination of essential topics and theorems assumes no background in logic. "Undoubtedly a major addition to the literature of mathematical logic." – *Bulletin of the American Mathematical Society.* 1978 edition. 592pp. 6 1/8 x 9 1/4. 0-486-46898-4

INTRODUCTION TO PROOF IN ABSTRACT MATHEMATICS, Andrew Wohlgemuth. This undergraduate text teaches students what constitutes an acceptable proof, and it develops their ability to do proofs of routine problems as well as those requiring creative insights. 1990 edition. 384pp. 6 1/2 x 9 1/4. 0-486-47854-8

FIRST COURSE IN MATHEMATICAL LOGIC, Patrick Suppes and Shirley Hill. Rigorous introduction is simple enough in presentation and context for wide range of students. Symbolizing sentences; logical inference; truth and validity; truth tables; terms, predicates, universal quantifiers; universal specification and laws of identity; more. 288pp. 5 3/8 x 8 1/2. 0-486-42259-3

Mathematics–Algebra and Calculus

VECTOR CALCULUS, Peter Baxandall and Hans Liebeck. This introductory text offers a rigorous, comprehensive treatment. Classical theorems of vector calculus are amply illustrated with figures, worked examples, physical applications, and exercises with hints and answers. 1986 edition. 560pp. 5 3/8 x 8 1/2. 0-486-46620-5

ADVANCED CALCULUS: An Introduction to Classical Analysis, Louis Brand. A course in analysis that focuses on the functions of a real variable, this text introduces the basic concepts in their simplest setting and illustrates its teachings with numerous examples, theorems, and proofs. 1955 edition. 592pp. 5 3/8 x 8 1/2. 0-486-44548-8

ADVANCED CALCULUS, Avner Friedman. Intended for students who have already completed a one-year course in elementary calculus, this two-part treatment advances from functions of one variable to those of several variables. Solutions. 1971 edition. 432pp. 5 3/8 x 8 1/2. 0-486-45795-8

METHODS OF MATHEMATICS APPLIED TO CALCULUS, PROBABILITY, AND STATISTICS, Richard W. Hamming. This 4-part treatment begins with algebra and analytic geometry and proceeds to an exploration of the calculus of algebraic functions and transcendental functions and applications. 1985 edition. Includes 310 figures and 18 tables. 880pp. 6 1/2 x 9 1/4. 0-486-43945-3

BASIC ALGEBRA I: Second Edition, Nathan Jacobson. A classic text and standard reference for a generation, this volume covers all undergraduate algebra topics, including groups, rings, modules, Galois theory, polynomials, linear algebra, and associative algebra. 1985 edition. 528pp. 6 1/8 x 9 1/4. 0-486-47189-6

BASIC ALGEBRA II: Second Edition, Nathan Jacobson. This classic text and standard reference comprises all subjects of a first-year graduate-level course, including in-depth coverage of groups and polynomials and extensive use of categories and functors. 1989 edition. 704pp. 6 1/8 x 9 1/4. 0-486-47187-X

CALCULUS: An Intuitive and Physical Approach (Second Edition), Morris Kline. Application-oriented introduction relates the subject as closely as possible to science with explorations of the derivative; differentiation and integration of the powers of x; theorems on differentiation, antidifferentiation; the chain rule; trigonometric functions; more. Examples. 1967 edition. 960pp. 6 1/2 x 9 1/4. 0-486-40453-6

ABSTRACT ALGEBRA AND SOLUTION BY RADICALS, John E. Maxfield and Margaret W. Maxfield. Accessible advanced undergraduate-level text starts with groups, rings, fields, and polynomials and advances to Galois theory, radicals and roots of unity, and solution by radicals. Numerous examples, illustrations, exercises, appendixes. 1971 edition. 224pp. 6 1/8 x 9 1/4. 0-486-47723-1

AN INTRODUCTION TO THE THEORY OF LINEAR SPACES, Georgi E. Shilov. Translated by Richard A. Silverman. Introductory treatment offers a clear exposition of algebra, geometry, and analysis as parts of an integrated whole rather than separate subjects. Numerous examples illustrate many different fields, and problems include hints or answers. 1961 edition. 320pp. 5 3/8 x 8 1/2. 0-486-63070-6

LINEAR ALGEBRA, Georgi E. Shilov. Covers determinants, linear spaces, systems of linear equations, linear functions of a vector argument, coordinate transformations, the canonical form of the matrix of a linear operator, bilinear and quadratic forms, and more. 387pp. 5 3/8 x 8 1/2. 0-486-63518-X

Mathematics–Probability and Statistics

BASIC PROBABILITY THEORY, Robert B. Ash. This text emphasizes the probabilistic way of thinking, rather than measure-theoretic concepts. Geared toward advanced undergraduates and graduate students, it features solutions to some of the problems. 1970 edition. 352pp. 5 3/8 x 8 1/2. 0-486-46628-0

PRINCIPLES OF STATISTICS, M. G. Bulmer. Concise description of classical statistics, from basic dice probabilities to modern regression analysis. Equal stress on theory and applications. Moderate difficulty; only basic calculus required. Includes problems with answers. 252pp. 5 5/8 x 8 1/4. 0-486-63760-3

OUTLINE OF BASIC STATISTICS: Dictionary and Formulas, John E. Freund and Frank J. Williams. Handy guide includes a 70-page outline of essential statistical formulas covering grouped and ungrouped data, finite populations, probability, and more, plus over 1,000 clear, concise definitions of statistical terms. 1966 edition. 208pp. 5 3/8 x 8 1/2. 0-486-47769-X

GOOD THINKING: The Foundations of Probability and Its Applications, Irving J. Good. This in-depth treatment of probability theory by a famous British statistician explores Keynesian principles and surveys such topics as Bayesian rationality, corroboration, hypothesis testing, and mathematical tools for induction and simplicity. 1983 edition. 352pp. 5 3/8 x 8 1/2. 0-486-47438-0

INTRODUCTION TO PROBABILITY THEORY WITH CONTEMPORARY APPLICATIONS, Lester L. Helms. Extensive discussions and clear examples, written in plain language, expose students to the rules and methods of probability. Exercises foster problem-solving skills, and all problems feature step-by-step solutions. 1997 edition. 368pp. 6 1/2 x 9 1/4. 0-486-47418-6

CHANCE, LUCK, AND STATISTICS, Horace C. Levinson. In simple, non-technical language, this volume explores the fundamentals governing chance and applies them to sports, government, and business. "Clear and lively ... remarkably accurate." – *Scientific Monthly*. 384pp. 5 3/8 x 8 1/2. 0-486-41997-5

FIFTY CHALLENGING PROBLEMS IN PROBABILITY WITH SOLUTIONS, Frederick Mosteller. Remarkable puzzlers, graded in difficulty, illustrate elementary and advanced aspects of probability. These problems were selected for originality, general interest, or because they demonstrate valuable techniques. Also includes detailed solutions. 88pp. 5 3/8 x 8 1/2. 0-486-65355-2

EXPERIMENTAL STATISTICS, Mary Gibbons Natrella. A handbook for those seeking engineering information and quantitative data for designing, developing, constructing, and testing equipment. Covers the planning of experiments, the analyzing of extreme-value data; and more. 1966 edition. Index. Includes 52 figures and 76 tables. 560pp. 8 3/8 x 11. 0-486-43937-2

STOCHASTIC MODELING: Analysis and Simulation, Barry L. Nelson. Coherent introduction to techniques also offers a guide to the mathematical, numerical, and simulation tools of systems analysis. Includes formulation of models, analysis, and interpretation of results. 1995 edition. 336pp. 6 1/8 x 9 1/4. 0-486-47770-3

INTRODUCTION TO BIOSTATISTICS: Second Edition, Robert R. Sokal and F. James Rohlf. Suitable for undergraduates with a minimal background in mathematics, this introduction ranges from descriptive statistics to fundamental distributions and the testing of hypotheses. Includes numerous worked-out problems and examples. 1987 edition. 384pp. 6 1/8 x 9 1/4. 0-486-46961-1

Mathematics–Geometry and Topology

PROBLEMS AND SOLUTIONS IN EUCLIDEAN GEOMETRY, M. N. Aref and William Wernick. Based on classical principles, this book is intended for a second course in Euclidean geometry and can be used as a refresher. More than 200 problems include hints and solutions. 1968 edition. 272pp. 5 3/8 x 8 1/2. 0-486-47720-7

TOPOLOGY OF 3-MANIFOLDS AND RELATED TOPICS, Edited by M. K. Fort, Jr. With a New Introduction by Daniel Silver. Summaries and full reports from a 1961 conference discuss decompositions and subsets of 3-space; n-manifolds; knot theory; the Poincaré conjecture; and periodic maps and isotopies. Familiarity with algebraic topology required. 1962 edition. 272pp. 6 1/8 x 9 1/4. 0-486-47753-3

POINT SET TOPOLOGY, Steven A. Gaal. Suitable for a complete course in topology, this text also functions as a self-contained treatment for independent study. Additional enrichment materials make it equally valuable as a reference. 1964 edition. 336pp. 5 3/8 x 8 1/2. 0-486-47222-1

INVITATION TO GEOMETRY, Z. A. Melzak. Intended for students of many different backgrounds with only a modest knowledge of mathematics, this text features self-contained chapters that can be adapted to several types of geometry courses. 1983 edition. 240pp. 5 3/8 x 8 1/2. 0-486-46626-4

TOPOLOGY AND GEOMETRY FOR PHYSICISTS, Charles Nash and Siddhartha Sen. Written by physicists for physics students, this text assumes no detailed background in topology or geometry. Topics include differential forms, homotopy, homology, cohomology, fiber bundles, connection and covariant derivatives, and Morse theory. 1983 edition. 320pp. 5 3/8 x 8 1/2. 0-486-47852-1

BEYOND GEOMETRY: Classic Papers from Riemann to Einstein, Edited with an Introduction and Notes by Peter Pesic. This is the only English-language collection of these 8 accessible essays. They trace seminal ideas about the foundations of geometry that led to Einstein's general theory of relativity. 224pp. 6 1/8 x 9 1/4. 0-486-45350-2

GEOMETRY FROM EUCLID TO KNOTS, Saul Stahl. This text provides a historical perspective on plane geometry and covers non-neutral Euclidean geometry, circles and regular polygons, projective geometry, symmetries, inversions, informal topology, and more. Includes 1,000 practice problems. Solutions available. 2003 edition. 480pp. 6 1/8 x 9 1/4. 0-486-47459-3

TOPOLOGICAL VECTOR SPACES, DISTRIBUTIONS AND KERNELS, François Trèves. Extending beyond the boundaries of Hilbert and Banach space theory, this text focuses on key aspects of functional analysis, particularly in regard to solving partial differential equations. 1967 edition. 592pp. 5 3/8 x 8 1/2. 0-486-45352-9

INTRODUCTION TO PROJECTIVE GEOMETRY, C. R. Wylie, Jr. This introductory volume offers strong reinforcement for its teachings, with detailed examples and numerous theorems, proofs, and exercises, plus complete answers to all odd-numbered end-of-chapter problems. 1970 edition. 576pp. 6 1/8 x 9 1/4. 0-486-46895-X

FOUNDATIONS OF GEOMETRY, C. R. Wylie, Jr. Geared toward students preparing to teach high school mathematics, this text explores the principles of Euclidean and non-Euclidean geometry and covers both generalities and specifics of the axiomatic method. 1964 edition. 352pp. 6 x 9. 0-486-47214-0

Mathematics–History

THE WORKS OF ARCHIMEDES, Archimedes. Translated by Sir Thomas Heath. Complete works of ancient geometer feature such topics as the famous problems of the ratio of the areas of a cylinder and an inscribed sphere; the properties of conoids, spheroids, and spirals; more. 326pp. 5 3/8 x 8 1/2. 0-486-42084-1

THE HISTORICAL ROOTS OF ELEMENTARY MATHEMATICS, Lucas N. H. Bunt, Phillip S. Jones, and Jack D. Bedient. Exciting, hands-on approach to understanding fundamental underpinnings of modern arithmetic, algebra, geometry and number systems examines their origins in early Egyptian, Babylonian, and Greek sources. 336pp. 5 3/8 x 8 1/2. 0-486-25563-8

THE THIRTEEN BOOKS OF EUCLID'S ELEMENTS, Euclid. Contains complete English text of all 13 books of the Elements plus critical apparatus analyzing each definition, postulate, and proposition in great detail. Covers textual and linguistic matters; mathematical analyses of Euclid's ideas; classical, medieval, Renaissance and modern commentators; refutations, supports, extrapolations, reinterpretations and historical notes. 995 figures. Total of 1,425pp. All books 5 3/8 x 8 1/2.
Vol. I: 443pp. 0-486-60088-2
Vol. II: 464pp. 0-486-60089-0
Vol. III: 546pp. 0-486-60090-4

A HISTORY OF GREEK MATHEMATICS, Sir Thomas Heath. This authoritative two-volume set that covers the essentials of mathematics and features every landmark innovation and every important figure, including Euclid, Apollonius, and others. 5 3/8 x 8 1/2.
Vol. I: 461pp. 0-486-24073-8
Vol. II: 597pp. 0-486-24074-6

A MANUAL OF GREEK MATHEMATICS, Sir Thomas L. Heath. This concise but thorough history encompasses the enduring contributions of the ancient Greek mathematicians whose works form the basis of most modern mathematics. Discusses Pythagorean arithmetic, Plato, Euclid, more. 1931 edition. 576pp. 5 3/8 x 8 1/2. 0-486-43231-9

CHINESE MATHEMATICS IN THE THIRTEENTH CENTURY, Ulrich Libbrecht. An exploration of the 13th-century mathematician Ch'in, this fascinating book combines what is known of the mathematician's life with a history of his only extant work, the Shu-shu chiu-chang. 1973 edition. 592pp. 5 3/8 x 8 1/2. 0-486-44619-0

PHILOSOPHY OF MATHEMATICS AND DEDUCTIVE STRUCTURE IN EUCLID'S ELEMENTS, Ian Mueller. This text provides an understanding of the classical Greek conception of mathematics as expressed in Euclid's Elements. It focuses on philosophical, foundational, and logical questions and features helpful appendixes. 400pp. 6 1/2 x 9 1/4. 0-486-45300-6

BEYOND GEOMETRY: Classic Papers from Riemann to Einstein, Edited with an Introduction and Notes by Peter Pesic. This is the only English-language collection of these 8 accessible essays. They trace seminal ideas about the foundations of geometry that led to Einstein's general theory of relativity. 224pp. 6 1/8 x 9 1/4. 0-486-45350-2

HISTORY OF MATHEMATICS, David E. Smith. Two-volume history – from Egyptian papyri and medieval maps to modern graphs and diagrams. Non-technical chronological survey with thousands of biographical notes, critical evaluations, and contemporary opinions on over 1,100 mathematicians. 5 3/8 x 8 1/2.
Vol. I: 618pp. 0-486-20429-4
Vol. II: 736pp. 0-486-20430-8

Physics

THEORETICAL NUCLEAR PHYSICS, John M. Blatt and Victor F. Weisskopf. An uncommonly clear and cogent investigation and correlation of key aspects of theoretical nuclear physics by leading experts: the nucleus, nuclear forces, nuclear spectroscopy, two-, three- and four-body problems, nuclear reactions, beta-decay and nuclear shell structure. 896pp. 5 3/8 x 8 1/2. 0-486-66827-4

QUANTUM THEORY, David Bohm. This advanced undergraduate-level text presents the quantum theory in terms of qualitative and imaginative concepts, followed by specific applications worked out in mathematical detail. 655pp. 5 3/8 x 8 1/2. 0-486-65969-0

ATOMIC PHYSICS AND HUMAN KNOWLEDGE, Niels Bohr. Articles and speeches by the Nobel Prize–winning physicist, dating from 1934 to 1958, offer philosophical explorations of the relevance of atomic physics to many areas of human endeavor. 1961 edition. 112pp. 5 3/8 x 8 1/2. 0-486-47928-5

COSMOLOGY, Hermann Bondi. A co-developer of the steady-state theory explores his conception of the expanding universe. This historic book was among the first to present cosmology as a separate branch of physics. 1961 edition. 192pp. 5 3/8 x 8 1/2. 0-486-47483-6

LECTURES ON QUANTUM MECHANICS, Paul A. M. Dirac. Four concise, brilliant lectures on mathematical methods in quantum mechanics from Nobel Prize-winning quantum pioneer build on idea of visualizing quantum theory through the use of classical mechanics. 96pp. 5 3/8 x 8 1/2. 0-486-41713-1

THE PRINCIPLE OF RELATIVITY, Albert Einstein and Frances A. Davis. Eleven papers that forged the general and special theories of relativity include seven papers by Einstein, two by Lorentz, and one each by Minkowski and Weyl. 1923 edition. 240pp. 5 3/8 x 8 1/2. 0-486-60081-5

PHYSICS OF WAVES, William C. Elmore and Mark A. Heald. Ideal as a classroom text or for individual study, this unique one-volume overview of classical wave theory covers wave phenomena of acoustics, optics, electromagnetic radiations, and more. 477pp. 5 3/8 x 8 1/2. 0-486-64926-1

THERMODYNAMICS, Enrico Fermi. In this classic of modern science, the Nobel Laureate presents a clear treatment of systems, the First and Second Laws of Thermodynamics, entropy, thermodynamic potentials, and much more. Calculus required. 160pp. 5 3/8 x 8 1/2. 0-486-60361-X

QUANTUM THEORY OF MANY-PARTICLE SYSTEMS, Alexander L. Fetter and John Dirk Walecka. Self-contained treatment of nonrelativistic many-particle systems discusses both formalism and applications in terms of ground-state (zero-temperature) formalism, finite-temperature formalism, canonical transformations, and applications to physical systems. 1971 edition. 640pp. 5 3/8 x 8 1/2. 0-486-42827-3

QUANTUM MECHANICS AND PATH INTEGRALS: Emended Edition, Richard P. Feynman and Albert R. Hibbs. Emended by Daniel F. Styer. The Nobel Prize–winning physicist presents unique insights into his theory and its applications. Feynman starts with fundamentals and advances to the perturbation method, quantum electrodynamics, and statistical mechanics. 1965 edition, emended in 2005. 384pp. 6 1/8 x 9 1/4. 0-486-47722-3

Physics

INTRODUCTION TO MODERN OPTICS, Grant R. Fowles. A complete basic undergraduate course in modern optics for students in physics, technology, and engineering. The first half deals with classical physical optics; the second, quantum nature of light. Solutions. 336pp. 5 3/8 x 8 1/2. 0-486-65957-7

THE QUANTUM THEORY OF RADIATION: Third Edition, W. Heitler. The first comprehensive treatment of quantum physics in any language, this classic introduction to basic theory remains highly recommended and widely used, both as a text and as a reference. 1954 edition. 464pp. 5 3/8 x 8 1/2. 0-486-64558-4

QUANTUM FIELD THEORY, Claude Itzykson and Jean-Bernard Zuber. This comprehensive text begins with the standard quantization of electrodynamics and perturbative renormalization, advancing to functional methods, relativistic bound states, broken symmetries, nonabelian gauge fields, and asymptotic behavior. 1980 edition. 752pp. 6 1/2 x 9 1/4. 0-486-44568-2

FOUNDATIONS OF POTENTIAL THERY, Oliver D. Kellogg. Introduction to fundamentals of potential functions covers the force of gravity, fields of force, potentials, harmonic functions, electric images and Green's function, sequences of harmonic functions, fundamental existence theorems, and much more. 400pp. 5 3/8 x 8 1/2. 0-486-60144-7

FUNDAMENTALS OF MATHEMATICAL PHYSICS, Edgar A. Kraut. Indispensable for students of modern physics, this text provides the necessary background in mathematics to study the concepts of electromagnetic theory and quantum mechanics. 1967 edition. 480pp. 6 1/2 x 9 1/4. 0-486-45809-1

GEOMETRY AND LIGHT: The Science of Invisibility, Ulf Leonhardt and Thomas Philbin. Suitable for advanced undergraduate and graduate students of engineering, physics, and mathematics and scientific researchers of all types, this is the first authoritative text on invisibility and the science behind it. More than 100 full-color illustrations, plus exercises with solutions. 2010 edition. 288pp. 7 x 9 1/4. 0-486-47693-6

QUANTUM MECHANICS: New Approaches to Selected Topics, Harry J. Lipkin. Acclaimed as "excellent" (*Nature*) and "very original and refreshing" (*Physics Today*), these studies examine the Mössbauer effect, many-body quantum mechanics, scattering theory, Feynman diagrams, and relativistic quantum mechanics. 1973 edition. 480pp. 5 3/8 x 8 1/2. 0-486-45893-8

THEORY OF HEAT, James Clerk Maxwell. This classic sets forth the fundamentals of thermodynamics and kinetic theory simply enough to be understood by beginners, yet with enough subtlety to appeal to more advanced readers, too. 352pp. 5 3/8 x 8 1/2. 0-486-41735-2

QUANTUM MECHANICS, Albert Messiah. Subjects include formalism and its interpretation, analysis of simple systems, symmetries and invariance, methods of approximation, elements of relativistic quantum mechanics, much more. "Strongly recommended." – *American Journal of Physics*. 1152pp. 5 3/8 x 8 1/2. 0-486-40924-4

RELATIVISTIC QUANTUM FIELDS, Charles Nash. This graduate-level text contains techniques for performing calculations in quantum field theory. It focuses chiefly on the dimensional method and the renormalization group methods. Additional topics include functional integration and differentiation. 1978 edition. 240pp. 5 3/8 x 8 1/2. 0-486-47752-5

Physics

MATHEMATICAL TOOLS FOR PHYSICS, James Nearing. Encouraging students' development of intuition, this original work begins with a review of basic mathematics and advances to infinite series, complex algebra, differential equations, Fourier series, and more. 2010 edition. 496pp. 6 1/8 x 9 1/4. 0-486-48212-X

TREATISE ON THERMODYNAMICS, Max Planck. Great classic, still one of the best introductions to thermodynamics. Fundamentals, first and second principles of thermodynamics, applications to special states of equilibrium, more. Numerous worked examples. 1917 edition. 297pp. 5 3/8 x 8. 0-486-66371-X

AN INTRODUCTION TO RELATIVISTIC QUANTUM FIELD THEORY, Silvan S. Schweber. Complete, systematic, and self-contained, this text introduces modern quantum field theory. "Combines thorough knowledge with a high degree of didactic ability and a delightful style." – *Mathematical Reviews*. 1961 edition. 928pp. 5 3/8 x 8 1/2. 0-486-44228-4

THE ELECTROMAGNETIC FIELD, Albert Shadowitz. Comprehensive undergraduate text covers basics of electric and magnetic fields, building up to electromagnetic theory. Related topics include relativity theory. Over 900 problems, some with solutions. 1975 edition. 768pp. 5 5/8 x 8 1/4. 0-486-65660-8

THE PRINCIPLES OF STATISTICAL MECHANICS, Richard C. Tolman. Definitive treatise offers a concise exposition of classical statistical mechanics and a thorough elucidation of quantum statistical mechanics, plus applications of statistical mechanics to thermodynamic behavior. 1930 edition. 704pp. 5 5/8 x 8 1/4. 0-486-63896-0

INTRODUCTION TO THE PHYSICS OF FLUIDS AND SOLIDS, James S. Trefil. This interesting, informative survey by a well-known science author ranges from classical physics and geophysical topics, from the rings of Saturn and the rotation of the galaxy to underground nuclear tests. 1975 edition. 320pp. 5 3/8 x 8 1/2. 0-486-47437-2

STATISTICAL PHYSICS, Gregory H. Wannier. Classic text combines thermodynamics, statistical mechanics, and kinetic theory in one unified presentation. Topics include equilibrium statistics of special systems, kinetic theory, transport coefficients, and fluctuations. Problems with solutions. 1966 edition. 532pp. 5 3/8 x 8 1/2. 0-486-65401-X

SPACE, TIME, MATTER, Hermann Weyl. Excellent introduction probes deeply into Euclidean space, Riemann's space, Einstein's general relativity, gravitational waves and energy, and laws of conservation. "A classic of physics." – *British Journal for Philosophy and Science*. 330pp. 5 3/8 x 8 1/2. 0-486-60267-2

RANDOM VIBRATIONS: Theory and Practice, Paul H. Wirsching, Thomas L. Paez and Keith Ortiz. Comprehensive text and reference covers topics in probability, statistics, and random processes, plus methods for analyzing and controlling random vibrations. Suitable for graduate students and mechanical, structural, and aerospace engineers. 1995 edition. 464pp. 5 3/8 x 8 1/2. 0-486-45015-5

PHYSICS OF SHOCK WAVES AND HIGH-TEMPERATURE HYDRO DYNAMIC PHENOMENA, Ya B. Zel'dovich and Yu P. Raizer. Physical, chemical processes in gases at high temperatures are focus of outstanding text, which combines material from gas dynamics, shock-wave theory, thermodynamics and statistical physics, other fields. 284 illustrations. 1966–1967 edition. 944pp. 6 1/8 x 9 1/4. 0-486-42002-7

Browse over 9,000 books at www.doverpublications.com